Isaac Hopkins

The Illustrated Australasian Bee Manual

And Complete Guide to Modern Bee Culture in the Southern...

Isaac Hopkins

The Illustrated Australasian Bee Manual
And Complete Guide to Modern Bee Culture in the Southern...

ISBN/EAN: 9783744792035

Printed in Europe, USA, Canada, Australia, Japan

Cover: Foto ©Andreas Hilbeck / pixelio.de

More available books at **www.hansebooks.com**

THE ILLUSTRATED

AUSTRALASIAN BEE MANUAL

AND COMPLETE

GUIDE TO MODERN BEE CULTURE

IN THE

SOUTHERN HEMISPHERE.

———

By ISAAC HOPKINS, MATAMATA, AUCKLAND, N.Z.

———

WITH WHICH IS INCORPORATED THE

"NEW ZEALAND BEE MANUAL"

GREATLY ENLARGED, REVISED, AND MOSTLY RE-WRITTEN BY THE AUTHOR,
ASSISTED BY

T. J. MULVANY, BAY VIEW APIARY, KATIKATI.

———

THIRD EDITION—FOURTH THOUSAND.

143 ILLUSTRATIONS.

———

Published by the Author.

H. H. HAYR & Co., Agents, High Street, Auckland, N.Z.
MDCCCLXXXVI.

Published by the Author.

H. H. HAYR & Co., Agents, High Street, Auckland, N.Z.
MDCCCLXXXVI.

PREFACE.

THE first edition of "The New Zealand Bee Manual," published in September, 1881, was a small book of some 150 pages, and was intended to convey, in a popular form, such information with regard to the modern system of bee-culture as might tend to awaken an interest in that pursuit among the settlers in these colonies, enable those who should desire to make a beginning to do so in the proper manner, and induce others, who were already working in the dark with candle-boxes and gin-cases, to discard such appliances and adopt a more rational method of bee-keeping. The best proof that the subject was beginning to attract attention will be found in the fact that the first edition was disposed of in thirteen months, and a second thousand had to be issued in October, 1882. That edition being also exhausted and a new one required, I felt bound to consider the greatly altered circumstances under which it would have to be published, and to endeavour to make it, as far as possible, suitable to those new conditions and to the advances made in the art of bee-culture up to the present day.

In the interval since the issue of the first edition, bee-culture has taken an established footing in New Zealand and the Australian colonies, the suitability of the climate and the flora being no longer a matter of speculation but one of experience. My duties as Editor of "The New Zealand and Australian Bee Journal" during the two years of its circulation, brought me into very agreeable communication with bee-keepers in all parts of these colonies, and the consequent interchange of views and experiences has enabled me to obtain an amount of information with regard to the condition and prospects of the industry in this part of the world which was previously entirely wanting. I came to the conclusion that there is a rapidly increasing class of people who are turning their attention to apiculture in Australasia, and who, whether as professed apiarists or as amateurs, would require a manual of bee-culture which, while giving in a general but not merely superficial way an epitome of our present knowledge upon all apicultural matters, scientific and theoretical, should also enter into full details upon all the practical points necessary for the guidance of a beginner in the art, and which should be specially prepared with reference to the seasons of the southern hemisphere, and to the flora and other local peculiarities of the Australasian colonies. A revision and enlargement of the original manual to meet these views has led to the re-casting and re-writing of all the chapters, and the introduction of so much new matter and additional illustrations as to constitute it in point of fact an entirely new work.

In carrying out this programme I have availed myself of the able co-operation of MR. T. J. MULVANY, of Bay View Apiary, Katikati, to whom I tender my sincere thanks for his valuable aid, without which, with the limited time at my disposal, I could not have undertaken the task. The share taken by MR. MULVANY in the new matter to be found in many of the following chapters will be recognised by the readers of " The New Zealand and Australian Bee Journal," to the columns of which he so ably contributed, during the two years of its existence; condensed extracts from several of his papers are here given in places where their introduction seemed to work in with the present plan.

I have also to acknowledge my indebtedness and tender my thanks to MR. C. FULLWOOD, of Brisbane, Queensland; MR. THOS. E. WILLIS, of Sydney, N.S.W.; MR. DAVID GLASS, of Ballarat, Victoria; MR. A. E. BONNEY, of Adelaide, South Australia; and MR. THOS. LLOYD HOOD, of Hobart, Tasmania, who so willingly and ably acceded to my request to furnish me with facts connected with apiculture in their respective colonies, and so aided me in giving much local information valuable to beginners in all parts of Australasia, and which must also prove interesting to apiarists working under very different circumstances in other parts of the world.

In endeavouring to place before the novice (whose necessities have been kept constantly in view in the preparation of this work) a clear picture of the rise, progress, and present condition of the art of bee-culture, I have made free use of all the standard works already pub-

lished in England, America, and Germany, as well as the bee journals
and technical periodicals of the same countries, extracting from each
what appeared to me to be the most essential to be impressed upon
the mind of the reader. In every case I have been careful not only to
specify the sources from which such extracts have been taken, but
also to give the quotations accurately in the words of the respective
authors. I feel convinced that no reader of this volume can fall into
the mistake of supposing that the quotations given from such writers
as HUBER, DZIERZON, LANGSTROTH, QUINBY, COOK, ROOT, CHESHIRE,
LUBBOCK, MÜLLER, and many others are meant to supersede the
necessity of studying the original works from which those quotations
are taken ; on the contrary, they cannot fail to excite the desire to do
so on the part of all who, after the general view here given, shall wish
seriously to pursue the study.

On all practical points of working detail I have given the practice
followed by myself, and with special regard to the experience of
myself and others in the peculiar circumstances of the Australasian
colonies. On many of these points the beginners in these countries
could find no reliable guide in any of the European or American works,
for although there are in both of those continents some honey-producing districts similarly situated, at least in point of climate, to the
semi-tropical parts of Australasia, yet it happens that all the standard
works on apiculture have had their origin in places situate in comparatively high latitudes, where the severe winters and the absence of
an evergreen flora tend to place the practice of bee-keeping upon
essentially different principles in many respects. Many subjects
indirectly connected with practical apiculture are also introduced in
this volume which I have not seen touched upon in any other work.
Among these I would direct special attention to Chapter XIX.

In order not to confuse the novice I have given, in all ordinary
matters, full details of such methods as I have adopted and consider
the best ; but in special cases, where there is some divergence of
opinion, I have also described the plans recommended by some of the
leading apiarists of the day, and in every case I have sought to bring
the accounts of improvements in the art down to the latest date.

I have spared no trouble or expense in procuring the numerous illustrations which appeared to me to be necessary for the completeness of the work ; and notwithstanding the necessary difference in price of the book as compared with the former editions, I trust it may be considered in that respect, as well as in regard to the style of bringing it out, as creditable to the publishing trade of New Zealand.

I. HOPKINS.

MATAMATA, AUCKLAND, N.Z.,
    JANUARY, 1886.

ERRATA.

Page 27 line 15—For *fasciaca*, read *fasciata*.
  ,,   72 ,,   1—After the word " which " insert " the eggs of."
  ,,   86 ,,   11—For execeted, read excreted.
  ,, 117 ,,   1—For fig. 39, read fig. 37.
  ,, 131 ,,   2 (from bottom)—For 2in. read 1¾in. and in
  ,,   ,, ,,   1   ,,   For 1¾in., read 2in.
  ,, 157 ,,   8   ,,   For " 1,800lbs. " read 1,450lbs.

CONTENTS.

CHAPTER I.

INTRODUCTORY AND HISTORICAL.

Antiquity of the Use of Honey—Origin of the Art of Bee-keeping—Modern Art of Bee-keeping—Bee-keeping in New Zealand and Australia—Introduction of the Black or German Bee into New Zealand—Introduction of Italian Bees—Improved System of Bee-keeping in New Zealand—Introduction of Bees and Bee Culture into Australia and Tasmania—Italian Bees in Queensland, in New South Wales, in Victoria, in South Australia, in Tasmania—Suitability of New Zealand and Australia for Apiculture—Climate—Native Flora—Importance of Apiculture as an Industry—Profits of Bee-keeping—Adaptation to Women—Advice to Beginners—State Aid to Apiculture—Bee Publications ... 1—24

CHAPTER II.

THE HONEY BEE, ITS VARIETIES AND DISTRIBUTION.

Apis Mellifica—Geographical Distribution of its Varieties—German or Black Bee—Italian or Ligurian Bee—Markings of pure Italian Bees—Hybrids—German, Italian, Cyprian, Syrian and Palestine Bees—Carniolans, Herzegovian, Dalmatian, Roumelian, and Hymettus Bees—Other Races of Bees—Native Bees of New Zealand and Australia ... ... 25—39

CHAPTER III.

INMATES OF THE HIVE, THEIR NATURAL HISTORY.

General Description—The Queen—The Drone—The Worker—Physiology and Anatomy of the Honey Bee—Classification of Species—General Characteristics of Structure—Nervous System

—Respiratory Organs—Air-Sacs of Bees—The Head—The Eyes
—The Mouth—The Antennæ—Senses of Hearing and Smelling—
The Wings—The Legs—The Honey Sac—The Sting—Reproductive Organs of Queen—Parthenogenesis—Dzierzon Theory —
Development from Egg to the Bee—Fertile Workers—Relation of
Bees to Flowers   ...   .   40—79

CHAPTER IV.

WHAT BEES COLLECT, AND WHAT THEY PRODUCE.

Honey—Nectar of Blossoms—Adulteration of Honey—Honey - Dew—
Wax—Bees Wasting Wax—Extravagant Waste of Wax—Method
of rendering Wax — Comb, how constructed—Advantages of
the hexagonal form of Cells—Pollen and Bee-Bread—Artificial
Pollen—Propolis   ...   .   80—98

CHAPTER V.

THE APIARY.

Location—General Arrangement—Shade—Water—Area of Ground—
Arrangement of Hives—Extracting House and Honey Store—
Workshop and Hive Store —Fumigating House — Stocking the
Apiary—Moving Hives—House Apiary ...   ...   99—110

CHAPTER VI.

HIVES, FRAMES, AND SECTION BOXES.

Movable Comb Hives—Choice of a Hive—An ideal Hive—Various
forms now in use—The Langstroth Hive—General Description
—Instructions for making—Body of Hive—Bottom Board—
Alighting Board — Cover – Half-Story Hives — Hive Cramp—
Nucleus Hive—Observatory Hive—Timber for Hive-making—
Painting Hives—Frames--Narrow or brood Frames—Broad, or
Section Frames—Half-story Frames—Frame Form—Number of
Frames to a Hive—Mats for covering Frames—Section Boxes—
Requisites of good Sections—To Make sections—One Piece
Sections—Putting dovetailed Sections together—Clinching the
Dovetails— Separators – Dispensing with Separators — Section
Racks and Cases—The Heddon Section Case—Honey Boards –
How to construct Honey Boards and Section Cases   111—144

CHAPTER VII.

THE HONEY EXTRACTOR, AND MANIPULATION OF EXTRACTED HONEY.

Strained or pressed and melted Honey—Invention of the Centrifugal Extractor — Two-comb Extractor— Single-comb Extractor — Six-comb Extractor—Preparing Combs for extracting—Uncapping Knife—Uncapping Can—Broken-Comb Basket—Manipulation of extracted Honey—Arrangement of Extracting House

145—157

CHAPTER VIII.

COMB-FOUNDATION.

History of the Invention—Advantages derived from its Use—Principal Points of good Foundation—Comb Foundation Machines—Drone Cell Foundation—Progress of Manufacture—To fasten Foundation in Frames—To Fasten Foundation in Sections—Wired Foundation —Wiring the Frames—Imbedding the Foundation — To secure straight Combs ... ... 158 – 170

CHAPTER IX.

MANIPULATION OF BEES, AND FEEDING.

Handling Bees—Bee Veil—Bee Gloves—Quieting Bees—Smokers— Fuel for Smokers—How to open a Hive—Comb Holders—Cures for Stings—Feeding—Feeding for Winter—Stimulative Feeding— What and When to feed—How to feed Syrup ... 171—182

CHAPTER X.

TRANSFERRING.

Implements Required—Wires and Clasps--Transferring Board — Driving—Fixing Combs in Frames—Mr. Heddon's new Practice —Open Driving ... ... 183—187

CHAPTER XI.

INCREASE OF STOCKS, NATURAL SWARMING, DIVIDING.

What Rate of Increase is desirable?—Circumstances which affect a Decision— Mode of attaining the Object—Natural Swarming—

Causes of Swarming—The Swarming Season—Symptoms of Swarming—Issue of the Swarm—Objections raised against Natural Swarming—Preparing for Swarms—Swarm Box—Taking and Hiving Swarms—Absconding Swarms—Clipping the Queen's Wing — Process of clipping — After-Swarms — Prevention of Swarming—Prevention of After-Swarming—Supplying the old Stock with Fertile Queen—Preventing Increase of Colonies—Dividing ... ... 188—207

CHAPTER XII.

QUEEN REARING.

Necessity of the Practice—A Word concerning Drones—Entrance Guards—How to secure choice Queen Cells—Forming Nuclei—How to insert Queen Cells—My Method of forming Nuclei and inserting Queen Cells—Mating Young Queens—Necessary Distance apart of different Races to insure pure Mating—Queen Nurseries—Introducing Queens—Direct Introduction of Queens—Introducing Virgin Queens—" Shipping " Queens—Shipping whole Colonies
208—230

CHAPTER XIII.

SURPLUS HONEY, MODE OF SECURING AND MARKETING.

Spring Management—Division Boards—Spreading Brood—Putting on surplus Boxes—Reversing Frames—Heddon's reversible Frames—Queen Excluder Honey Board—Taking surplus Honey—Ripening Honey—Comb Baskets—Marketing Honey—Extracted Honey—Comb Honey—Benefits of co-operation ... 231—243

CHAPTER XIV.

WINTERING, UNITING.

Difference in Character of Winter Seasons—Australasian Winters—Precautions for Wintering—Reducing interior Room in Hive—Providing Food—Inner Covering of Frames—Reducing Entrance—Securing Covers—Providing Winter Forage—Ventilation—Chaff Hives—To make a Chaff Hive—Providing Space above Frames in Winter—Uniting weak and queenless Colonies—Uniting Swarms ... ... 244—251

CHAPTER XV.

ROBBER BEES.

Causes of Robbing—How to know Robber Bees—Precautions to be observed—Bee Tents—How to stop Robbing ... 252—255

CHAPTER XVI.

DISEASES OF BEES.

Dysentery and its Prevention—Bacillus alvei (foul brood)—Symptoms of Bacillus alvei—Investigations of Mr. Frank Cheshire—Bacillus alvei under the Microscope—The Cheshire Cure—The Salicylic Acid Remedy—Bacillus Gaytoni—Other Diseases of Bees—Arrenotokia—Spray Diffuser ... . 256—268

CHAPTER XVII.

ENEMIES OF BEES.

Australasian Exemptions—Spiders—Mice—Ants—Bee-hawk (*libellula*) —The Bee or Wax-moth—Tinea cereana—Damage to Combs— Remedies—Bee-mite—A Caution to Importers of Bees—Fumigating Combs... ... ... 269—277

CHAPTER XVIII.

BEE FORAGE.

Ordinary Sources—Native Flora of New Zealand—Native Flora of Australia - -New South Wales—South Australia — Victoria — Queensland—Tasmania—Eucalypti and Acacias in New Zealand— European Plants and Trees—American Plants and Trees—Basswood—Sages — Horse Mint—Figwort—Golden Rods — Asters— Spider Plant—Mellilot Clover—Giant Mignonette—Duration of the Honey Season—Flight of Bees—Over-stocking ... 278—296

CHAPTER XIX.

APICULTURE IN RELATION TO AGRICULTURE.

Are Bees Trespassers ?—Beneficial Influence of Bees on Agriculture— Can bees harm the Soil or the Crops ?—Saccharine Matter of Plants not derived from the Soil—Derived from the Atmosphere and Rain Water—Nectar of Plants intended to attract Insects— Sometimes thrown off as superfluous—Superfluous Nectar evaporated if not taken by Insects—Question as to grazing Stock— Quantity of Honey furnished by pasture Lands—Proportion possibly consumed by Stock—Bee-keeping as a Branch of Farming ... . 297—307

CHAPTER XX.

USES OF HONEY FOR FOOD, DRINKS, MANUFACTURES AND MEDICINE.

Honey as Food—For Domestic Cookery—For Preserves—Moderation in Use—Deleterious Honey –Fermented Drinks—Manufactures— Honey as Medicine—Recipes—Honey Cakes—Mead —Metheglin— Miodomel — Fruit Wine with Honey — Light Beverages — Medicinal   ...   .   308—319

CHAPTER XXI.

CALENDAR, BEEKEEPERS' AXIOMS.

Variability of Seasons—Use of Meteorological Observations—Calendar, adapted to Auckland, New Zealand—Bee-keepers' Axioms
320—321

GLOSSARY...   ...   ...   326—330

INDEX   ...   ...   331—335

LIST OF ILLUSTRATIONS.

FRONTISPIECE : Burwood Apiary, Matamata.

FIG.		PAGE
1.	German or Black Queen	28
2.	Italian Queen	29
3.	Abdomen of Italian Worker	32
4.	The Queen Bee	41
5.	The Drone	42
6.	The Worker	42
7.	Wings of a bee	46
8.	Nervous system of the bee	47
9.	Trachea, magnified	48
10.	Respiratory organs of the bee	49
11.	Air sacs and ovaries of queen	50
12.	Air sacs of worker	50
13.	Head of worker bee	51
14.	Hooklets of wings	56
15.	Hind leg of bee, showing pollen basket	57
16.	Anterior leg of worker, magnified	58
17.	Bee sting, magnified	60
18.	Ovaries of queen	63
19.	Queen's egg under the microscope	69
20.	From the egg to the bee	70
21.	Worker nymph and larva, in comb	71
22.	Worker larvæ and queen cells	72
23.	Queen cells built over worker cells	74
24.	Salvia officinalis, young flower visited by a bee	77
25.	Ditto, stamens and anthers	78
26.	Ditto, older flower, with pistil developed	78
27.	Under side of abdomen of worker bee, showing wax pockets and wax scales	88
28.	The Gerster wax extractor	91
29.	Jones' wax extractor	92

FIG.		PAGE
30.	Worker cells and drone cells	93
31.	Hexagonal cells	94
32.	Circular cells	94
33.	Pollen grains under the microscope	96
34.	Water bottle	102
35.	Arrangement of hives	103
36.	House apiary for 20 hives	110
37.	Movable comb-hive	111
38.	Two-storied Langstroth hive	116
39.	End of hive (inside view)	119
40.	Ditto (outside view)	119
41.	Side of hive (inside view)	120
42.	Showing how the different parts of the hive go together	121
43.	Bottom board	122
44.	Alighting board	122
45.	End of cover (inside view)	123
46.	Side of cover, ditto	123
47.	Ridge board of cover	123
48.	Roof board of cover	124
49.	Cover complete	124
50.	Hive cramp	125
51.	Nucleus hive, with cover and mat	127
52.	Observatory hive	128
53.	Narrow or brood frame	131
54.	Broad or section frame, with sections and tin separators	131
55.	Half-story frame, with sections and tin separator	132
56.	Frame form or gauge	133
57.	One-pound section box, with starter of comb-foundation	135
58.	One piece section	136
59.	Section cramp and form	137
60.	Prize section rack	139
61.	The Heddon section case	140

xvi.
LIST OF ILLUSTRATIONS.

FIG.   PAGE

62. Heddon's honey board .. 142
63. Honey extractor 147
64. Framework for two-comb extractor 148
65. "Little Wonder" 149
66. Reversible six-comb honey-extractor, with one basket detached 150
67. Cowan's automatic basket.. 151
68. Root's uncapping knife .. 152
69. Bingham and Hetherington knife, with cap-catcher .. 152
70. Dadant's uncapping can ... 153
71. Broken-comb basket.. .. 154
72. Cross-section of extracting house, with view of extractor, strainer, and tanks 155
73. Ground plan of ditto.. .. 156
74. Comb foundation 158
75. A. I. Root's 10in. roller machine 161
76. The Given press 162
77. Gauge for trimming foundation 164
78. Comb-foundation board . 165
79. Mode of fastening in frame 165
80. Wax-smelter 166
81. Parker's comb-lever.. .. 167
82. Wired frame 168
83. Wiring board 169
84. Easterday's wire imbedder 169
84A. Device for securing straight combs 170
85. Wire-cloth bee-veil .. 172
86. Tarlatan bee-veil .. . 172
87. Clark's cold-blast smoker .. 174
88. Bingham's direct-draught smoker 174
89. Comb-holder 176
90. "Simplicity" comb-holder.. 177
91. "Simplicity" feeder .. . 181
92. Grey's entrance feeder .. 182
93. Transferring wires and claps 183
94. Transferring board . .. 184
95. Pieces of comb transferred to frame 185
96. Jones' entrance guard .. 210
97. Alley's drone excluder, drone and queen trap .. 210

FIG.   PAGE

98. Comb containing eggs .. 211
99. Frame for raising queen cells on.. 212
100. Frame of queen cells . 214
101. Comb stand 216
102. Inserted queen cell, from which queen has emerged 217
103. Alley's queen nursery 221
104. Alley's queen nursery cage 221
105. Alley's introducing cage 223
106. Queen shipping cage .. 227
107. Benton's shipping cage for two queens 227
108. Division board 232
109. Hive with division boards.. 233
110. Heddon's reversible frame.. 237
111. Queen excluder 238
112. Bliss' sun evaporator.. .. 239
113. Comb basket 240
114. Shipping crate and show case for comb-honey 243
115. Brickell's chaff-hive .. 249
116. Hill's device 250
117. Bee tent 255
118. Bacilli 260
119. Bacilli spores 260
120. Healthy juices of larva .. 261
121. B. alvei (last stage) 261
122. B. alvei (early stage).. .. 262
123. B. alvei (late stage) 262
124. Spray diffuser 268
125. Bee-hawk (*Libellula*) 271
126. Bee-moth.. 272
127. Male and female bee-moth 273
128. Silken tube of bee-moth larva 273
129. Silken tube in comb 274
130. Larvæ of bee-moth . .. 274
131. Comb destroyed by bee-moth larvæ 275
132. American linden or bass-wood 288
133. Horse-mint of Texas.. .. 289
134. Figwort 290
135. Three varieties of golden rods 291
136. Aster .. 292
137. Spider plant 293
138. Melilot clover 294
139. Giant mignonette 295

CHAPTER I.

INTRODUCTORY AND HISTORICAL.

IN the whole range of created objects presented to our contemplation in the study of what we familiarly call NATURE, from the inconceivably great systems of inanimate matter rolling in infinite space to the inconceivably small but animated forms revealed by the microscope, there is probably no class more calculated to excite our wonder and admiration than that of *Insects;* and of all the different kinds of insects there is none more interesting as an object of study, and none that can be made more useful and profitable to man, than the Honey Bee. Its history is as old as that of the human race; its product, honey, was recognised in the earliest ages as a most desirable, almost an indispensable, addition to the food of man : and yet it is only now, some 3400 years after its first authentic historical mention, that we are beginning to realise the full economic importance of that product and to avail ourselves fully of the bounty of Providence, evidenced not only in its production, but also in the endowment of the bee with those wonderful instincts which render its collection so easy.

ANTIQUITY OF THE USE OF HONEY.

A certain proportion of saccharine matter in the food of man appears to be essential for his sustenance in a healthy condition, and previous to the comparatively modern invention of preparing sugar from vegetable juices, the only form in which such saccharine matter was attainable in a concentrated state was that of honey. The temperate or semi-tropical climate of that part of the globe which formed " the cradle of the human race " was most favourable to the spontaneous spreading of the honey-bee and the collection of surplus honey in its natural hives or nests. These would be built in the hollows of trees, in the clefts and under the ledges of rocks, as they are at the present day in such climates, and their stores would soon be discovered by men engaged in the grazing of flocks and herds in a very thinly populated land. It is not, therefore, surprising that in the Scriptures of the Old Testament, the earliest written records of the human race, we find frequent reference made to honey as a thing universally known and intimately connected with the comforts of man. The name is said to be *ghoneg* in the original Hebrew, signifying "delight," evidently the root of the German word "honig," which easily becomes "honey" in English. The name is used generally in the ancient Scriptures in combination with that of milk, the most universal of all foods, to form the Oriental metaphor denoting abundance—" a land flowing with milk and honey" being the words used in nearly twenty passages of those writers, from Moses down to the prophet Ezekiel, to describe the country promised to the descendants of Jacob. In the non-historical parts of Scripture, the Prophecies, the Psalms of David, the Song of Solomon, Proverbs, and the book of Sirach, the words "honey" and "honey-comb" are always used as the types of everything good and wholesome as well as sweet; in the last mentioned book (which, though its canonical value is a matter of dispute, may be safely quoted in this respect) it is distinctly mentioned as one of the *necessaries* of life. In the historical portions it is first mentioned as one of the choice articles sent as a present by Jacob to the ruler of the Egyptians when his sons went to that people to obtain a supply of corn during a time of scarcity, about 3600 years ago. Some 700 years later King Jeroboam sent a "cruse of honey" with other presents to propitiate the prophet Ahijah. A curious case is mentioned about Samson (in the twelfth century before our Christian era) finding " a swarm of bees and honey in the carcass of the lion " which he had killed some time before. In explanation of this strange sort of bee-hive, we are told that in the climate of Palestine, in some hot seasons, dead bodies are often so quickly dried up that they become like mummies and remain a long time undecayed, so that a swarm of bees might well select the inside of a dried-up lion's body (supposing it to have been disembowelled) to build in.* Somewhat later in date a circumstance is related of

---

* Possibly some such case may have given rise to the extraordinary theory put forward by Virgil that bees were generated in the decaying entrails of a bullock!

Jonathan, with a part of Saul's army, entering a wood and finding "honey on the ground." "When the people came into the wood, behold the honey dropped," and Jonathan refreshed himself by "putting forth the end of the rod that was in his hand and dipping it in a honey-comb and putting his hand to his mouth." This is very interesting as showing so clearly how honey was then commonly obtained. About the year 1023 B.C. honey is mentioned as one of the things supplied by friendly hands for the refreshment of David and his followers when "they were hungry and weary and thirsty in the wilderness;" and three centuries later it is enumerated amongst the things of which tithes were to be paid to the Levite priests by order of King Hezekiah. Finally, it is mentioned in the Prophecy of Ezekiel, when describing the ancient commerce of Tyre, as an article of commerce sent to that port from Palestine.

ORIGIN OF THE ART OF BEE-KEEPING.

Those passages relating to honey in the writings of the Old Testament are quite sufficient to prove the great antiquity of its use, but they give us no grounds for looking upon the patriarchs and the early inhabitants of the earth as *bee-keepers ;* on the contrary, there is ample evidence afforded that, at the time referred to, honey was obtained from the natural haunts of the bees—in the forests and rocky pasture lands—just as it may be obtained at the present day in the bush districts of warm climates, and especially in parts of India, where the bees build not so much in the hollows of trees as in the open air in the branches, and under ledges of rock on the sides of hills. The climate of Palestine, Assyria, and Egypt is quite suited to the natural propagation of bees in the woods and "wildernesses" on the borders of the Arabian desert, and the nomadic life of the shepherds and cattle-herds afforded the best opportunities for tracing the bees to their haunts and collecting the wild honey. We may then fairly conclude that such were the sources from which honey was ordinarily obtained by the inhabitants of those Eastern countries, and we have no reason to suppose that they practised any art of bee-keeping, or knew anything about a system of providing bees with artificial dwellings and inducing them to gather honey and to store it in a manner more convenient to man. We must suppose that the haunts of the wild bees, when found, had to be pillaged with the aid of smoke or sulphur fumes, and that the operation was not always an agreeable one we may conclude from the way in which bees are mentioned in the few passages of Scripture where they are incidentally alluded to, as in Deut. i. 44, where Moses, recapitulating all that had happened to the Israelites during their migration, tells them, "The Amorites came out against you and chased you as bees do." And in Psalm cxviii. 13, "They (the heathens) encompassed me about like bees."

Amongst the Western nations the civilised Greeks had unquestionably practised the art of bee-keeping at a very early period. The laws of Solon, 600 years B.C., contain regulations as to the distances apart at which bee-hives may be kept; and both Greeks and Romans wrote and sang about bees and bee-keeping from the times of Homer down to those of Aristotle, Virgil, Palladius, Pliny, and Columella. It is very probable that the Romans first introduced the practice into Palestine. The term "wild" honey is never met with in the ancient Scriptures, simply because *all* honey deserved that name in those times; but the Evangelists Matthew and Mark, who wrote when Palestine had been for nearly a century virtually a Roman Province, both use the term "locusts and wild honey." We may conclude that at that time the people were accustomed to keep bees in artificial hives, and they would naturally make a distinction between honey so obtained and that gathered by " wild " bees in the " wildernesses " or unfrequented places.

When Alexander carried his conquests into India, in the fourth century B.C., he found honey so plentiful there that he imposed a tribute payable in honey and wax. The Romans, at a much later period, levied a tribute of 200,000 lbs. of wax yearly upon Corsica, and the countries of the "barbarians" outside the limits of the Roman Empire in Europe were known to produce (and certainly without any art of bee-keeping) large quantities of both honey and wax. In the early part of the third century, when the Goths were gradually migrating towards the Roman Provinces, Gibbon mentions that when they took possession of the present Russian district of the Ukraine,* " The plenty of game and fish, the innumerable bee-

* This is the part of Russia from which the largest quantities of honey are obtained at the present day.

hives deposited in the hollows of old trees and in the cavities of rocks, and forming, even in that rude age, a valuable branch of commerce, . . . all displayed the liberality of nature and tempted the industry of man." The same writer tells us that in the time of Constautine the Great (A.D. 306 to 337) the people of Chersonesus (the present Crimea) were supplied from the Roman Provinces of the East, with corn and manufactures "which they purchased with their only productions, salt, *wax* and hides." The ambassadors of Theodosius II. to Attila, king of the Huns, when travelling through part of the country now called Hungary (about A.D. 450) "received from the contiguous villages a plentiful supply of provisions," amongst which is noted "*mead* instead of wine." But however primitive may have been the mode of obtaining honey in those unsettled countries, great progress, both in the art of bee-keeping and in mercautile dealings in honey and wax, must have been made in the civilised provinces, as it is mentioned, on the authority of a writer named Synesius, that when the Goths, under Alaric, sacked the city of Athens, A.D. 396, that city "was at that time less famous for its schools of philosophy than for its trade in honey."

In the seventh century the Emperor Heraclius raised a sort of forced loan from the churches at Constantinople to meet some war expenses, and on that occasion it is related that barrels of honey (ostensibly) packed away among the church stores were found to be really filled with gold. This anecdote serves to indicate how extensively honey was used, and how it was kept in those times. About the same time, when Persia was overrun by the Saracen Caliph, after the great battle of Nehavend, the fugitive general of the Persians was stopped and overtaken "in a crowd of camels and mules *laden with honey*," an incident which, as Gibbon remarks, "however slight or singular, will denote the luxurious impedimenta of an Oriental army." It is also related that Mahomet, who was very temperate and sparing in his diet, "delighted in the taste of milk and honey;" and that this taste was general among the Arabs we may conclude from the circumstance mentioned by Gibbon, that with them "the perfection of language out-stripped the refinements of manners, and their speech could diversify the *fourscore* names of honey."

In the earliest history of the Russian people, in the ninth and tenth centuries, we find mentioned among the chief articles of their trade, " the spoils of their bee-hives and the hides of their cattle," and " their native commodities of furs, wax, and hydromel ;" and a Greek historian, describing the state of Britain at the time of the visit of the Greek Emperor Manuel (about 1400), says : " The land is overspread with towns and villages ; though destitute of vines, and not abounding in fruit trees, it is fertile in wheat and barley, in *honey* and wool."

The true history of the rise and progress of the art of bee-keeping amongst the Greeks and Romans, and its extension over Europe during the middle ages, is as yet unwritten, but there can be no doubt that amongst the Northern nations the use of honey became with time more and more a matter of necessity, much of their fermented liquors being prepared from it, and the more northern the positions, and the more severe the winter seasons, the more essential it became to domesticate the bees, or use artificial means for preserving them during the winter months.

The primitive system of bee-keeping adopted in the earliest period of Greek civilisation seems to have been followed with little change or improvement by the Romans and the nations which rose upon the ruins of that empire, and to have been handed down from father to son almost unaltered until the close of the last century. In the first half of the present century some important improvements were introduced into England, especially by Thomas Nutt, a self-instructed apiarist, who was one of the first to condemn and abolish the barbarous custom of destroying the bees with sulphur, and to invent and practice a more rational and humane method of taking the surplus honey in separate boxes and bell-glasses. Since the middle of the seventeenth century much attention had been paid to the natural history of the bee and other insects by Von Swammerdam in Holland, Maraldi in Italy, Réaumur, Lepeletier and Latreille in France, Bonnet in Switzerland, Linnæus in Sweden, and by Dr. John Hunter and Dr. Bevan in England ; but it is to the researches and discoveries of Huber and Dzierzon that we are indebted for that knowledge of the physiology of the honey-bee which has led to those great practical improvements which may be said to constitute the

MODERN ART OF BEE-KEEPING.

This may be considered to have commenced with the second half of the present century, although the most important strides in the progress of the honey industry have only taken place within the last twenty years. In the year 1845, the results of Dzierzon's investigations were first made known in the *Eichstadt Bienen-Zeitung*, and in 1848 his book on the "Theory and Practice of Bee Culture" was published at the instance of the Prussian Government. Not many years afterwards, Langstroth's work on "The Hive and Honey Bee," and Quinby's "Mysteries of Bee-keeping Explained," appeared almost simultaneously in America. All these men had been working independently for some twenty years, studying the habits of the bee, and inventing a hive and a system which should enable the apiarist to control the working of his bees, and to obtain the largest amount of surplus honey without injury to them. They all attained a very high degree of success, and they bestowed the knowledge of their successful labours upon the public nearly at the same time. All their works have great and independent merits, and must always remain as classics in bee literature. To Dzierzon must be allowed the merit of having so completely worked out and supplemented Huber's theory with regard to the physiology of the bee, and also the priority at least in the publication of his system of bar-hives. Langstroth and Quinby both produced frame-hives, simpler and more practical than that of Dzierzon, and each of them have their advocates to the present day. Subsequently the invention of the honey extractor, of comb-foundation, and a number of ingenious implements and appliances, have led to a complete revolution in the practice of bee-keeping, and helped to raise it to the rank of an important national industry which can no longer be neglected in any country possessing the natural capabilities for its establishment.

BEE-KEEPING IN NEW ZEALAND AND AUSTRALIA.

None of the countries of the New World, of North or South America, or of Australasia, were found, when first discovered, to possess any variety of the true honey-bee (*Apis mellifica*); a necessary preliminary, therefore, to the practice of bee-culture in any of those regions was the introduction of bees from the old country, an operation which was attended with far greater difficulties even forty or fifty years ago than in these days of rapid steam navigation.

INTRODUCTION OF THE BLACK OR GERMAN BEE INTO NEW ZEALAND.

The first bees introduced into New Zealand are said to have arrived in the ship *Westminster*, in the early part of 1840. These bees belonged to Lady Hobson, wife of the first Governor. and were watched over on board the vessel by Mr. McElwaine, the Governor's gardener. They were landed at the Bay of Islands. Mr. William Mason, who was, at the period above mentioned, Government Architect and Inspector of Public Works, told me that he distinctly recollected seeing the bees on board the ship, and that they were in straw hives, which were wrapped in blankets. He believed they remained at the Bay when the Government party left to establish the seat of government on the Waitemata, now the city of Auckland.

Dieffenbach, in his "Travels in New Zealand," mentions having seen (in December, 1840) a hive of bees, thriving remarkably well, with the Rev. Richard Taylor at Waimate, but says "the bees had been introduced into New Zealand from New South Wales." This may be an error. It is not improbable that the hives referred to may have been stocked with some of Lady Hobson's bees, but it is also quite possible that they may have been brought from New South Wales where they had been first introduced in 1822.

For the introduction of bees into this colony we are also indebted to the late Rev. William Charles Cotton, and to Mrs. Allom, mother of our respected and esteemed fellow-citizen, A. J. Allom, Esq., of Parawai. With regard to Mr. Cotton's success, I quote the following from the *British Bee Journal* of January 1st, 1880 :—

"In 1841 Mr. Cotton became chaplain to the late Bishop of New Zealand, Dr. Solwyn, with whom he embarked on board the *Tomatin* at Plymouth, on the 30th December of that year. On the voyage out, and subsequently, Mr. Cotton rendered the Bishop much assistance in translating the Bible into the native tongue.

"Mr. Cotton took with him four stocks of bees ; and many marvellous stories are told of his mastery over his favourites on ship board. He was very successful in the introduction of the cultivation of bee-keeping in his adopted country, and in 1848 he produced his 'Manual for New Zealand Bee-keepers,' published at Wellington, New Zealand. Before the introduction of the honey bee into New Zealand, they had to send over to England every year for the white clover seed (*Trifolium repens*), as it did not seed freely there, but by the agency of the bees they are now able to export it. New Zealand is such a good country for bees, that Mr. Cotton told me, one stock had increased to *twenty-six* in one year. The natives call the bee the white man's fly."

Mrs. Allom, the lady before referred to, some time in 1842 (as I am informed by Mr. Allom), sent some colonies of bees to Nelson and Wellington ; those sent to Nelson were consigned to Captain Wakefield, the then head of that settlement, and reached their destination safely, while those forwarded to Wellington died before arrival. This lady's claim has never before, as far as I know or can ascertain, been recognised except by the Society for the Encouragement of Arts, Adelphi, London. That Society awarded her the Silver Isis Medal in 1845, for her "communication respecting her successful introduction of bees to New Zealand."

From the bees thus introduced in the years 1840 to 1842 have sprung the whole of the black stock of New Zealand.

INTRODUCTION OF ITALIAN BEES INTO NEW ZEALAND.

Previous to the year 1880 several unsuccessful attempts were made to introduce Ligurian bees into this colony. I believe the Honourable Thomas Russell, C.M.G., spent a large sum of money to secure this object, but in vain. The hot weather encountered in the passage from America to this country, and also the imperfect knowledge as to the best mode of packing bees to travel long distances, acted as almost insuperable barriers to their introduction. By these repeated failures, however, apiarists gained knowledge, and as a result, in September, 1880, two splendid colonies of Ligurians were landed in Auckland—one consigned to the Acclimatisation Society, Christchurch, the other to Mr. J. H. Harrison, Coromandel. Too much praise cannot be given to Captain Cargill, who took charge of the little creatures from the moment they were shipped and personally attended to all their wants on the passage across. These hives came from Los Angelos County, California, and were procured by Mr. R. J. Creighton, the

New Zealand Government representative, to whom much praise is due. This consignment, owing to the method of packing, having been so successful, Messrs. Hopkins and Clark, of the Parawai Apiary, took steps to procure some colonies, and two were received in due course from Ventura County, California. These, too, were received in splendid condition, thanks again to the care taken of them by Captain Cargill. Following upon this I obtained from America two other consignments, in all twenty nuclei and two full colonies. An event of considerable importance in the history of bee-keeping in New Zealand was the first successful importation of queens direct from Italy. After some correspondence with Mr. Fullwood, of Brisbane, I decided to give the matter a trial, and the result was that four out of eight queens shipped at Naples by Mr. Chas. Bianconcini on 10th of November, 1883, arrived in good condition at the Matamata Apiary on the 11th of January, 1884. Another shipment was made later in the same year, when six out of twelve queens arrived alive. Since the first importations numbers of Italian queens have been reared and distributed over the colony ; fresh importations have been made by other parties, and the greater number of New Zealand apiaries are now being Italianised. These bees flourish splendidly in this country, and will, I am quite sure, eventually replace with profit the German or common black bee. A full account of the Ligurian bee is given in another chapter.

IMPROVED SYSTEM OF BEE-KEEPING IN NEW ZEALAND.

Till within the last five or six years bee-keeping here was, with a few exceptions, in a very backward state. The hives in general use were composed of old gin cases, candle boxes, and in fact any wooden material in the shape of a case that was handy to the bee-keeper when his colonies happened to swarm. As a rule, no preparations were made for the swarming season, and it was not until the swarm was in the air that the need of a spare hive was realised. These boxes in some cases have been so neglected that they have actually fallen to pieces through age, and the bees left exposed to the weather. The sulphur pit has, I am sorry to say, not been unknown here, and it is in use even at the present day. In a German work on bees the following epitaph is given, which, as Langstroth remarks, might be properly placed over every pit

of brimstoned bees, as a brand of disgrace to those who practise
this horrid system :—

HERE RESTS,

CUT OFF FROM USEFUL LABOUR,

A COLONY OF

INDUSTRIOUS BEES,

BASELY MURDERED

BY ITS

UNGRATEFUL AND IGNORANT OWNER.

But this most barbarous and cruel practice is fast passing
away, through the efforts of more enlightened aud humane
bee-keepers.

Amongst those who have done good service in this direction
is Dr. Irving, of Cauterbury, who, soon after his arrival there
in 1879, took steps to put bee-culture in the South Island on
a proper footing. To do this, he placed a modern hive, containing a colony of bees, in the Public Gardens at Christchurch,
and occasionally delivered lectures, with experiments, to those
interested in bee-keeping. He has also written many interesting and valuable articles on bee-culture in the *Canterbury
Times.* He was elected first president of the Christchurch Bee-
keepers' Association, which he was mainly instrumental in
founding.

About the same time, with the object of giving information
to our bee-keeping settlers, I wrote a series of articles upon
bee-culture, which appeared in our local papers, and which
created such a large amount of interest and produced so many
requisitions to me to publish them in book form, that I was
induced, in the year 1881, to publish the first edition of this
Manual. The extent of the newly-awakened interest in the
improved system of culture was shown by the fact that a new
edition of the work was required within thirteen months ; and
that being now exhausted, I am led to lay before the public
the third edition in its present revised and greatly enlarged
form. In July, 1883, the *New Zealand and Australian Bee
Journal* was started by Mr. J. C. Firth, under my editorship,
I having in the meantime entered into arrangements with that
gentleman for the establishment aud working of extensive

apiaries on the Matamata estate. The Journal was, for two years, widely circulated and ably supported by the literary contributions of enthusiastic and successful apiarists in New Zealand and the Australian colonies. It was, for reasons given at the time, incorporated with the *New Zealand Farmer, Bee and Poultry Journal*, in June, 1885.

Attractive displays of honey and of apiarian appliances have been made at the last two annual shows of the Auckland Horticultural Society. The New Zealand Bee-keepers' Association was formally constituted on the 7th of August, 1884, and held its first annual meeting this year. It forms an admirable centre point for the combined action of all New Zealand bee-keepers in their endeavours to promote the general interests of the industry in the colony. The Auckland Provincial Bee-keepers' Association has been in operation since February, 1884, and its proceedings are likely to help effectively in advancing the new industry, especially in the Waikato district. Other local associations are about being formed, and it is hoped that the example will be followed wherever there is a sufficient number of apiarists living within such a distance of each other, or of their common centre, as may render their regular periodical meetings practicable. A great many persons in different parts of the country have already taken up bee-keeping with the intention of making it their sole or principal occupation; many others have commenced to practise the improved system of culture on a small scale, for their own gratification and the supply of honey for their own households. The numbers of both these descriptions of apiarists are increasing every day. The production of honey in the Auckland province alone is calculated to have exceeded eighty tons last season. As a further proof of the progress of the industry, we may take the number of hives and other implements sent out by that well-known firm of hive-makers, Messrs. Bagnall Bros. & Co., of Turua, Auckland, since 1879. In the year mentioned, I arranged with them to cut my hives, etc., at their saw-mill, and in 1882 they took over my supply business. Since then the firm has sent hives and all other apiarian implements to every part of Australasia, and they are fairly entitled to be called the premier hive-makers of these colonies. In response to some inquiries I made concerning the number of hives, etc., they had supplied since first commencing the business, Messrs. Bagnall Bros. & Co.

kindly sent me the following :—" The following figures are as near correct as possible of the number of hives, extractors, and smokers we have supplied : Hives, 4,500 ; extractors (single and double), 300 ; smokers, 750. These are the principal items : sections are a very large item ; last season we sent out about 100,000." Of comb-foundation, since I first commenced making it, I have supplied nearly eight tons. I think we shall not be very far out if we allow a like number of hives and half as much comb-foundation as being home-made and supplied from other sources. Presuming this to be correct, we have, then, about 9,000 hives and twelve tons of comb-foundation distributed through Australasia—not at all a bad showing for so young an industry. On the whole, there is the gratifying prospect that New Zealand and Australia, before many years have elapsed, shall have taken an important station among honey-producing countries.

INTRODUCTION OF BEES AND BEE-CULTURE INTO AUSTRÁLIA AND TASMANIA.

The black or German bee was introduced into New South Wales in 1822. The following extract is from Haydn's Dictionary of Dates, for which I am indebted to the kindness of a correspondent in Sydney :—

" Bees were first imported by Captain Wallis, in the ship *Isabella*, into Sydney, in April, 1822, and from these original hives the stocks were propagated into the interior by the colonists."

Mr. Thos. Lloyd Hood, of Hobart, has very kindly furnished me with the following information concerning the introduction of bees and state of apiculture in Tasmania. He says :—

" Bees were first introduced into Tasmania by Dr. Wilson, R.N., in the ship *Catherine Stewart Forbes*, in the year 1831. Great interest was taken in their arrival, and there was a general expression of gratitude to Dr. Wilson for the disinterested benefit he had conferred on the colony at considerable trouble and cost to himself.

" Bee-keeping here is carried on on the most primitive principles, frame hives and other appliances are only known by repute. Bees are generally kept in some handy sized boxes (gin cases, etc.), and at the end of summer these boxes are lifted, and the heavy ones mercilessly put over the ' sulphur pit ;' or by the more merciful bee-keepers, the bees are driven into another box, and so on from year to year. Very great interest is now being taken in the improved system and modern appliances since I introduced them last year, and I hear of many who intend taking up bee-culture as a commercial industry."

I have not been able to obtain any information as to the introduction of the German bee into South Australia, Victoria, or Queensland. Probably the importation may have been made from New South Wales or Tasmania, and not direct from Europe. To Mr. C. Fullwood, of Brisbane, I am indebted for much information as to the progress of bee-culture in Queensland. The following extract from a communication of his in the first number of the *New Zealand and Australian Bee Journal* gives a graphic description of bee-culture under the old regime :—

"Some years ago large quantities of bees were kept by farmers and others in a very primitive fashion, and the bush resounded with the hum of the 'busy bee.' Timber getters, wood carters, and aborigines frequently secured large quantities of honey from hollow trees; both the black bee and stingless bee, peculiar to Australia, were found almost everywhere. Gin cases, tea, or any kind of rough boxes were appropriated to bee use, and such is the climate, and the yield of honey so regular, that bees appeared to thrive everywhere, and in any kind of hive, so long as they had a cover under which to build their comb and rear their brood. No skill was demanded in their management. Given a swarm—put it in a box, on a stand, under a 'sheet of bark; then look out for swarms in a few weeks; and, after a while, turn up the box, cut out some honey, or drive the bees into another box to go through the process of building and storing, to be again despoiled in like manner.

"No thought about the destruction of brood, waste of honey and wax; no care about the queens. Would not know a queen from a drone, or their value in the hive. What matter if a few boxes (stock) perish? Such was the natural increase by swarming that a few losses were of no consequence.

"Anybody could keep bees who had courage enough to rob them. The aborigines knew how to do it. With a tomahawk and fire-stick they would attack the 'white-fellow sugar bag,' and driving the bees with smoke, deprived them of their honey. 'Pettigrew's old Irishman' was not required here to teach the Australian aborigines how to rob the bees by means of smoke.

"A few years ago, however, a great change came over the land. A moth, unknown previously, commenced its ravages. The bees succumbed before it, and were rapidly swept away. Farmers owning from fifty to two hundred stocks lost all. The bees in the bush gave way also before the terrible onslaught, leaving the invader all but master of the field. Only a very few individuals, by dint of determined persevering watchfulness and care, managed to save a few stocks amid the general devastation.

"Bee-keeping naturally came to be viewed as a very precarious, risky, and unprofitable business; and, although it has its charms for many, there are but two or three persons in the colony who have any number of stocks, or who attempt bee-keeping as a means of obtaining an income."

ITALIAN BEES IN QUEENSLAND.

In order to remedy this state of things Mr. Fullwood very properly determined to introduce Italian bees, which are known to defend themselves more effectually than the German bee against the inroads of moths, ants, and other enemies. In the year 1880 he brought five queens with himself from Liverpool to Melbourne, and thence to Brisbane. In 1882 he got twelve queens sent direct from Charles Bianconcini of Bologna, and of these five arrived alive; and again in 1883 he got a second consignment of twelve, of which seven arrived safely. These spirited efforts appear to have been crowned with the success they deserve. The Italian bees seem to be quite able for the moths, and honey in abundance can be gathered by them in Queensland. Owing to Mr. Fullwood's enterprise and example, a number of people are now turning their attention to bee-keeping, and I have no doubt that in a comparatively short time, Queensland will show to the front as a honey-producing country.

ITALIAN BEES IN NEW SOUTH WALES.

It is stated by Dr. Gerstaecker, that four stocks of Ligurian bees were shipped in England by Mr. I. W. Woodbury, in September, 1862, and that they arrived safely in Australia, after a passage of seventy-nine days. It does not appear, however, that these stocks succeeded and propagated their race, any more than a colony which Mr. Angus Mackay, the present editor of the *Town and Country Journal* in Sydney, subsequently brought with him to Brisbane, at great expense, from America. Mr. S. McDonnell, of Sydney, imported two colonies from America in 1880, and succeeded in raising stock from them; and later Mr. Abrams, a German bee-master, brought some colonies with him from Italy in 1883, settled in Paramatta, and having succeeded in rearing a pure race from his queens, started an apiary for the Italian Bee-Farming Company, of which he is the manager, and Mr. McDonnell secretary.

ITALIAN BEES IN VICTORIA.

In Victoria, we are told that the late Mr. Edward Wilson had a stock of Ligurians sent out to him in 1862, by Messrs. Neighbour and Sons; but I am informed that no successful attempt had been made to establish the race there until quite recently, when, in the latter part of 1884, Mr. Herman Naveau, of Hamilton, obtained some of those bees from Queensland, and has had great success with them.

ITALIAN BEES IN SOUTH AUSTRALIA.

In South Australia, as Mr. Bonney informs me, the Chamber of Manufactures imported a colony of Italian bees from Mr. Fullwood, of Brisbane, in December, 1883, and succeeded in establishing them on Kangaroo Island, where they are doing remarkably well. Mr. Bonney himself has since successfully imported queens direct from Italy, a parcel of twelve from Bologna, to his order, having arrived safely in September, 1884, at Adelaide. He states that " around Adelaide, bee-keeping is now all the rage, very many persons taking it up as an amusement, while a few are making it a means of livelihood." Much credit is due to this gentleman for the trouble he has taken to place apiculture on a proper footing in South Australia.

ITALIAN BEES IN TASMANIA.

To Mr. Thos. Lloyd Hood, of Hobart, the gentleman already referred to, belongs the credit of being the first person to introduce Italian bees into Tasmania. They arrived at Hobart from New South Wales in the s.s. *Flora*, Captain Bennison, on the 4th October, 1884. Mr. Hood, writing in May, 1885, informs me that he has had great success with them. " Though kept in the city they increased the first season to five strong colonies and two rather weak ones. Most of the young queens are hybrids."

SUITABILITY OF NEW ZEALAND AND AUSTRALIA FOR APICULTURE.

Any person who had a practical knowledge of apiculture, who had witnessed the results obtained from its improved scientific practice in Europe and America, and who afterwards visited the New Zealand and Australian colonies, could not fail to be struck with the advantages they offer for the prosecution of the honey industry. First, as regards

CLIMATE.

The influence of climate upon the operations of the bee-keeper is of a two-fold nature : first, as it affects the bee itself, especially the condition of the insect during the winter season ; and, secondly, as it is favourable or otherwise to that class of vegetation which affords forage for the bee and a flow of nectar in the honey season. Looking to the old country, we shall find that all those portions of Southern Europe, Asia Minor, and Phœnicia which constitute the ancient home of the honey-bee lie between the isothermal lines of 41° and 59° mean winter temperature, the medium line of 50° passing through or close to all the localities most celebrated both in ancient and modern times for the quality of its honey. The same places lie between the summer isothermals of 68° and 77°. California in North America and Chili in South America, both rich honey-producing countries, have a mean winter temperature of 50° to 51° and a summer temperature of 67° to 68°. The colony of Victoria in Australia and the province of Auckland in New Zealand have exactly the same mean temperature as these last mentioned countries, both in winter and summer ; and the *whole* of the New Zealand islands, as well as nearly all the Australian colonies south of Queensland, lie between the lines of 41° and 59° mean winter temperature, exactly as in the case of the most favoured honey countries in the northern hemisphere. Queensland and some northern portions of New South Wales and South Australia have a winter temperature several degrees warmer than, and a tropical summer nearly equal to, that of Egypt and Syria. The rainfall in most of these colonies is amply sufficient for a luxuriant vegetation. In the most southern parts of New Zealand and Tasmania bees can fly about and even gather some honey and pollen all through the winter ; and in some of the Australian colonies they can even gather *surplus* honey all the year round. When we remember the trouble, risk, and expense that has to be incurred in wintering bees in many parts of Europe and America, where they have to be confined in cellars for three or even for five or six months of the year in a state of semi or complete torpor, we can appreciate the advantages for the purposes of bee-keeping of a climate such as we enjoy in these colonies.

NATIVE FLORA.

This is a matter to be dealt with more fully in the chapter on bee forage; it is only necessary to mention here, in elucidation of this part of our subject, that the indigenous trees are nearly all honey bearers. This is abundantly proved by the amount of honey sometimes taken from colonies wild in the bush. It is quite a frequent occurrence to take from 100 to 200 lbs. and very often more from these hives. This honey could not, in many cases, have been gathered from any other source than the bush, for as the colonies are sometimes found eight or ten miles from any cultivation, and as the bee does not usually exceed from one mile and a half to two miles' radius in its flight, it follows that honey obtained from the hives mentioned must have come exclusively from the indigenous flora. If further proof be required, may we not find it in the fact that the bee has so quickly and universally spread over New Zealand and other parts of Australasia from a few colonies?

The apiarist has not only the benefit of a splendid native flora, but the climate being so well adapted for the growth of all honey-producing plants of the old world, he is especially favoured in this country.

In Australia the native acacias and eucalypti are especially valuable for bee forage, varying as they do in their times of blossoming, so that some of them are available at almost any season of the year. These trees also grow rapidly and thrive well when introduced into New Zealand.

IMPORTANCE OF APICULTURE AS AN INDUSTRY.

The degree to which the production of honey may be developed in a comparatively short time will be best illustrated by the case of the United States of America. Professor Cook, in the last edition of his *Manual of the Apiary*, gives the following picture of the present state of the industry:—" An excellent authority places the number of colonies of bees in the United States in 1881 at 3,000,000, and the honey production of the year at more than 200,000,000lbs; the production for that year was not up to the average, and yet the cash value of the year's honey crop exceeds thirty millions of dollars." G. W. Mead and Co., of San Francisco, in their annual review issued a few months ago, state that the total output of California honey, comb and extracted, for 1884, aggregated nearly the enormous total of 9,000,000 lbs. It has no parallel in any part of the world. These appear to be enormous results, and yet the apiarists of America still speak and act as men convinced that their industry is scarcely out of its infancy as yet, and who see no prospect of a sudden or early check to the progressive increase either of production or consumption.

The use of manufactured sugar has now for many generations almost entirely supplanted that of honey, which could not be, under the old system, produced in sufficient quantities or at a sufficiently low cost to compete with the new sweet. But sugar, although so nearly the same in chemical constitution, *is not honey*, and never can take the place of that delicious product as an agreeable and wholesome addition to the food of man. Modern apiculture, which renders possible an enormously increased production of honey at a greatly reduced cost, cannot fail to lead to its general use again, not of course to the exclusion of sugar, but upon a scale which would have been quite impossible in former times.

PROFITS OF BEE-KEEPING.

This is a question with reference to which it is necessary to guard against false or exaggerated views. It must be recollected that all industries require the combination in certain proportions of three elements—capital, labour, and skill. Some afford a ready and safe investment for the first ; others require an immense quantity of the second ; and others again are chiefly dependent upon the exercise of the third. The honey industry especially may be reckoned of the latter sort. An apiary cannot, it is true, be established without a certain expenditure of capital, nor worked without some labour ; but both these factors are small as compared with the value of the personal care and attention of the skilled apiarist, upon which the question of profit *or loss* mainly depends, and the profits of a successful apiary are rarely indeed more than sufficient to fairly remunerate the time and skill so applied. Bee-keeping is therefore not to be looked upon as a *profitable investment for large capital*, or as a large employer of labour, but as a fair field, and certainly a fairly remunerative one, for the industry, skill and perseverance of him who lays himself out to be not only a bee-keeper but a *bee master.*

The question is continually asked—" What is the average yearly production of honey, and what the average profit from each hive ?"   The answer must be, the former depends practically upon the skill of the apiarist (within certain limits of course), and the latter mainly upon his commercial intelligence.   It is easy to show what results are attained in some cases, but it is dangerous to apply these results as a measure of success or failure in our own case.   Such results as 300, 400, or even 500 lbs. of extracted honey from one hive in a good season are not unknown nor even very rare.   An average of 200 lbs. per hive may be often attained under favourable circumstances and good management ; but 100 lbs. of extracted, or 60 lbs. of comb honey per hive may be nearer to the mark of what a prudent apiarist will look forward to obtaining, and any one who can show such results as the *average of a number of successive seasons,* may fairly count himself a successful bee-keeper, and his location a favoured one. It must, however, be understood that it is a rule, with perhaps no exceptions, that the larger the apiary the lower the average production per hive ; so that supposing 100 lbs. of extracted honey to be a fair average through an apiary of 100 hives I would consider 75lbs. a good one per hive for 250 hives under the same conditions.

ADAPTATION TO WOMEN.

There is a feature in this industry which, I think, especially recommends it to notice, viz., its adaptation to women.   In both England and America, at the present time, some of the most successful apiarists are ladies, and several of the most extensive bee-keepers in America are assisted by their wives and daughters.   Professor Cook states that Mrs. L. B. Baker, of Landsing, Michigan, who has kept bees very successfully for four years, read an admirable paper before the Michigan convention of bee-keepers, in which she said :—

" But I can say, having tried both (referring to boarding-house-keeping and bee culture), I give bee-keeping the preference, as more profitable, healthful, independent, and enjoyable.   I find the labours of the apiary more endurable than working over a stove, and more pleasant and conducive to health. I believe that many of our delicate and invalid ladies would find renewed vigour in body and mind in the labours and recreation of the apiary. My own experience of the apiary is that it is a source of interest and enjoyment far exceeding my anticipations."

Although apiculture offers as good an opening to people of either sex as can be found amongst ordinary industries, I do not mean to say that it is a "royal road" to wealth, or that it is suitable to every person who thinks proper to engage in it; but we have ample proof that it has been the means of many people of both sexes regaining their health and strength, and so enabling them to earn a respectable livelihood when they were almost incapable of undertaking any other employment. One notable instance in this respect I can quote in the person of Mrs. L. Harrison, of Illinois, now one of the most successful lady apiarists and writers on bee matters known. This lady was at one time told by her physician that she could not live; but, as she herself states, "apiculture did for her what the physicians could not do—restored her to health, and gave her such vigour that she has been able to work a large apiary for years."

ADVICE TO BEGINNERS.

Let me impress upon the minds of those about to embark in the culture of bees the fact that success in this industry, as in all others, can only be obtained by tact, patience, and perseverance. As the Rev. L. L. Langstroth says :—"There is no royal road to profitable bee-keeping; and while large profits can be realised by careful and experienced bee-keepers, those who are otherwise will be almost sure to find their outlay result only in vexatious losses. An apiary neglected or mismanaged is worse than a farm overgrown by weeds or exhausted by ignorant tillage ; for the land, by prudent management, may again be made fertile, but the bees when once destroyed are a total loss."

It would be injudicious for an inexperienced person to start with a large number of colonies, not more than four or five, for under modern management these could be increased very rapidly after he had acquired skill and experience. I would recommend beginners to procure good stocks or early swarms to start with, from some reliable person in his immediate neighbourhood if possible. For the sake of economy get black bees, and after gaining some little experience procure either an Italian queen or a nucleus colony of Italian bees, and Italianise the black stocks, according to directions given in another chapter. By adopting this plan much expense and risk will be saved at the commencement. The beginner should of course adopt all the latest improvements in bee-culture, and, if possible, visit an apiary where the modern system of management has been introduced. Let me also impress on him the necessity and advantage of only having one kind and size of hive throughout the apiary. The reason of this is so obvious that it needs no further comment. Let him also remember that "practice makes perfect;" that no matter how fully any book may enter on a given subject, yet without experience the reader, practically, will be like a ship without a rudder, for it is only by practical experience that we gain lasting knowledge and success. He should become a member of the nearest bee-keepers' association and a subscriber to a local bee journal, which will keep him posted up in everything relating to the progress of the industry in these colonies, and serve as a means of communication between him and his fellow bee-keepers. He should follow the simple instructions given in this Manual, and avoid trying new experiments until he feels that he is master of the rudiments of the art; he may *afterwards* with advantage study *all* that has been written on the subject of apiculture, and form his own judgment on points (not a few) where he finds that "the doctors differ."

STATE AID TO APICULTURE.

Germany and other continental states have long felt it to be one of the duties of a paternal government to promote the diffusion of a knowledge of the principles of bee-culture by means of suitable publications and by placing at the disposal of agricultural societies an annual contribution in aid of their objects. The United Sates of America have, at their Agricultural Colleges, professors of entomology and of bee-culture, who give both theoretical and practical instruction to the students; and in England a movement has been for some time in progress, which has now received the sanction of the Education Department, to place among the "extra subjects," the optional study of which is provided for by the Education Act, the branch of " practical scientific bee-keeping, the natural history of the honey-bee, and the fertilization of flowers by insects generally." This is a step which might well be followed by our own Education Boards. It is to be hoped that our Colonial Governments will show themselves alive to the importance of encouraging the study of scientific bee-culture. The industry itself is not one which calls for state aid in the shape of subsidies or of protective duties, and it is therefore all the more deserving of all the indirect assistance which an enlightened Government may find opportunities of extending to it.

BEE PUBLICATIONS.

The following are some of the best works on modern apiculture :—

"LANGSTROTH ON THE HIVE AND HONEY BEE."—This work is considered by some to be the best bee book in the English language. I can, from my own knowledge, bear testimony to its excellence and usefulness.

"COOK'S MANUAL OF THE APIARY."—This is unquestionably a first-class work, and is fully up with the times as far as regards apiculture and the physiology of the honey-bee.

"THE A B C. OF BEE-CULTURE."—This work is in the form of an Encyclopædia, containing all the latest information relative to this matter, and is both plain and practical. By A. J. Root.

"QUINBY'S NEW BEE-KEEPING."—This is a very useful work, written by one of America's most practical bee-keepers.

"KING'S NEW BEE-KEEPERS' TEXT BOOK."—A capital work, kept well up with the times. Has had an enormous sale.

"BEES AND HONEY," by T. G. Newman.—This is a sprightly little book, well illustrated, and contains a large amount of practical information.

"ALLEY'S HAND-BOOK."—'New method of queen-rearing,' by one of the most experienced queen-breeders living.

"BLESSED BEES" and "PHIN'S DICTIONARY OF PRACTICAL APICULTURE" are well worthy of a place in every bee-keeper's library.

The foregoing are all American publications; the following are some by English authors :—

"BEVAN ON THE HONEY-BEE."—This book is especially devoted to the natural history, anatomy, and physiology of the honey-bee, and would be very valuable to the student.

'HUNTER'S MANUAL."—A good work ; specially written for British bee-keepers.

"THE APIARY, OR BEES, BEE HIVES, AND BEE CULTURE," by Alfred Neighbour.—This is, I believe, the best work now published in England—quite up with the times.

"COWAN'S BRITISH BEE-KEEPERS' GUIDE BOOK."—Written by a thoroughly practical beekeeper. Cheap, and first-class.

PERIODICALS (*Weeklies*). — "AMERICAN BEE JOURNAL," "KANSAS BEEKEEPER," and "CANADIAN BEE JOURNAL." (*Bi-Monthlies*)—"BRITISH BEE JOURNAL," and "GLEANINGS IN BEE CULTURE." (*Monthlies*)—"AMERICAN BEE JOURNAL," "BEEKEEPERS' MAGAZINE," "AMERICAN APICULTURIST," THE BEE-KEEPER'S GUIDE," and "NEW ZEALAND FARMER, BEE AND POULTRY JOURNAL."

AXIOM.

"BEES GORGED WITH HONEY NEVER VOLUNTEER AN ATTACK."

Langstroth.

CHAPTER II.

THE HONEY BEE: ITS VARIETIES AND DISTRIBUTION.

THERE are many species of the genus Apis, or Bee, but only one which stores honey in such a manner as to be practically useful to man, and which Linnæus distinguished by the name

APIS MELLIFICA.

The particular variety of this species known to Linnæus was the Black, or German bee. Since the beginning of the present century, other varieties were observed and described by Spinola and others, and were classed at first as distinct species. In the year 1862, Dr. A. Gerstaecker, of Berlin, first published the results of his investigations upon the

"GEOGRAPHICAL DISTRIBUTION OF THE HONEY-BEE AND ITS VARIETIES,"

from which I take the following condensed extracts. He says that up to within some ten years of the time when he was writing, bee-keepers knew only one sort of honey-bee—that which had been reared for ages—the *Apis mellifica* of Linnæus; but they then (in 1862) distinguished the German from the Italian bee. The latter had, in fact, been noticed in the beginning of this century, by Spinola and by Latreille, as a separate species of the genus Apis, and was named by the former zoologist, *Apis ligustica;* nevertheless it proved to be only a coloured variety of the same species; the size, as well as the structural peculiarities of the insect, being the same in every respect, and the two sorts admitting of 'cross-breeding to any extent whereas, if they belonged to different species, the off-spring would, in all probability, consist of unprolific hybrids. The knowledge of the practical apiarist was, at all events, then confined to these two varieties of the honey-bee, and they were supposed to be indigenous almost exclusively to Europe, the northern coast of Africa being their supposed boundary on the south, and the coast of Asia Minor on the east. When Dr. Gerstaecker, however, undertook his investigations, he obtained samples of a large number of varieties mentioned in the works of Fabricius, Latreille, Lepeletier, and others, as being found in various parts of Africa and of Asia, north of the Himalayas, and subjected them to a minute examination, comparatively, with each other, and with the two varieties already known in Europe. He soon satisfied himself beyond all doubt that they were all merely varieties of the one species, the *Apis mellifica*, differing only in colour and size—all capable of being cross-bred, and of being utilised by the apiarist. He also found that this one species, represented by many different varieties, was spread over a vastly larger area than had been supposed, comprising nearly the whole of Europe (up to 60°, or even 64° north latitude in some places), the whole continent of Africa, and the whole of Asia Minor, Syria, Persia, and other portions of Asia north of the Himalayan range, up to eastern Siberia and China.

Out of the numerous varieties brought under review, six have been selected as being of sufficient importance to be separately dealt with. These, with their distinguishing marks and the regions to which they seem particularly to belong, are classified and described as follows :—

1. The single coloured, dark, northern, or German bee (*Apis mellifica* of Linnæus), found in the whole of north and middle Europe, and also in the south of France, south of Spain, Portugal, a few parts of Italy, in Dalmatia, Greece, at the Crimea, and along the coast of Asia Minor, including the adjacent islands. It is also found on the African continent, at Algiers, Guinea, and at the Cape of Good Hope, to which latter place, however, it was probably introduced direct from Europe. As a very slight sub-variety of the same may be noted the Hymettus bee (*Apis cecropia*), differing only in being slightly smaller and more hairy, often also showing reddish spots on the sides of the second abdominal ring. This bee is found in the south of Spain, as well as in Greece, and even, in some isolated cases, in Germany.

2. The Italian bee (*Apis ligustica* of Spinola), of equal size with the German bee, but with golden yellow colour on the first three abdominal rings, whilst the back plate (of the

thorax) is of a dark colour. It was first noticed by Spinola as being peculiar to all parts of Liguria. Its first or original habitat was difficult to be ascertained in 1862, as during the previous ten years it had been artificially distributed to many new places. Although to be found in various parts of Italy, it is by no means general in that country. Besides the province of Liguria, the southern slopes of the Tyrolese and Swiss Alps would appear to have been its original home.

3. The Italian bee, with yellow back plate—otherwise of the same size and colour as the last. It is found in southern France, Dalmatia, Banat, at Sicily, and in the Crimea, in the islands and on the coast of Asia Minor, and in the Caucasus, and in many of those places in common, partly with the Italian (No. 2), and partly with the German bee.

4. The Egyptian bee (*Apis fasciaca* of Latreille). It is nearly one-third smaller than the German or Italian bee, its body coloured like the latter, and the back plate also yellow ; the hair of the chest and body whitish. Its proper habitat is Egypt, Arabia, and Syria, but it is found, with scarcely any observable difference, on the northern slopes of the Himalayas and in China. It was introduced into Germany in 1863, by the Acclimatisation Society of Berlin, and thence into England in 1865.

5. The specific African bee (*Apis Adansonii* of Latreille) is of the same size and colour as the last, but differs in the greyish-yellow colour of the hair on the chest and body. It is spread over the whole African continent, with the exception of Algiers and Egypt, from Abyssinia and Senegambia to the Cape of Good Hope.

6. The remarkable black Madagascar bee (*Apis unicolor* of Latreille) is something smaller than the German bee, all dark coloured, and its hairs black. It is confined to Madagascar and the Mauritius.

With reference to the countries of the New World, North and South America, and Australasia, Dr. Gerstaecker asserts that in none of them were any species of the genus Apis found until they had been imported from Europe. He gives the dates of importation into Florida, North America, as 1763 ; thence to Kentucky in 1780, and to New York in 1793 ; into Brazil, South America, in 1845, Rio Grande in 1853, and to Buenos Ayres (from Chili) in 1852. Into Mexico and central America generally, the bee appears to have been introduced at an early period by the Spaniards, and probably spread itself thence to the districts of Venezuela, Peru, and Chili, in South America. Its introduction into Australia and New Zealand has been noticed in the preceding chapter. With regard to its first importation into North America, Dr. J. P. H. Brown, an eminent American apiarist, in a paper read by him at the National Convention in 1881, says, "The Black or German bee was introduced, it is believed, into Pennsylvania from Germany, about the year 1627." It certainly appears very probable that William Penn's followers would have endeavoured to introduce bees from England, if not from Germany, as soon as they began to settle down in their new home ; nevertheless it is very likely that in the severe winter climate of Pennsylvania and New York the bees would not spontaneously wander far from the human settlements, and that it was only when they got fairly established in the favourable climate of Florida, as mentioned by Dr. Gerstaecker, that they began to spread themselves westwards in advance of the new settlements.

GERMAN, OR BLACK BEE.

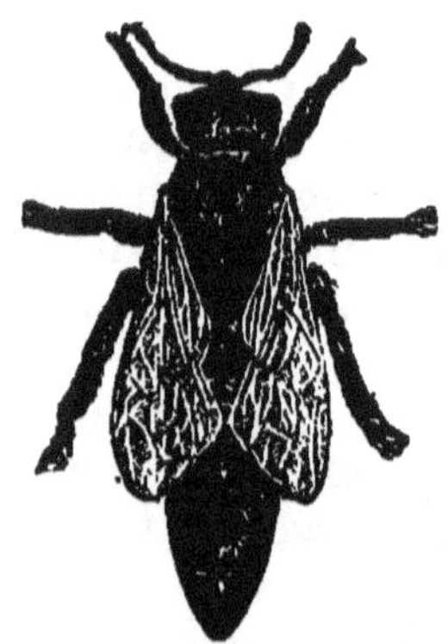

Fig. 1.—BLACK QUEEN.

Neither of the names, German or Black, is a correct designation of this variety ; for, as Dr. Gerstaecker has shown, it was by no means confined originally to Germany, and its prevailing colour is more brown than black ; but these are the names by which it is now universally known.

This variety has held undisputed sway in the north and west of Europe for a couple of thousand years at least, and has been the pioneer in culling the sweets of all the countries of the New World. Wherever Europeans have colonised, there may be found this little insect. It is now being rapidly superseded by the Italian race, but it has still some faithful admirers, and in more than one respect it is admitted by all to hold the advantage over its Italian rival. I shall compare its qualities with those of the Italian further on. However we may admit the superior beauty, as well as the more useful qualities, of the new races, we cannot avoid feeling a sort of regret for the extinction of our old favourites.

ITALIAN, OR LIGURIAN BEE.

Fig. 2.—ITALIAN QUEEN.

The Italian bee was evidently known to Aristotle and Virgil. The latter writer refers to it in the following lines :—

" These gaily bright their radiant scales unfold,
Spangled with equal spots, and dropped with gold "

Although known so well to these ancient writers, very little notice appears to have been taken of this variety till quite modern times, when, in the beginning of the present century, the Marquis de Spinola described it as being distinct from the common bee, and gave it the name of " Ligurian," after a province in Northern Italy, where it was first discovered. This district being very mountainous, and the Alps intervening between it and Northern Europe, it is in a manner isolated, which will no doubt account in some measure for so little having been known of this bee, and, as some think, has helped to develop a distinct variety.

It was introduced into Germany in 1853, by Dr. Dzierzon, who, on the twenty-fifth anniversary of its introduction, published in a German bee paper a very interesting account of it, speaking in very high terms of its superior quality.

It was introduced into England in 1859, by Mr. Neighbour, and into America by Messrs. Wagner and Colvin the same year. Wherever it has been introduced, its superiority over the common bee in many respects has been always acknowledged. In America it has become so universal, that it is said to be impossible to find a pure black bee in some of the States ; and I have no doubt, now it has become established in Australasia, that, notwithstanding the fears of a few individuals, it will be found to deserve all that has been said regarding its superiority over the black bee.

With regard to the many excellent qualities it is said to possess, I will first quote Cook ; he being a professor of entomology, as well as an apiarist, his opinion should be doubly valuable. He says :—

"The Italians certainly possess the following points of superiority:—

" 1. They possess longer tongues, and so can gather from flowers that are useless to the black bee. How much value hangs upon this structural peculiarity I am unable to state. I have frequently seen Italians working on red clover. I never saw a black bee thus employed. It is easy to see that this might be at certain times and seasons a very material aid. How much of the superior storing qualities of the Italians is due to this lengthened ligula, I am unable to say.

"2. They are more active, and with the same opportunities will collect a good deal more honey. This is a matter of observation, which I have tested over and over again ; yet I will give the figures of another. Mr. Doolittle secured from two colonies 309 lb. and 301 lb. respectively of *box honey* during one season. These surprising figures, the best he could give, were from his best Italian stocks. Similar testimony comes from Klien and Dzierzon, over the sea, and from most of our apiarists.

"3. They work earlier and later. This is not only true of the day, but of the season. On cool days in spring I have seen the dandelions swarming with Italians, while not a black bee was to be seen. On May 7th, 1877, I walked less than half a mile, and counted sixty-eight bees gathering from dandelions, yet only two were black bees. This might be considered an undesirable feature, as tending to spring dwindling ; yet, with proper management, I consider this no objection, but a great advantage

"4. They are far better to protect their hives against robbers. Robbers who attempt to plunder Italians of their hard-earned stores soon find that they have dared to 'beard the lion in his den.' This is so patent, that even the advocates of black bees are ready to concede it.

"5. They are almost proof against the ravages of the bee moth's larva. This is almost universally conceded.

"6. The queens are decidedly more prolific. This is probably in part due to the greater and more constant activity of the workers. This is observable at all seasons, but more especially when building up in the spring. No one who will take the pains to note the increase of brood will long remain in doubt on this point.

"7. They are less apt to breed in winter, when it is desirable to have the bees very quiet. This refers to cold climates.

"8. The queen is more readily found, which is a great advantage. In the various manipulations of the apiary, it is frequently found desirable to find the queen. In full colonies, I would rather find three Italian queens than one black one. Where time is money this becomes a matter of much importance.

"9. The bees are more disposed to adhere to the comb while being handled, which some might regard as a doubtful compliment, though I consider it a desirable quality.

"10. They are, in my judgment, less liable to rob other bees. They will find honey when the blacks gather none, and the time for robbing is when there is no gathering. This may explain the above peculiarity.

"11. And, in my estimation, a sufficient ground for preference, did it stand alone, the Italian bees are *far more amiable*. Years ago I got rid of my black bees because they were so cross. Two years ago I got two or three colonies, that my students might see the difference, but to my regret; for, as we removed the honey in the autumn, they seemed perfectly furious, like demons seeking whom they might devour, and this, too, despite the smoker, while the far more numerous Italians were safely handled without smoke. The experiment at least satisfied a large class of students as to superiority. Mr. Quinby speaks in his book of their being cross, and Captain Hetherington tells me that if not much handled they are more cross than the blacks. From my own experience I cannot understand this. Hybrids are even more cross than the pure blacks, but otherwise are nearly as desirable as the pure Italians. I have kept these two races side by side for years, I have studied them most carefully, and I feel sure that none of the above eleven points of excellence is too strongly stated."

Having now had over four years' experience of Italian bees, I can fully endorse nearly all that Professor Cook says of them, though I am not convinced as regards his third point. I have often seen black bees out in the morning, when not an Italian was stirring; at other times they were about equal in this respect. Neither can I admit that black bees never work on red clover, as I have frequently myself of late seen them do so. And although it is quite true, and a decided advantage, that Italians and hybrids defend their hives better than black bees against the attacks of robbers, I cannot acquit them of a propensity to act as robbers themselves.   I have found hybrids, at least, as bad as the black bees in this respect.

MARKINGS OF PURE ITALIA N BEES.

In describing the markings of pure Italian bees, all writers agree that the workers should have three yellow bands on the abdomen.

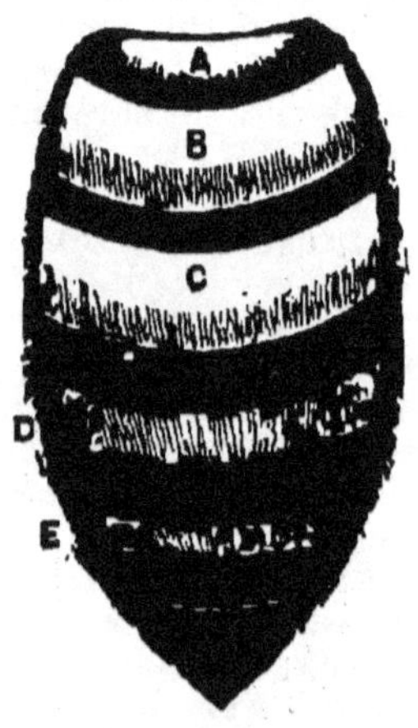

Fig. 3.—ABDOMEN OF ITALIAN WORKER BEE.

A B C, Fig. 3, represent the three yellow semi-transparent bands ; D E, and the shaded parts of A B C, are rows of greyish hairs. A strain of Italian bees having these rows of hair unusually developed have been sold in America under the name of Albinos. Now, with some of the worker bees bred from queens which I have imported direct from Italy, although they have the three bands, the first next to the thorax (A, Fig. 3) is so narrow that it cannot be seen unless closely inspected, while others show all three bands plainly. In a personal conversation with Mr. Fullwood, of Brisbane, on this subject, he told me he had frequently noticed the same difference in the markings of his imported bees. I mention this the more particularly, because some persons, after reading the description of pure Italian bees in other works, where they are stated to be of a bright colour, with the three yellow bands plainly visible, are apt to think their bees are not pure unless they answer that description. The bees from the queens I imported from America were, as a rule, much lighter and handsomer than those that came direct from Italy, and I account for this by their having been bred for lightness of colour rather than for honey-gathering qualities, though I have no doubt the two objects may be attained in the same bee. It is worthy of note that some of the most experienced bee keepers of America prefer a strain of dark leather-coloured bees to the lighter ones. My own opinion is that the test of purity is uniformity of the markings on the whole of the worker-bees of a colony, whether the three bands be plainly visible or not. The lightest coloured and most handsome variety of the Italian bee is to be found in the Swiss-Alpine districts, from which place the most of the English importers now obtain their supplies.

HYBRIDS—GERMAN-ITALIAN.

Much has been said for and against the cultivation of hybrids. My own experience leads me to believe that, as honey-gatherers and for hardiness, they are far superior to either the German or Italian race pure, but as regards docility they would be nowhere in a comparison. The first cross between an Italian queen and a black drone produces, I believe, the best workers. Should, however, any person prefer a greater degree of gentleness in his bees to a larger production of honey, I would advise him not to keep hybrids longer than he can possibly help.

In briefly stating what I consider to be the superior qualities of each sort, Germans, Italians, and hybrids, as compared with each other, I shall first take the Germans, or black bees. Without a doubt, for raising comb-honey they beat both Italians and hybrids. First, they will take to the section boxes sooner than the others; second, they leave a slight air space between the honey and the capping of the cells, which preserves the brightness of the cappings and gives to comb-honey that nice white appearance which is so much admired. On the contrary, the Italians, and in a less degree the hybrids, allow little or no air space, consequently the comb has a dark, damp look, on account of the proximity of the honey to the cappings. Italians are superior to the Germans—

first, in being better honey-gatherers ; second, in possessing longer tongues ; third, in being more prolific ; and fourth, in being more gentle, though, if once aroused, I believe them to be as vicious as hybrids. Hybrids I have found best of all for honey-gathering and for hardiness. As to prolificness, I think they are about equal to Italians. To sum up, I would place the three sorts in the following order for the different qualities required :—As honey-gatherers—Hybrids, Italians, Germans ; for gentleness, Italians, Germans, Hybrids ; for prolificness, Italians and Hybrids equal, Germans; for hardiness, Hybrids, Italians and Germans I have seen little difference between ; for protecting their hives against robbers, Italians, Hybrids, Germans ; for comb-honey raising, Germans, Hybrids, Italians.

CYPRIAN, SYRIAN, AND PALESTINE BEES.

The first of these varieties is a native of the Island of Cyprus. The name " Syrian " is now confined to a race of bees coming from the part of Syria north of the mountain range which extends from the Mediterranean at Mount Carmel eastward to the Jordan, while those coming from the south of that range, although still in Syria, are called " Palestine" or " Holy Land" bees. The first two differ very little from each other ; they have the yellow bands of the Italian, with which race they are probably nearly related, but have also more or less yellow on the thorax. They are evidently those comprised by Dr. Gerstaecker under one head, No. 3, which he mentions as being found on the coast of Asia Minor and the adjacent islands, as well as in other places. The third sort, or " Palestine bee," is as evidently the No. 4, or Egyptian bee of Gerstaecker, which he says inhabits Egypt, Arabia, and Syria.

Mr. D. A. Jones, of Ontario, Canada, one of the most extensive and enterprising apiarists in the world, paid a visit to Europe and the East in 1879, in search of a superior race of bees, believing that such existed somewhere in those parts. He was accompanied by an experienced entomologist and bee master, Mr. Benton. These gentlemen, after visiting Cyprus, established a queen-rearing apiary there, consisting of about 100 colonies. Mr. Jones also procured some bees from Syria and Palestine ; shortly after which he returned to Canada with a number of these bees, leaving Mr. Benton in charge of the Cyprus apiary.

The connection between these gentlemen has since been severed, and Mr. Benton has established apiaries at Beyrout, Syria; Larnaca, Cyprus; and Munich, Germany, where he at present resides.

Opinions are still much divided as to the positive and relative value of the different races of Eastern bees. Mr. Benton, in a circular he is now (July, 1885) issuing, says: " After five years' experience I am of opinion that the first rank should be given to Cyprian bees as the best bees, all things considered, yet cultivated." He gives them credit for all the good qualities of the Italian bees, but in a much greater degree, and with regard to their stinging propensities, which has been the principal objection urged against their general cultivation, he says: " The claim that the Cyprian bees are possessed of such great stinging propensities as to make them nearly unmanageable I have not found well based; indeed, in common with many others who have carefully tested them, I prefer to manipulate Cyprians rather than Italians, and find that, while getting no more stings from them, I can get on much faster with the work."

Syrian bees Mr. Benton considers nearly equal in every respect to Cyprians. The former vary slightly more in their markings, and are not quite so active as the latter, but in other respects they are about equal. He says, " Syrian bees are never to be confounded with Palestine bees." Of the latter race he does not speak very assuringly, and remarks, " Though Palestine bees possess some valuable qualities common to Cyprians and Syrians, still, on account of their bad temper and poor wintering qualities, I cannot recommend them for general introduction." He concludes that for the experienced—above all, for the professional—beekeeper either of the two sorts, Cyprians or Syrians, is most to be recommended. For those who suffer much from bee stings, or who " haven't steady enough nerves to manipulate Cyprians or Syrian bees," the variety most to be recommended is the

CARNIOLAN.

These bees take their name from the Austrian province of Carniola, a part of the ancient Illyria, to the east of the Carnic Alps, and on the upper part of the river Save, the great southern tributary of the Danube. Attention was first called to the qualities of this race, as we are informed by the *British Bee Journal*, by Mr. Edouard Cori, of ·Bohemia, who calls it the Carniolan or the Ukraine bee. Now the Ukraine is a Russian province on the Dnieper River, more than a thousand miles east of Carniola, with the whole of Hungary, Roumania, etc., lying between. It appears strange, then, that the two names should be connected in this way; but it will be found that Dr. Gerstaecker has described the German bee as being found also at Dalmatia (a little south of Carniola), and at the Crimea (a little south of the Ukraine); and it is evident from all accounts that the Carniolan is only a very slight variety of the German bee, although Mr. Benton implies that it is not. A writer in the *British Bee Journal* describes it so, and says that " great difficulty is experienced in keeping the Carniolan bee pure, from its resemblance to the common black bee, which renders it difficult to distinguish an hybrid of the two." They are greatly praised for their gentleness and other good qualities. Mr. Benton says they are " the *gentlest* of bees," that " their gentleness casts that of the gentlest Italians all in the shade ;" that they are even more prolific than the Italians, and are equal to them in honey-gathering qualities and in sticking to their combs and defending their hives (when not queenless); while they " are equal to the black bees in comb-building, disposition to enter boxes," etc. Their faults are, considerable disposition to swarm, the same tendency to rob which the black bees show, and that, when queenless, they do not defend their hives as well as the other Eastern bees. Mr. Marshall, the manager of Mr. Neighbour's apiary, says he prefers them to Ligurians, that—

"They are hardier, and therefore more suitable to our changeable climate. They breed as quickly, are very quiet, and a practised bee-keeper can handle them without smoke or veil, and he very rarely gets a sting. Some of our bee-masters give them the character of being much given to swarming, but I have not found them more disposed than Ligurians or blacks in that respect. If you want a business bee, get a good English queen mated with a Carniolan drone ; the combination of the two races makes a really useful bee. They are the silver bees, as the Ligurians are called the golden bees."

The description of this bee given by a writer in the *British Bee Journal* is as follows :—

"In outward appearance the Carniolan bee is slightly larger than the Italian, and not so slender in shape. It is, in fact, a larger bee—probably the largest domesticated bee. The entire body is of a rich dark brown, almost approaching to black. The golden rings of the Italian are wanting, but each rim of the abdomen is clearly marked by whitish-grey hairs, which render it distinct from any other known race ; and these hairs being longer and brighter than those of the Italian, give the bee a silvery-bright appearance which is very pleasing to the eye. . . . The Carniolan queen is a larger bee, broader in the thorax, and especially in the upper part of the abdomen, than the Italian or black queen."

HERZEGOVIAN, DALMATIAN, ROUMELIAN, AND HYMETTUS BEES

Mr. Benton has lately introduced queens under these four names. The three first mentioned would seem to be, like the Cyprians and Syrians, only slight varieties of the sort comprised by Gerstaecker under head No. 3. The Hymettus bee is, however, specially mentioned by him as a slight variety of the German or black bee, and already known by the name of *Apis cecropia*. This bee has, at all events, the advantage of a classic reputation. Gibbon mentions, on the authority of Geoponica, that " the ancients, or at least the Athenians, believed that all the bees in the world had been propagated from Mount Hymettus." The qualities of all these four varieties (if they are such) have yet to be tried. Mr. Benton wrote of them in November, 1883—

"Of the four races mentioned last, I have only tried practically the Hymettus, or Greek bees (also called Cecropian or Attic bees). They are prolific, good honey-gatherers, quite cross, but can be managed with plenty of smoke. Herzegovian and Dalmatian bees I know by reputation, and am thus safe in calling them superior to common bees and to Italians. Of Roumelian bees I know nothing ; but as I have an opportunity to get some of them next spring, and having reason to hope they may have good qualities, I shall try them."

OTHER RACES OF BEES.

There are three sorts of bees mentioned by Dr. Gerstaecker as being " indigenous to India and the adjacent islands," *Apis dorsata, A. indica*, and *A. sirialis* of Fabricius ; all, no doubt, being varieties of the *A. mellifica*. The Indian Government lately published the results of some inquiries they had instituted concerning the " popular treatment of bees in India," from which it would appear that there are plenty of bees and honey in the hilly districts both in the north and south of India. Mr. Morgan, Deputy Conservator of Forests, reports that "only one kind of bee, the *Apis indica*, is capable of domestication, and that only in hilly districts, not in the plains." The larger sort of bees, which they call "large cliff bees" (building in cliffs, under projecting ledges of rock) are represented as so ferocious in habit, and furnished with such formidable stings, as to be dangerous to both men and beasts coming within their neighbourhood. A circumstantial account of a bad case of stinging by these bees appeared in a recent issue of the *American Bee Journal*, taken from the London *Lancet*, which called forth the following editorial remarks : "We do not think we want any of these bees in America. The Cyprians are bad enough ; but for these bees of India (*Apis indica*), as well as their more irascible cousins (*Apis dorsata*), we have no use. Let them stay where they are." Mr. John Douglas, of the Indian Telegraph Department, says, "A swarm of these bees has been known to put a regiment of cavalry to flight, and innumerable are the instances in which man and beast have fallen victims to their unrelenting animosity; " yet he proposes the domestication of this " great tiger honey fly " (as it is called in parts of the country) as the "first question for Indian apiculturists !" Mr. Benton has been making efforts to import the *Apis dorsata* from Ceylon ; but if they are not very different in disposition from the *Apis indica*, we in Australasia may echo the words of the *American Bee Journal*, " Let them stay where they are ! "

NATIVE BEES OF NEW ZEALAND AND AUSTRALIA.

Notwithstanding the assertion of Dr. Gerstaecker, there are indigenous bees both in New Zealand and Australia. The small bees indigenous to New Zealand, I believe, belong to the species *Apis tregona*. It makes its nest in the ground, by boring a small hole from the surface, about two inches in depth; holes then branch off in different directions : these branch holes extend two or three inches, and at the bottom of each is deposited a mixture of honey and pollen, in which the eggs are laid. I dug up several nests last season, and found brood in different stages, but there only appeared to have been one egg deposited in each compartment. It is of no service to the apiarist.

Mr. F. A. Joyner, of North Adelaide, S. A., at the suggestion of Mr. Bonney, very kindly sent me lately some specimens of native bees, accompanied by the following remarks :—

"I have observed them pretty closely for some time past, and find that they gather honey and pollen, are very swift in their movements, have particularly long stings and proboscis, and as late as last evening (April 16th, 1885), for the first time discovered that they exist in swarms. I was unfortunately unable to take the swarm, as the bees were disturbed before I reached them, and I was afterwards unable to find their new alighting place. It appeared an average-sized swarm, and moved similarly to our black bees, with the exception of moving much swifter."

I submitted the bees, which had been very much broken to Mr. T. J. Mulvany, for microscopical examination, and'he very kindly supplied me with the following information concerning their structure :—

"The rings on the abdomen, which are alternately black and silver-grey, are very handsome, and the three 'subcostal cells' in the front wings, as well as the hooklets in the under wings, beautifully developed. The wings, compared with those of a common bee, measure as 16 to 19, and as 11 to 13, which shows them to be 0·84 of the length ; from this, and from appearance of head, legs, and abdomen, I take the live insect to be more than four-fifths of the size of the black bee and therefore larger than the Egyptian or the Palestine bee. The hind legs have the pollen basket and the long hairs on the 'basal tarsus,' the front legs the peculiar spur at the knee joints, and both have a coating of silver-white hair on the outer side of the 'tibia.' The head is very handsome ; the compound eyes wide apart, with golden hairy forehead between. The mouth organs appear to me to be remarkable ; the mandibles are horny, with sharp double points like teeth. I should think this bee could bite as well as sting. The maxillæ are also stiff, and, at least in the dry state, look like the beak of a bird. I could also see the 'labial palpi,' but not the tongue itself. Altogether, to my unpractised eye, it looks more like a variety of the *Apis mellifica* than a different species. It would be very desirable to get some more specimens in a better state of preservation, and, if possible, to take a swarm and try them in a hive."

From what I have heard of the wild or native bees in other parts of Australia, I take them all to be the same as those described.

CHAPTER III.

INMATES OF THE HIVE--THEIR NATURAL HISTORY.

THE honey-bee is, above all things, gregarious in its habits. As Langstroth remarks, "It can flourish only when associated in large numbers as a colony. In a solitary state a single bee is almost as helpless as a new-born child, being paralysed by the chill of a cold summer's night." This is true; but it is not alone for the sake of mutual warmth that bees aggregate; their nature compels them to form a sort of republic (or, if rather a monarchy, then certainly a *very limited* one), which presents the peculiar feature that all the active citizens are, as we shall see further on, females, who are doomed to a life of celibacy as well as of toil, while the head of the community is, in the strictest sense of the word, the mother of her whole people; and although they support, for a time, a "pampered aristocracy" of idle males, they use very little ceremony in getting rid of them as soon as there appears to be no further chance of their presence being required.

GENERAL DESCRIPTION.

Every hive in a normal working condition, during the swarming season, will be found to contain bees of the three different kinds, the characteristics and relative sizes of which are shown in the illustrations which follow. First, one bee only of the peculiar form which denotes the *queen* or *mother bee;* secondly, a few hundreds (sometimes more than a thousand) of large bees, called *drones;* and thirdly, many thousands of the smaller kind, called *workers,* which are the common bees to be seen on blossoms, as neither the queen bee nor the drones gather honey or work outside the hive.

The queen is indispensable to the prosperity of the hive. She is the only perfectly developed female, and lays all the eggs, of which she can, on occasions, produce two to three thousand in twenty-four hours. Without her the colony would soon dwindle down and die out, or be attacked and killed for the sake of its stores, as, after being deprived of their queen, the workers generally (unless they are in a position to rear a new one, as will be seen further on) lose the disposition to defend themselves and their home. The queen is not provided with the special organisation which enables the workers to gather honey and pollen and to secrete wax. She is furnished with a sting, which, however, she very rarely uses, except in a struggle with a rival queen. When she has been once impregnated, and has taken her place in a hive, she never leaves it except to accompany a swarm. Her term of life may extend

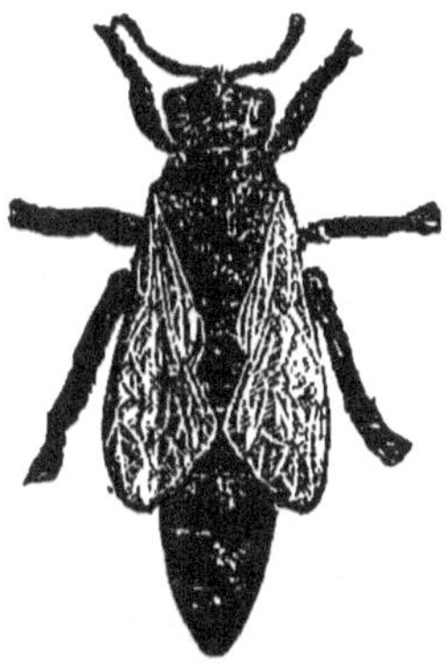

Fig. 4.--THE QUEEN.

to four years at least, and during that time she may lay many hundreds of thousands of eggs; but she is considered to be in her prime in the second year, and is seldom very prolific after the third. She can be easily distinguished from the other bees, and be recognised even by the most inexperienced from the following description :—Her body is not so bulky as that of a drone, though longer; it is considerably more tapering than that of either drone or worker; her wings are much shorter in proportion than those of the other bees; the under part of her body is of a lighter and the upper of a darker colour than the worker's; her movements are generally slow and matronly, and indeed she looks every inch a queen.

The drones, or male bees, are much stouter than either the queen or workers, although their bodies are not so long as that of the queen. They are neither furnished with a sting nor a suitable proboscis for gathering honey, no baskets on their legs for carrying pollen, and no pouches on their

Fig 5.—THE DRONE.

abdomens for secreting wax, so that they are physically incapable of doing the ordinary work of the hive. Their office is to impregnate the young queens, but very few have the chance of doing so; those that have, die immediately afterwards, and the rest are usually destroyed by the workers at the end of the swarming season, having by this time become an incumbrance only.

Fig. 6.—THE WORKER.

The worker bees, the smallest in size, constitute the bulk of the population of the hive. A good-sized swarm should contain *at least* twenty thousand,* and a well-stocked hive, during the

* About 4000 ordinary bees weigh one pound, so that a 5 lb. swarm contains about 20,000. Good swarms, however, sometimes weigh 7 lb. to 10 lb.

full working season, will have twice, and sometimes three times, that number of workers. They are all females, but not fully developed as regards their sexual organization—they are incapable of being impregnated by the drones ; but in some rare cases their ovaries are sufficiently developed to admit of their laying eggs, which, however, as will be shown later, are unfertilised, and produce only drones. On the other hand, these workers are specially provided with the means of successfully prosecuting their useful labours. They have a wonderfully constructed tongue, or proboscis, which enables them to suck or lap up the liquid sweet from the nectaries of blossoms, and to store it in a " honey sac," which is, in fact, a first or extra stomach, from which they can again disgorge it at will into the cells of their combs. Their hinder legs are provided with a hollow, or " basket," for carrying pollen, which they are enabled, by the use of their front legs and their proboscis, to work up into little pellets, and pack in these receptacles. They have the power of secreting wax in small scales under the folds of the abdominal rings of their body, and they are furnished with a sting to protect themselves and their stores, and of which they make effective use when provoked. They perform all the work both inside and outside the hive ; collect the materials for honey, beebread, and propolis ; carry water, secrete the wax, build the combs, nurse and feed the young brood, ventilate the hive, and stand guard at the entrance when it is necessary to keep out intruders. Although division of labour is beautifully exemplified in the economy of the hive, still there are not separate classes of worker bees (as was at one time supposed) to perform the different sorts of work; on the contrary, every worker bee is capable of doing all these things, and they take their turns accordingly. " One *bee* in *her* time plays many parts." The young bees are employed on " home duty " for the first week or two ; they then take their turn of outdoor work, and are gradually worn out in the service. Their term of life is short, varying from only six or seven weeks in the busiest working season to nearly as many months after that busy time is past.

PHYSIOLOGY AND ANATOMY OF THE HONEY-BEE.

I shall now devote some space in the endeavour to place before the reader, as concisely as practicable, and with the aid of illustrations, a clear view of the more important facts relating to the physical structure and functional peculiarities of these wonderful insects. For the present advanced state of our knowledge on these points we are mainly indebted to the investigations of Huber and Dzierzon, which have been successfully followed up by the skilful dissections and microscopic examinations more recently made by Professor Cook in America, and by Mr. F. R. Cheshire in England. A familiar acquaintance with these facts may be said to be indispensable to all earnest apiarists, not only because the system of modern scientific bee-keeping is based upon the knowledge so obtained, but also because the close observation of the habits of the bee, and of the operations performed in the hive, which constitute the great charm of the bee-keeper's occupation, can only be really effective and satisfactory when guided by the light of those brilliant discoveries. A writer upon the fine arts, when pointing out the necessity of a knowledge of anatomy to the draughtsman of the human form, has remarked that "no one can see things as they *are*, unless he knows how they *ought* to be." This is perfectly true with all of us in our observation of the works of nature. As long as we are uninstructed, we "have eyes and cannot see." The bee-keeper who shall have acquired some knowledge of the physiological peculiarities of the honey-bee, and its "relation to flowers," will ever afterwards view its every movement and all the phenomena of the hive with new eyes; he will take an entirely new interest in the various structural features of the honey-bearing plants and their blossoms; he will have obtained at least some inkling of the Divine *intention* in these varieties of form, and in the simple act of watching a bee's visit to a flower, he will perceive, what would otherwise have escaped his notice, how beautifully in the different cases the instinct of the insect and the structure of the flower combine to attain the object of the former—the collection of the nectar and the pollen—and the intended beneficial effect upon the latter—its cross fertilization.

CLASSIFICATION OF SPECIES.

All the different *races* of the honey-bee with which we are as yet well acquainted are (as mentioned in Chapter II.) only varieties of the one species, the *Apis mellifica*; that is to say, they differ only in size, colour, and perhaps in the greater or less development of some organs; but none of them present any marked distinction in structure or in habits; they therefore admit of cross-breeding, producing fertile crosses or hybrids which may be continued as a new variety, or re-crossed with other varieties. This would not be the case if they belonged to different species. They belong further to the genus *Apis*, the family *Apidæ*, the order *Hymenoptera*, sub-class *Hexapoda*, and class *Insecta*.

GENERAL CHARACTERISTICS OF STRUCTURE.

As characteristic of their class and sub-class, the body of these insects is divided into three parts—the head, the thorax, and the abdomen, which are connected by small and hollow ligaments; they have six legs, attached to the thorax, and they breathe air through a system of tubes to be described further on. As belonging to the order *Hymenoptera*, they have four membranous wings attached to the thorax, of which the two foremost cover the hinder ones when at rest; also a proboscis or tongue by which they can suck or lap, and strong jaws for biting. The family *Apidæ*, according to Professor Cook,

"Includes not only the hive-bee but all insects which feed their helpless young or larvæ entirely on pollen, or honey and pollen. . . The larvæ of all insects of this family are maggot-like—wrinkled, footless, tapering at both ends. . . . They are helpless, and thus, all during their babyhood—the larvæ state—the time when all insects are most ravenous, and the only time when many insects take food, the time when all growth in size, except such enlargement as is required by egg-development occurs, these infant bees have to be fed by their mothers or elder sisters. They have a mouth with soft lips and weak jaws, yet it is doubtful if all or much of their food is taken in at that opening. There is some reason to believe that they, like many maggots, such as the Hessian fly larvæ, absorb much of their food through the body walls. From the mouth leads the intestine, which has no anal opening, so there are no excreta other than gas and vapour. What commendation for their food, *all* capable of nourishment, and thus all assimilated!"

The genus *Apis*, to which the species *A. mellifica* belongs, is characterised chiefly by slight peculiarities in the legs and wings. All bees of this genus have *no* tibial spurs (stiff spines

at the end of the tibia or long joint of the leg) on their posterior
legs, and as to their wings, "they have three cubital or subcostal cells—the second row from the costal or anterior edge—
on the front or primary wings."   Here we cannot help
remarking how wonderful are those minute rules of organic
structure.   Who that examined for the first time the wing of
a bee, and compared it with that of any other insect of the
same order, would imagine that the differences in the subdivisions of the membrane, which have all the appearances
of chance formation, and which are probably not *precisely* the
same in any two examples, are yet so characteristic in their
arrangement as to afford an easy means of distinguishing the

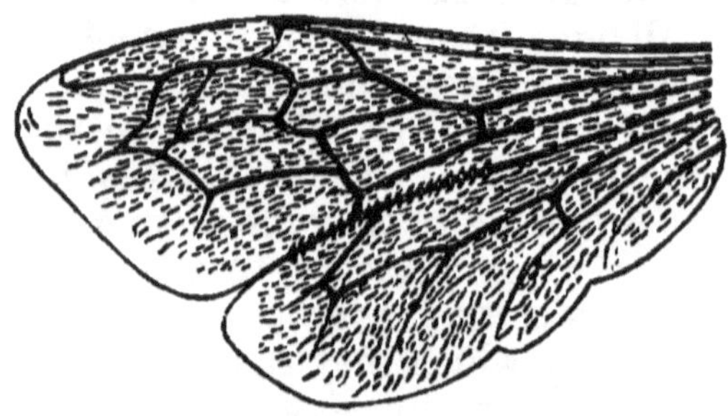

Fig. 7.—WINGS OF A BEE.

genus to which the insect belongs !   Professor Cook gives the
following further details as marks of this whole genus :—

" On the inner side of the posterior basal tarsus, opposite the pollen
baskets, in the neuters or workers, are rows of hair (Fig. 15) which
are probably used in collecting pollen.   In the males, which do no
work except to fertilise the queens, the large compound eyes meet
above, crowding the three simple eyes below, while in the workers and
queens the simple eyes, called ocelli, are above, and the compound
eyes wide apart.   The drones and queens have weak jaws, with a
rudimentary tooth, short tongues, and no pollen baskets, though they
have the broad tibia and wide basal tarsus."

NERVOUS SYSTEM.

Coming now to the special anatomy and physiology of the
Apis mellifica, it may be well, in the first place, to show the
general arrangement of the nervous system as depicted by Mr.
F. R. Cheshire in his admirable "Diagrams on the Anatomy and

Physiology of the Honey-Bee," of which I shall present a few examples for the benefit of the reader, as an incentive to a more thorough study of the whole subject. This illustration speaks for itself, the black spots representing the *ganglia* or nerve-centres, from which issue the minor systems of nerves appropriated to particular uses, " the great masses positioning

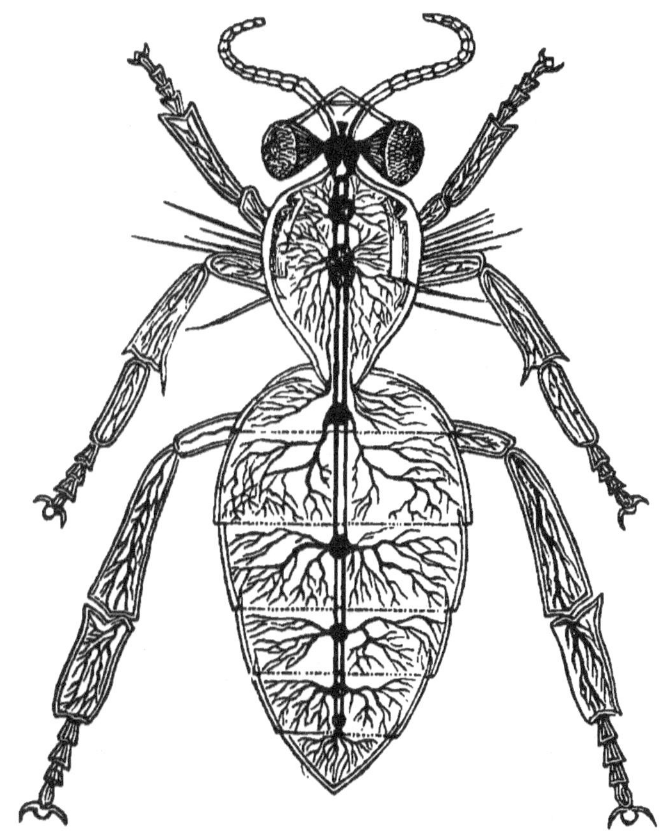

Fig. 8.—NERVOUS SYSTEM OF THE BEE.

themselves," as expressed by Mr. Cheshire, " at the roots of the wings and articulations of the legs, to supply stimulus to the organs of locomotion. The ganglia energising the compound eyes are also large."

RESPIRATORY ORGANS.

It has been already mentioned, as one of the main characteristics of the class *Insecta*, that they breathe air through a peculiar system of tubes. These tubes are called trachea, and are of a very beautiful formation. They are thus illustrated and described by Mr. Cheshire :—

Fig. 9.—TRACHEA, MAGNIFIED.

b. Elastic Spiral of Trachea.

"The tubes consist of two membranes, between which lies the elastic spiral thread, which prevents the closing of the tube through movement. In the same intervening space the fluids of the insect become aerated, so that the purpose of the lungs of the higher animals is answered." The outer openings of these tubes, in the sides of the insect, are called spiracles. The bee has fourteen of these spiracles, two on each side of the thorax, and five on each side of the abdomen. The trachea expand into large lung-like sacs in the abdomen of the bee, as shown in the following illustration, where *a* represents one of the air sacs, *b b b* the spiracles, and *c c* some of the trachea.

In this engraving details of only one side are shown, and only one air sac on that side.

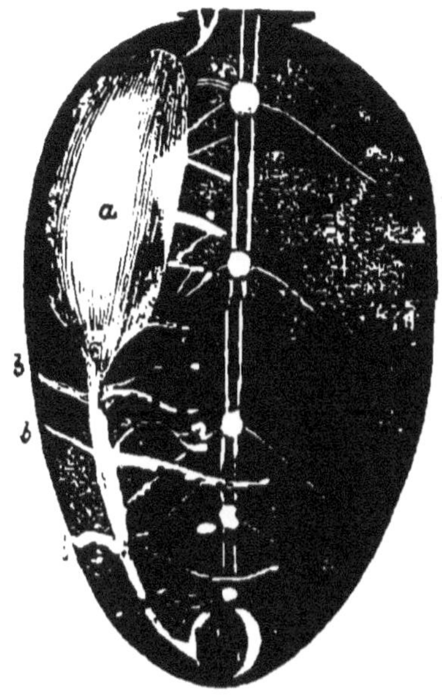

Fig. 10.—RESPIRATORY ORGANS OF THE BEE.

AIR SACS OF BEE.

There are however, in the worker bee, two sacs on each side a large and a small one ; and, what is very remarkable, the larger sac is in fact the undeveloped ovary of the insect, and in the queen bee is replaced by the ovary proper, so that she possesses only one small air sac on each side. The large air sac of the worker is only distended during the time of flight, and we may notice in this substitution of a valuable auxiliary to the flight and carrying power of the worker bee, in place of an organ not required by her, a beautiful adaptation of means to the end. The queen we know is not required to fly far or often, and then not to carry any loads of honey or pollen—indeed it is a well-known fact that she cannot fly far, when her ovaries are filled with eggs, and the smaller air sacs therefore are sufficient for her purpose. The following diagrams show (in Fig. 12) the arrangement of the large and small air sacs in the case of the worker, and it will be seen at a glance that they take the places of the smaller air sacs *a*, and of the ovary *b*, in the case of the queen (Fig. 11).

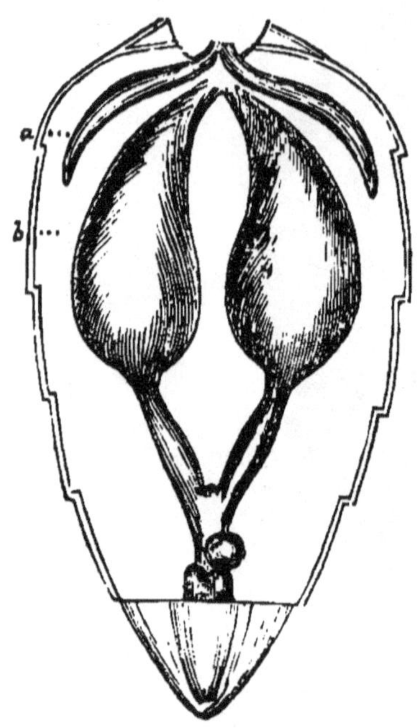

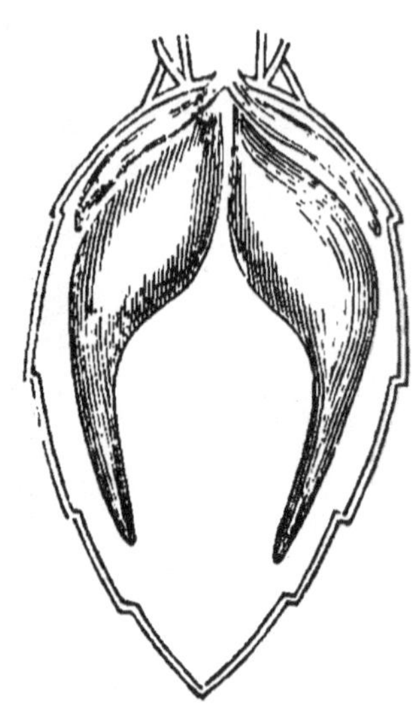

Fig. 11.—AIR SACS AND OVARIES OF QUEEN.

Fig. 12.—AIR SACS OF THE WORKER.

a. Air sack ; b. Ovary.

Every person accustomed to bees must frequently have observed that some of the workers, when returning from the field, remain for some time on the alighting-board before entering the hive, and that during that time the rings of the abdomen are in constant motion. These bees are simply breathing themselves after a long and tiresome flight. Mr. Cheshire remarks :—

"The constant elongation and contraction of the abdomen of the bee's body has for its object the ejection of air which has become carbonised and the drawing in of fresh supplies. The spiracles admit of being closed voluntarily. When the bee is in flight with the air sacs filled, if the spiracle be closed and the abdomen contracted, the fæces are extruded. This explains why bees never soil their hives, except when in a dysenteric condition, and why mating only occurs on the wing."

From this sketch of the general features of the bee's structure I shall now proceed to notice the separate parts more particularly appertaining to the *Apis mellifica*, commencing with

THE HEAD.

Within the small limits of a bee's head there are contained several important organs, some of them of a very complex nature. These are—the compound eyes; the simple eyes, or stemmata; the mouth and its appendages; and the antennæ. The following engraving shows a front view (on a greatly magnified scale) of a worker bee's head :—

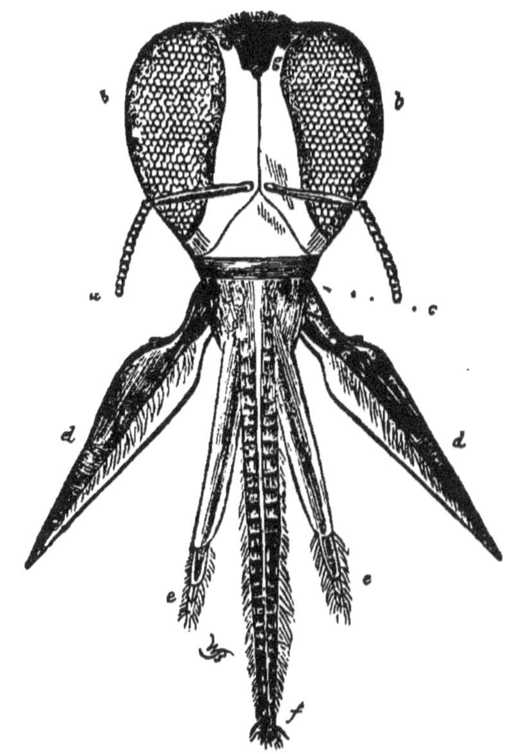

Fig. 13.—HEAD OF WORKER BEE.

a. Antennæ; b. Compound eyes; c. Jaws; d. Maxillæ; e. Lateral palpi; f. Ligula, or tongue; g. Stemmata.

THE EYES.

In this illustration the compound eyes are shown on the right and left hand side, at *b b*, and the simple eyes between, on the top of the forehead, in a triangular position, at *g*. In the drone the compound eyes meet together at top, and the simple eyes are forced down to near the middle of the forehead. The compound eyes occupy a great portion of the head; they form on each side an oval lobe, convexly rounded, with a brown, horny surface, which is divided into an immense number of hexagonal facets, looking, when magnified, very like a honeycomb, and each of which facets is in fact the surface of a separate eye. There are supposed to be about 3,500 in each compound eye of a bee. These are immovable, and each has a very limited range of vision, but from the way in which they are placed the bee must be able to see with them in a great range without turning its head. These eyes are not supposed to be capable of adjustment for different distances, but to be chiefly useful for distant vision, while the small simple eyes, or stemmata, in front of the head, are most probably intended for seeing objects near at hand. Our knowledge about the bee's sense of vision, as well as its other senses of hearing, taste, and smelling, is still very imperfect. Sir John Lubbock, after making very careful experiments, is clearly of opinion that bees can distinguish different colours easily, and that they have a partiality for blue. It has been frequently remarked that they seem to discern objects better at a great distance than when near at hand. They fly homewards from any distant point in very direct lines, apparently guided by remote landmarks, but they frequently knock against persons or things, as if they had not perceived them at a little distance; and if they happen to alight ever so little to one side of the entrance to their hive, they are just as likely to go to the wrong side as to the right one in looking for it.

THE MOUTH.

The mouth consists of an upper lip (*labrum*) and under lip (*labium*), and of two pairs of jaws, the upper pair short and horny, called the mandibles, shown at *c* in the figure, and the lower pair long and more membranous, called the maxillæ, and

shown at *d d*. The under-lip, or labium (not seen in the figure)
has a broad base, called the mentum, which forms the floor of
the mouth, and at the same time the root of the tongue, or
ligula, *f*, and of the labial palpi, *e e*. The ligula itself is sometimes called a proboscis, a name which is, in most minds, associated with the trunk of the elephant, which is a hollow tubular
prolongation of the nasal organ. This is no ways analogous with
the proboscis of insects in general, and of the honey-bee in
particular; the latter is a prolongation of the labium, as above
mentioned, which organ is capable of being pushed forward and
drawn back into the under cavity of the head, carrying the
ligula of course with it. The labial palpi, together with the
maxillæ, appear to be brought into use to assist the ligula, or
tongue, in conveying nectar from the flowers to the honey sac
of the bee. The end of the tongue is furnished with a spoon-
shaped hollow on the under side, which opens into a capillary
tube on the upper side, covered with whorls of hair, as is also
the end of the ligula. When the bee is sipping, the liquid
enters the capillary tube, and the tongue is drawn back by
muscles at the base into what Herman Müller terms a suctorial
apparatus formed by the labial palpi and maxillæ. In his
valuable work on "The Fertilisation of Flowers," he beautifully describes the above process as follows :—

"When the bee is sucking honey which is only just within her reach,
all the movable joints of its suction apparatus, cardines, the chitinous
retractors at the base of the mentum, laminæ (maxillæ), labial palpi,
and tongue, are fully extended, except that the two proximal joints
of the labial palpi are closely applied to the tongue below, and the
laminæ to the mentum and hinder part of the tongue above. But as
soon as the whorls of hair at the point of the tongue are wet with
honey, the bees, by rotating the retractors, draw back the mentum,
and with it the tongue, so far that the laminæ now reach as far forward as the labial palpi; and now labial palpi and laminæ together,
lying close upon the tongue, and overlapping at their sides, form a
tube, out of which only a part of the tongue protrudes. But almost
simultaneously with these movements, the bee draws back the basal
part of its tongue into the hollow end of the mentum, and so draws
the tip of the tongue, moist with honey, into the tube, where the
honey is sucked in by an enlargement of the foregut, known as the
sucking stomach, whose action is signified externally by a swelling of
the abdomen."

Doubts have been expressed as to whether the bee empties
the contents of its honey sac into the cells through its proboscis

or simply through its mouth; but the statement of Muller, that when gathering pollen from some kinds of flowers the bee ejects a little honey on the anthers through its suction tube—which in another part of his work he calls the " proboscis " for shortness—would incline us to suppose that the honey may be ejected into the cells in the same manner.

The maxillæ, or so-called lower jaws, form the under sheath of the ligula and palpi when at rest, and the whole organ is then folded under the lower part of the head.

THE ANTENNÆ.

The antennæ, or "feelers," as they are commonly and not inappropriately called, are very sensitive organs of touch-sensations, and, beyond all doubt, of vital importance to the insect. Huber tried experiments with queens deprived of the antennæ, and found that the loss of one was not very injurious; but when both were gone, the bee became apparently delirious, avoided the worker bees, dropped her eggs at random about the hive, and rushed towards the opening, as if to escape. Having introduced a second queen, similarly mutilated, it was found that they had both lost their natural instinct for a combat, and met several times without exhibiting the smallest resentment. The worker bees did not seem to distinguish their own mutilated queen from the strange one, and both were left to do as they liked ; but when Huber introduced a third unmutilated queen, the workers seized her, bit her, and confined her so closely that she could hardly move. When he removed this last and one of the others, and left one fertile but mutilated queen in charge of the hive, she left it, and tried to fly away, but being unable, she fell and died on the ground. Mr. Harris mentions also that worker bees, if deprived of the antennæ, and allowed to fly, become incapable of recognising their own hive again, and are hopelessly lost as to their whereabouts. Huber tried other experiments by dividing the bees of one stock by two fine wire gratings placed so far apart that the antennæ of the bees on each side could not meet, and afterwards removing one of the gratings, so that they could touch each other ; and from the results of these experiments he drew the conclusion that they could actually *communicate intelligence* to each other by means of their antennæ. There is, at all events, much yet to be learned with regard to the true nature of these organs.

SENSES OF HEARING AND SMELLING.

Closely connected with the conjectures as to the uses of the antennæ are those which many naturalists have made as to the organs of hearing and smelling in the honey-bee. That they are possessed of a keen sense of smell, as well as of taste is indisputable ; and that they are capable of hearing seems very probable, though still doubtful ; but through what organ these senses may be worked upon is only matter of conjecture. Huber suggested that the antennæ might be organs of smell as well as touch ; Lehmann and Cuvier considered that the spiracles used for the purposes of respiration were also the means by which the sense of smell was exercised ; while Kirby and Spence incline to the belief that the organ lies somewhere in or near the mouth. We can only assert with certainty that bees have a very keen sense of smell, that they are attracted by the odour of flowers, honey, etc., and rendered furious by disagreeable odours, especially by the smell of their own sting-poison. As to their sense of hearing, it seems hard to believe that they are unconscious of such sounds as their own humming, so varied according to circumstances, or to the calls so distinctly made by young queens, and which appear to us to exercise such an influence on their conduct ; still it is true that Sir J. Lubbock, who has tried so many experiments on the hearing of bees, with musical instruments, dog whistles, shrill pipes, etc., seems to have satisfied himself that no noise he could make, either harmonious or discordant, was capable of making any impression on them, or of disturbing them in the least.

THE WINGS.

It has been already stated that the honey-bee, like all insects of the order *Hymenoptera,* is provided with four membranous wings, springing from the thorax, of which the foremost, or primary wings, slide over and cover the hinder ones when at rest. This arrangement is of importance, as enabling the bee to enter without difficulty the narrow cells of the comb in order to stow honey and bee-bread, to feed the larvæ, etc., as well as to work its way into the small corolla-tubes of some flowers and so reach the nectaries at bottom. A pair of these wings has been shown in Fig. 7 to illustrate the arrangement of the cells or subdivisions of the membrane, upon referring to which it will be seen that the two wings, fully expanded, look as if they were stitched together for a certain length. This apparent stitching is a line of about twenty hooklets upon the front edge of the under wing, of which a greatly magnified view is given below :—

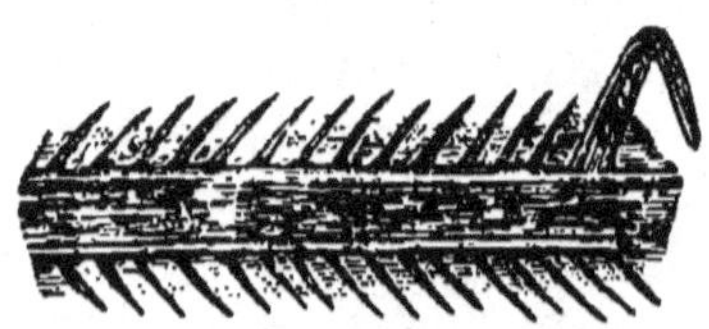

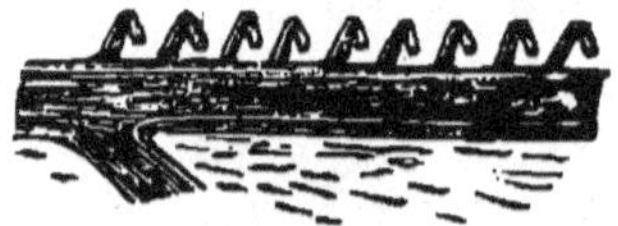

Fig. 14.—HOOKLETS OF WINGS.

Nine of the hooklets are shown in the lower diagram, while the upper one shows, on a still larger scale, the last hooklet of the row, and a line of strong bristly hairs, which furnishes the margin of the wing where the hooklets cease.

Mr. Cheshire explains that

"These four wings, though individually small, collectively present sufficient surface for a rapid flight, which is greatly aided by a beautiful arrangement for locking the two pairs into one. The front wing is folded under at its posterior edge, and as the wing is advanced to bring it into position for flying this fold catches into a line of hooks from twenty to twenty-three in number, which turn upwards from the front edge of the back wing."

Can anything more admirable than this arrangement be imagined ?

THE LEGS.

The legs, six in number, spring, like the wings, from the thorax, which is the chief seat of muscular power. It is, indeed, one mass of muscles, with the exception of the narrow passage of the œsophagus or gullet connecting the mouth with the honey sac and stomach. Each leg has four principal joints, the coxa, trochanter, femur, and tibia, and five smaller joints, called tarsi, terminating in a two-hooked claw. The coxa and trochanter are short and broad joints, the former working with a ball and socket movement in the so-called coxal cavity in the body ring; the femur represents the thigh, the tibia the leg, and the tarsi the foot joints of the higher animals. In the honey-bee the first of the tarsi is nearly as large as the tibia, to which it is attached; it is called the basal tarsus, and in the posterior legs of the worker bee it and the tibia are widened out and hollowed on the under side so as to form the "pollen basket" already mentioned at page 42, and as shown in the following engraving :—

Fig. 15.—HIND LEG OF BEE, SHOWING POLLEN BASKET.

The front legs of the workers have also a very peculiar formation, shown in the next engraving. Under what may be termed the knee joint there is a cavity, *c*, in the tibia, and a spur or finger, *b*, on the femur joint, which can be pressed over the cavity or opened at the will of the bee.

Most modern writers* describe this apparatus as performing an important part in the gathering of pollen, as follows :— When a bee is about to transfer the pollen she has gathered to her pollen baskets, she places her tongue in the cavities of both legs, closes the blades, and then withdraws it, leaving the pollen adhering to the sides of her knees. It is then worked

* Especially Mr. A. I. Root, in his " A B C of Bee Culture " it is not mentioned by Langstroth or Quinby.

up into a small pellet by her tongue and feet, and placed by the second pair in the spoon-shaped hollows, or baskets, on the third or last pair of legs, and neatly patted down.

Fig. 16.—ANTERIOR LEG OF WORKER, MAGNIFIED.

It must, however, be observed that Professor Cook, one of the best authorities, writes very cautiously on this point. He says :—

" For several years this has caused speculation among my students, and has attracted the attention of observing apiarists. Some have supposed that it aided bees in reaching deeper down into tubular flowers ; others, that it was used in scraping off pollen, and still others, that it enabled bees to hold on when clustering. The first two suggestions may be correct, though other honey and pollen-gathering bees do not possess it. The latter function is performed by the claws at the end of the tarsi."

This throws a doubt upon the matter, and we must be cautious not to assert as a fact anything that is not already universally admitted to be such, or that we cannot decisively prove by our own investigation. I have often watched bees gathering pollen, and *thought* I could observe the process of scraping the tongue, or something very like it. But it must be admitted that the movements of the bee on such occasions are so amazingly rapid that it would be difficult to say there could be no mistake as to the operation performed.

Muller says :—

"In collecting pollen, hive-bees and humble-bees use their mouth-parts in two different ways to moisten it, according as it is the fixed pollen of entomophilous flowers, or the loose, easily-scattered pollen of anemophilous flowers. In the former case (*e.g.*, when *Apis mellifica* collects pollen on *Salix*) the bee has its suctorial apparatus completely folded down, bringing the mouth opening, which lies between the mandibles and the labrum, close over the pollen. The bee ejects a little honey on the pollen, and then takes it up by means of its tarsal brushes and places it in the baskets on the tibiæ of its hind legs ; it often makes use of its mandibles to free the pollen before moistening it with honey. In the latter case, which I have observed in *Plantago lanceolata* . . . the bee, hovering over the flower, ejects a little honey upon the anthers from its suction-tube which is fully extended, but completely sheaths the tongue. . . . Since hive-bees and bumble-bees on entomophilous flowers suck honey with outstretched proboscis, and collect pollen with it folded up, and on anemophilous flowers collect pollen only, it follows that they can never suck honey and gather pollen simultaneously ; they must always do first one and then the other, and since the pollen has to be moistened with honey, the act of sucking must always be first."

THE HONEY SAC.

This is merely a widening of the œsophagus, forming a first stomach, in the anterior part of the abdomen, a sort of ante-chamber to the true stomach, which is very different in shape, and which is followed by the intestines leading to the anus, or vent. Everything passing from the mouth to the stomach must go through the honey sac, but the bee has the power of retaining the nectar in this sac, and afterwards disgorging it through the mouth, without letting it enter the true stomach at all. Connected with the œsophagus, in front of the honey sac, there are important glands in the head and in the front part of the abdomen, which secrete the so-called salivary juice, which, as Professor Cook states, " aids in kneading wax, etc., as already described. It also probably aids in modifying the sugar while the nectar is in the bee's stomach." This would account partly for the difference observable between honey and other merely saccharine matter.

THE STING.

The sting of the worker bee is a very complicated organ, as will be seen by a study of the following engraving, taken from Root's " A B C of Bee Culture."

In the general view of the sting, I, is the double gland which secretes the poison ; A, the cylindrical reservoir in which the poison is collected from the glands, and from which it is transmitted through hollows in the spears or lancets to the wound ; B, the two barbed lancets ; and D, the third spear or awl, usually styled the sheath, in which the other two partly slide when at work. In the cross section (greatly enlarged) of the lancets, at the point D, it will be seen how the two hollow lancets, A and B, slide on ribs or guides in the concave side of the so-called sheath, D. They have tubes, F and G, through which, as well as through the tube E, formed between the three parts of the sting, the poisonous fluid is transmitted. There is a hollow, C, in the awl or sheath, D, but it is only for strength and lightness, and is not open either above or below. In the

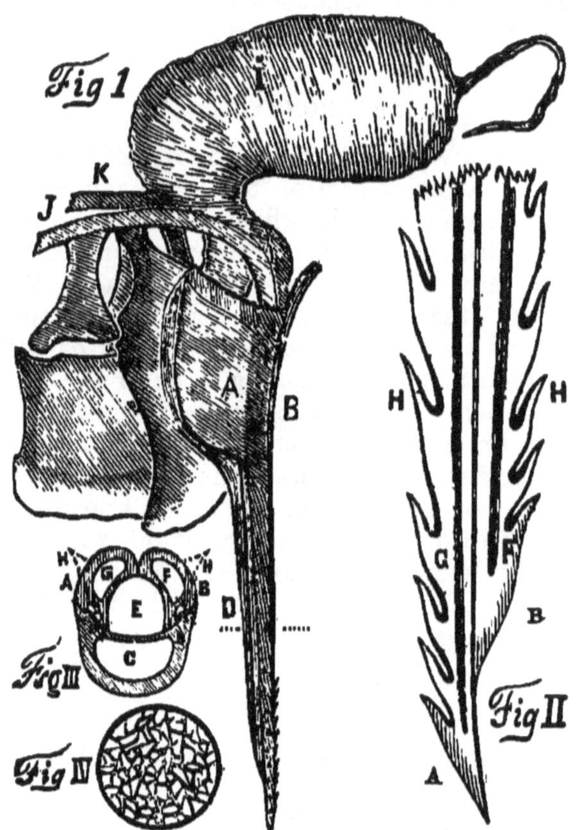

Fig. 17.—THE BEE STING.

I. *Bee sting, magnified.*
II. *One of the barbed lancets.*
III. *Cross section of lancets at D.*
IV. *Drop of the poison, crystallised.*

barbed lancets, the end of one of which is shown, greatly magnified, there are grooves, G, to fit on the ribs of the sheath, and the poison, which is conveyed down the hollow tube inside of each, finds vent by small side openings to the barbs at H H. It appears that when the wound is first pierced by the smooth and highly polished point of the awl D, a sliding motion is communicated to the barbed lancets by the muscles shown at J and K, and the poison is *pumped* into the wound through the centre cavity E; the barbed lancets are then driven in by alternate motions, and at the same time the centre cavity is closed by valves at the root of the sting, and the poison is forced through the tubes in the hollow lancets, and through the side openings near the barbs. The barbs having once penetrated any tough material, such as the human skin, cannot be withdrawn by a direct pull. The bee, if left to itself, will gradually work round and round until it screws out the sting, but if it be abruptly shaken or brushed off, the whole sting is torn out of its body and left behind. In that case the muscles will continue to work and to force poison into the wound for some time, if the sting be not carefully extracted, which should be done without squeezing the poison reservoirs at its base. The body of a bee that had been dead for hours has been known to sting in that way. The injury occasioned to a bee by the tearing out of its sting must be very severe, and it has been generally supposed that they must die immediately afterwards. Sir John Lubbock, however, in his work on " Ants, Bees, and Wasps," says : " Though bees that have stung and lost their sting always perish, they do not die immediately, and in the meantime they show little sign of suffering from the terrible injury." He mentions having seen a bee after losing its sting, remain twenty minutes on the floor-board, enter the hive, return in an hour, feed quietly on some honey, and again return to the hive. Mr. A. I. Root says he has kept bees some time in confinement after being so injured, " and could not see but they flew off just as well as bees that had not lost their sting." He even inclines to think they may live and gather honey afterwards.

Recent researches by the French naturalist, M. G. Carlet, show that the two glands secrete two different sorts of liquids, the combined action of which makes the poison so virulent. In the translation of M. Carlet's paper, given in the *American Apiculturist* of December, 1884, it is stated that although the stinging of a fly by a bee causes the instantaneous death of the former, yet if a fly be inoculated with the product of one of the two venomous glands of the bee, it will not die for a long time ; the successive inoculations of the same fly with the products of the two glands causes death in a very short time after the second inoculation. The conclusions come to are—" 1. The poison of the hymenoptera is always acid. 2. It is composed of a mixture of two liquids, one strongly acid, the other feebly alkaline, and acts only when both liquids are present. 3. These are produced by two special glands that may be called the acid gland and the alkaline gland. 4. These two glands both expel their contents at the base of the throat from which the sting darts out."

REPRODUCTIVE ORGANS OF THE QUEEN.

The most important organs of the queen-bee—themselves forming perhaps one of the most wonderful objects of nature, and of which the very accurate knowledge which we now possess, owing to the patient researches of many naturalists, has done more than aught else for the progress of scientific bee-culture—are her ovaries and the parts attached thereto, which are illustrated in the following engraving.

The two fig-shaped bodies are the ovaries, which are multi-tubular, there being more than a hundred tubes (called the ovigian tubes) in the two ovaries of a queen bee. In these tubes the eggs grow and develop themselves until they are fit to be deposited. Each ovary has a separate oviduct at bottom, through which the eggs pass for some distance, until the two join in one common oviduct leading to the vulva, or vent, through which the eggs are ultimately deposited. A little below the junction of the passages from the two ovaries, and on the outside of the common oviduct, is a small globular body, shown on the right hand side in the engraving. This is a hollow vessel, called the spermatheca, of which much has to be said. More than two hundred years ago Swammerdam published an excellent illustration of the ovaries of a queen bee, showing the spermatheca, but he conjectured that it secreted a fluid for sticking the eggs to the bottom of the cells in the comb. In his time but little was known of what went on within the hive. It was no doubt assumed by many that every single egg laid by the queen required to be fertilised by a separate act of the drone, while Swammerdam himself conceived the idea that no copulation was necessary, but that some gaseous emanations from the body of the drone produced fecundation by penetrating the body of the queen. About a hundred years later great advances were made in the knowledge of the physiology of the bee. It is said that Jansha, apiarist to the Empress Maria Theresa of Austria, discovered the fact that young queens have to leave the hive to meet the drones ; but it is to the labours of Huber in 1787 and following years, and com-

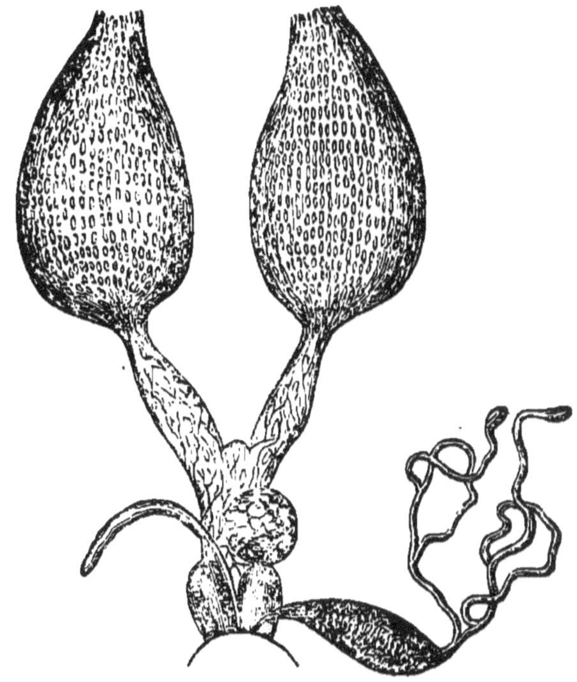

Fig. 18.—OVARIES OF QUEEN.

municated in his letters addressed to Bonnet in the years 1789 to 1791, that we owe the first knowledge of the following main facts :—1. That the queen bee is truly oviparous; that what she deposits is a true egg, which takes three days to produce a living maggot or larva—(even the great Bonnet was inclined up to that time to believe that a minute worm, and not an egg, was produced by the queen). 2. That the queen must be impregnated by the drone in order to become fertile. 3. That copulation is accomplished outside the hive and while on the wing high in the air. 4. That one impregnation was sufficient to fertilise all the eggs laid by the queens subsequently for two years at least, perhaps for life. 5. But that if the act of impregnation was delayed beyond the twenty-first day of the queen's life, her eggs would afterwards produce only drones. Huber also proved that queens could be reared from the larvæ of worker eggs, and also that in some rare cases workers were able to lay eggs, which, however, could only produce drones. He investigated other matters of the greatest importance to the science of bee-culture, and has been gratefully designated THE PRINCE OF APICULTURISTS by Langstroth. He failed, however, to discover the secrets of the spermatheca, and remained under the false impression that the fertilisation of the eggs took place in the ovaries and that there were two sorts of eggs, one sort to produce workers and queens, the other to produce drones, and that they occupied separate portions of the ovaries. His cotemporary, Schirach, who also contributed much to apiarian science, supposed that one branch of the ovaries contained the one sort and the second branch the other sort of fertilised eggs. In this state the science remained for some sixty years. Langstroth says it is now ascertained that Posel, in a work published at Munich in 1784—therefore previous to the experiments of Huber—" describes the spermatheca and its contents and the use of the latter in impregnating the passing egg;" and also that "years ago the celebrated Dr. John Hunter and others supposed that there must be a permanent receptacle for the male sperm opening into the oviduct." Nothing certain was known, however, until 1845, when the brilliant discoveries of Dzierzon led to the promulgation of the THEORY which bears his name, and especially to the doctrine of

PARTHENOGENESIS.

On this point Professor Cook says :—

" This strange anomaly—development of the eggs without impregnation—was discovered and proved by Dzierzon in 1845. Dr. Dzierzon, who as a student of practical and scientific apiculture must rank with the great Huber, is a Roman Catholic priest of Carlsmarkt, Germany. This doctrine—called Parthenogenesis, which means produced from a virgin—is still doubted by some quite able bee-keepers, though the proofs are irrefragable."

No wonder that people were slow to accept this wonderful doctrine. Von Berlepsch, in his exposition of the Dzierzon Theory, says :—

"From time immemorial naturalists have regarded as universally true the doctrine that no living creature can be developed from the egg of a female without male impregnation. And when occasionally exceptional cases were adduced the men of science treated the statements with contempt, or endeavoured to impugn their force or validity by assuming that the observers were either incompetent or careless."

Dr. Dzierzon's discoveries accordingly were received with incredulity and sometimes with derision ; but *magna est veritas, et prevalebit !* Dr. Dzierzon was assisted in proving his case by such scientists as Professors Leuckart and Von Siebold of Munich, and by the Baron von Berlepsch, the author of the celebrated "Apistical Letters."

Von Siebold "demonstrated clearly, that not only do living larvæ occasionally issue from a portion of the unimpregnated eggs of the silkworm, and develop as moths—some male, others female ; but that in various species of butterflies the virgin females regularly lay eggs which, not partially only and occasionally, but uniformly and without exception, produce females."

Prof. Leuckart subsequently noticed a still greater number of exceptions, and says :—

"There can be no doubt that parthenogenesis exists far more extensively among insects than is now known or anticipated."

And Von Berlepsch adds :—

"This exception is found also among bees ; with this difference, however, that among them *all* the eggs which remain unimpregnated invariably develop as males, and those which are impregnated invariably develop as females, and that the impregnation of the egg determines its feminine sexuality. Consequently, in the case of bees, not only is every egg susceptible of development, though unimpregnated, but masculinity *pre-exists* therein, which (marvellous indeed !) is transformed into feminity by impregnation with the male sperm."

THE DZIERZON THEORY.

Space will not admit of going into the details of observations and experiments by which the case has been proved. I shall only add the thirteen "propositions" of the Dzierzon theory, which are now accepted as correct in almost every particular, and each one of which is fully discussed in the excellent work of the Baron von Berlepsch : —

"1. A colony of bees, in its normal condition, consists of three characteristically different kinds of individuals—the queen, the workers, and (at certain periods) the drones.

"2. In the normal condition of a colony, the queen is the only perfect female present in the hive, and lays all the eggs found therein. These eggs are male and female. From the former proceed the drones; from the latter, if laid in narrow cells, proceed the workers, or undeveloped females; and from them also, if laid in wider, acorn-shaped, and vertically suspended, so-called royal cells, lavishly supplied with a peculiar pabulum or jelly, proceed the queens.

"3. The queen possesses the ability to lay male or female eggs at pleasure, as the particular cells she is at the time supplying may require.

"4. In order to become qualified to lay both male and female eggs, the queen must be fecundated by a drone, or male bee.

"5. The fecundation of the queen is always effected outside of the hive, in the open air, and while on the wing. Consequently, in order to become fully fertile, that is, capable of laying both male and female eggs, the queen must leave her hive at least once.

"6. In the act of copulation, the genitalia of the drone enter the vulva of the queen, are there retained, and the drone simultaneously perishes.

"7. The fecundation of the queen, once accomplished, is efficacious during her life, or so long as she remains healthy and vigorous; and when once become fertile, she never afterwards leaves her hive, except when accompanying a swarm.

"8. The ovaries of the queen are not impregnated in copulation; but a small vesicle, or sac, which is situated near the termination of the oviduct, and communicating therewith, becomes charged with the semen of the drone.

"9. All eggs germinated in the ovary of the queen develop as males, unless impregnated by the male sperm while passing the mouth of the seminal sac or spermatheca when descending the oviduct. If they be thus impregnated in their downward passage (which impregnation the queen can effect or omit at her pleasure) they develop as females.

"10. If a queen remain unfecundated she ordinarily does not lay eggs. Still exceptional cases do sometimes occur, and the eggs then laid produce drones only.

"11. If, in consequence of superannuation, the contents of the spermatheca of a fecundated queen become exhausted; or, if from enervation or accident, she lose the power of using the muscles connected with that organ, so as to be unable to impregnate the passing egg, she will thenceforward lay drone eggs only, if she lay at all.

"12. As some unfecundated queens occasionally lay drone eggs, so also in queenless colonies, no longer having the requisite means of securing a queen, common workers are sometimes found that lay eggs, from which drones only proceed. These workers are likewise unfecundated, and the eggs are uniformly laid by some individual bee, regarded and treated more or less by her companions as their queen.

"13. So long as a fertile queen is present in the hive, the bees do not tolerate a fertile worker. Nor do they tolerate one while cherishing the hope of being able to rear a queen. In rare instances, however, exceptional cases occur. Fertile workers are sometimes found in the hive immediately after the death or removal of the queen, and even in the presence of a young queen, so long as she has not herself become fertile."

The foregoing enunciation of the Dzierzon theory is now generally accepted, with scarcely any modification, as the basis of the modern science of Apiculture. As regards Proposition 5, some doubts are still entertained as to whether it may not be *possible* that fecundation may be, in some rare instances, accomplished within the hive. Professor Cook indeed states a case as having come under his own observation, where a "queen whose wing was clipped just as she came from the cell, and the entrance to whose hive was guarded by perforated zinc so that she could not get out, was impregnated, and proved an excellent queen." He adds, "so it seems more than possible that mating in confinement may yet become practicable." Certainly it has not been found practicable as yet, and many additional authentic cases must be recorded before it can even be admitted that there was *no possibility* of a mistake in the isolated case referred to.

Attempts have also recently been made to effect artificial fecundation of the queen larva, and thus produce a queen capable, when she first issues from the cells, of laying both male and female eggs. Some cases of success in this delicate operation have even been asserted, but no satisfactory practical results have been as yet attained. It is the opinion of some who are well qualified to judge, that ultimate success in this direction may be *possible*, and no doubt the most searching investigation and the most careful experiments will be made until certainty shall be attained on the point. Should it ever be found really practicable to regulate cross-breeding in such a certain manner, it would undoubtedly open quite a new era in queen rearing and in the propagation of peculiar races of bees.

The statement in Propositions 3 and 9, that the queen "can effect or omit at pleasure" the impregnation of the egg when passing the spermatheca, has also been open to much doubt until the recent investigations made by Mr. F. R. Cheshire, which were published in the *British Bee Journal* in the months of November and December, 1884, and which seem to decide that question in the most satisfactory way, and beyond all doubt. Referring to fig. 18, the reader will bear in mind that it is now ascertained, as set forth in Proposition 8, that the result of the act of fecundation by the drone is simply to fill the spermatheca with a glandular secretion containing the infinitely small spermatozoa, one of which must be introduced into the egg through a small opening at its lower end, called the micropyle, in order to change its nature from male to female. The eggs so developed in the ovaries are now understood to be all male eggs; if they pass through the oviduct unaltered, they will produce drones; if they receive a spermatozoon into the micropyle while passing the spermatheca on the downward passage, they develop into workers or queens, *i.e.*, into worker eggs. The question heretofore has been, how can the queen control the fecundation of the egg? It was for a long time supposed that when the queen inserted her abdomen into the narrow worker cells, in order to deposit her egg, the pressure from the sides of the cell was sufficient to open the passage from the spermatheca, and lead to the impregnation of the passing egg, whereas in a drone cell there was supposed to be no pressure on the body, and that therefore an unimpregnated egg was laid in it. With regard to the queen cells, in order to support this theory, it had to be assumed that the queen did not lay direct in them, but that the workers supplied these cells with impregnated eggs from the worker cells. Mr. Cheshire, on the contrary, proves that the queen *does* lay eggs in the queen cells, and further, that no outside pressure, even if leading to the death of the insect, could force open the spermatheca; that, on the other hand, she has perfect control over the impregnation of the eggs, and can lay male or female eggs when and where she pleases. He has discovered and described the beautiful valves by which the passage from the spermatheca into the oviduct are guarded, the muscles by which the queen can open and close them at will, and the manner in which the egg passing down from the ovary is allowed to receive one of the spermatozoa through the opening of the micropyle; and this leads him to the very natural reflection—

" What children we must feel ourselves, how utterly baffled and confounded by the reflection that this tiny spermatozoon, eight or ten millions of which the queen may carry in her microscopic spermatheca, has about it—somehow and somewhere—that which shall determine, not sex merely, but all distinctions of species, such as the external form of the body, the length and modelling of the tongue, the arrangement of the pigment cells, the colour of the covering plates, the tint of every hair, and the general temper and disposition of the resulting insect, besides a thousand other peculiarities! We speak of what the microscope *has* revealed, and without gainsaying it is surpassingly wonderful; but how little have we found, compared with what lies behind!"

And in another place he has a beautifully expressed passage which we gladly copy as an appropriate conclusion to our slight survey of the anatomy and physiology of the bee.

" Our bees," he concludes, " are miracles of creative skill, which to a better insight but thinly veils the Worker whose understanding is infinite; and we are not in our weakest moments when He, ' clearly seen by the things that are made,' draws us to bow the head and worship."

DEVELOPMENT FROM THE EGG TO THE BEE.

Having now come to understand the manner in which the egg, whether male or female, comes to be laid, we may examine the egg itself, and the way in which the germ it contains becomes developed into the full-grown insect.

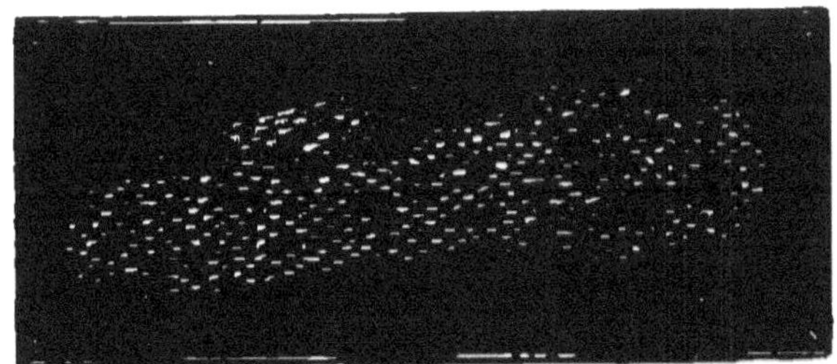

Fig. 19.—QUEEN'S EGG UNDER THE MICROSCOPE.

The egg, when laid in the cell, requires a tolerably sharp sight to distinguish it as it lies at the bottom, attached by one end to the comb by means of some glutinous fluid with which

it is coated. It is very small, and not round or oval like a
bird's egg, but long, like a small worm or maggot. It is, however, a true egg, and presents, when greatly magnified, the
appearance shown.

It appears covered with a sort of delicate network, which is,
in fact, its shell, and it has a yolk and surrounding white, or
albumen, like all eggs of birds or reptiles. When deposited in
a worker cell, it remains unchanged in outward appearance for
three days, when the larva first appears as a minute worm, and
goes through the stages of development shown in the following
figure ; the numbers underneath denoting the age, in days,
from the laying of the egg.

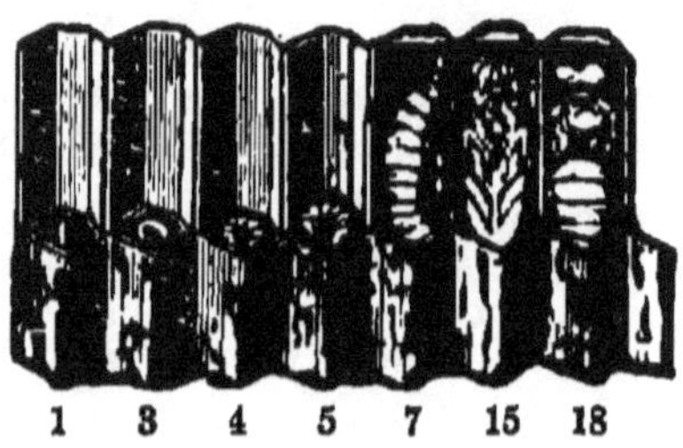

Fig. 20.—FROM THE EGG TO THE BEE.

The larva, when it emerges from the egg, is fed by the
workers, which act as nurses, with a mixture of bee-bread,
honey, and water, the two first-mentioned materials having
undergone a partial digestion in the stomachs of the bees, and
been converted into a species of chyle. Whether the water is
mixed with the food so prepared, or is required for the process
of digestion to prepare it, certain it is that during the breeding
time great numbers of bees are to be seen imbibing water, and
bringing it to the hive. This process of feeding the larvæ
continues five days for the workers and six and a half days for
the drones, and the cells are then capped with a mixture of
wax and pollen, which forms a safe covering for the cells, but
is sufficiently porous to admit the air necessary for the life of
the larva and pupa, or nymph, during its period of metamorphosis. As soon as the cell is closed, the grub begins to spin a
web or cocoon round itself ; this spinning goes on for thirty-six
hours, when the cocoon is complete, and then ensues a period
of rest, or apparent rest, and subsequent metamorphosis, during
which time a wonderful transformation is going on from hour

to hour. This includes the pupa or nymph period, and lasts altogether thirteen days for workers and fourteen and a half for drones ; and at length, on the twenty-second day from the laying of the egg in the former, or on the twenty-fifth day in the latter case, the fully formed bee cuts through the capping of the cell with its mandibles, and emerges complete in every respect, and ready, *without any previous training, education, or experience*, to fulfil its functions, to execute all the delicate operations, and to observe those rules of conduct which appear to us (and justly) to be such marvels of intelligence, ingenuity, dexterity, and even foresight. It is true that the actions of these insects, from the moment they break through the covering of their cells, are evidently prompted and guided by such intelligence and foresight—so indeed was the action of the grub in spinning its own cocoon ;—but is it not absurd to attribute the consequent results to any exercise of a reasoning faculty in

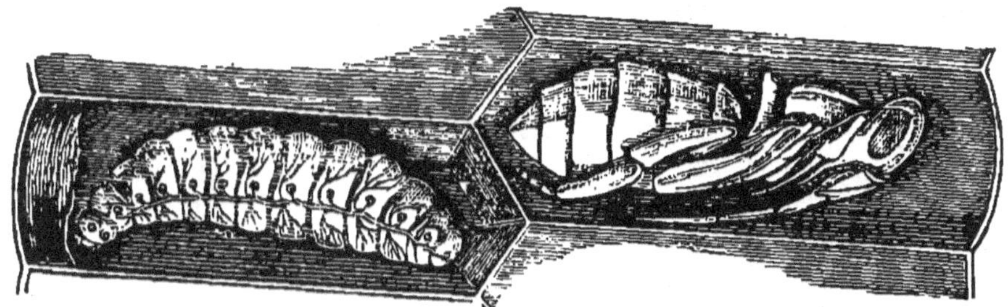

Fig. 21.—WORKER NYMPH AND LARVA, IN COMB.

the insect ? Even if we suppose it endowed at once with the reasoning powers of man himself, would it not require a long period of experience, or education, or both, before it could be capable of building a cell or seeking for and bringing home a load of honey or of pollen ? It is therefore a mistake to talk of the intelligence or ingenuity of the bee ; we have here to deal evidently with *instinct*, which is simply the exercise, on the part of the insect, of an *intelligence not its own*, and which, to make use again of Mr. Cheshire's most appropriate words, "but thinly veils the Worker whose understanding is infinite."

The foregoing illustration (Fig. 21) shows very clearly, at about three times the natural size, the larva when just closed in its cell, and before spinning its cocoon, and the pupa, or nymph, when nearly developed, with the exception of the wings.

The cells in which queen, or perfect female bees are laid and developed differ widely from those of the workers and drones : in the natural state, they are only built in the swarming season, or in cases where the colony has become queenless ; in the former case the cells are laid out for the purpose on the under side or on the edges of the comb, as shown in the following engraving, which exhibits, on an enlarged scale, the top view of a number of worker cells, with the egg and larva in the different stages of development up to the time of capping the cells (in the line marked *a*); a section of a queen cell (*b*), showing the larva and a supply of the royal jelly, and a similar one completed and closed (at *c*).

Fig. 22.—WORKER LARVÆ AND QUEEN CELLS.

Langstroth, in describing the queen cells, says :—

"These cells somewhat resemble a small pea-nut, and are about an inch deep and one-third of an inch in diameter. Being very thick, they require much wax for their construction. They are seldom seen in a perfect state after the swarming season, as the bees, after the queen has hatched, cut them down to the shape of a small acorn cup."

The material of which these cells are composed is not pure wax; there is much pollen mixed with it. The outside surface is uneven and indented like the sides of a thimble. The number built at one time varies much, according to circumstances— sometimes only two or three, but ordinarily not less than five, and sometimes more than a dozen. They are built to hang as nearly vertical as possible, the broader end uppermost, and gradually narrowing towards the point below. The queen lays her egg in the cell when it is about half built; after three days, as in the case of workers and drones, the larva is hatched ; the workers then feed the larva for five days with an abundance of so-called "royal jelly," which appears to be much the same as that fed to worker larvæ, only perhaps more carefully prepared. Mr. Root says on this point :—

"It has also been said that the queens receive the very finest, most perfectly digested, and concentrated food that they (the workers) can prepare. This I can readily believe, for the royal jelly has a very rich taste—something between cream, quince jelly, and honey, with a slightly tart and rank, strong milky taste, that is quite sickening if much of it be taken. I am inclined to think that the same food that is given the young larvæ at first will form royal jelly, if left exposed to the air, as it is in the broad open queen cells."

This food is deposited in a considerable quantity in the queen cell before it is closed (see *b* in last figure). The queen larva takes only one day to spin its cocoon, as it only covers the upper half of the body. This fact was observed by Huber, and commented upon in the following manner :—

"The worms both of workers and males fabricate *complete* cocoons in their cells ; that is, close at both ends and surrounding the whole body. The royal larvæ, on the other hand, spin imperfect cocoons, open behind, and enveloping only the head, thorax, and first ring of the abdomen. The discovery of this difference, which at first may seem trifling, has been the source of extreme pleasure to me, for it evidently demonstrates the admirable art with which nature connects the various characteristics of the industry of bees. . . . . Of several royal nymphs in a hive, the first transformed attacks the rest, and stings them to death " (by biting through the side of the cell, and inserting her sting through the hole so made). "But were these nymphs enveloped in a complete cocoon, she could not accomplish this. Why? Because the silk is of so close a texture, the sting could not penetrate, or if it did, the barbs would be retained by the meshes of the cocoon, and the queen, unable to retract it, would become the victim of her own fury. Thus, that the queen might destroy her rivals, it was necessary the last rings of the body should remain uncovered ; therefore the royal nymphs must only form imperfect cocoons. You will observe that the last rings alone should be exposed, for a sting can penetrate no other part. The head and thorax are protected by connected shelly plates, which it cannot pierce."

The transformations of the queen larva are completed in seven days from the closing of the cell, so that on the sixteenth day from the laying of the egg (six days shorter than the period for the worker, and nine days shorter than that for the drone) the fully developed queen emerges from the cell.

The only other matter to be noticed in this place is the exceptional development of a queen bee from an egg or young larva originally laid in a worker cell. This takes place in abnormal cases only where the hive has suddenly become queenless. As soon as their loss is discovered by the workers, they proceed to build queen cells over worker eggs, or over larvæ not more than three days old. They select a cell for the purpose, with the egg or very young larva in it; they break down the partitions of the adjoining cells, and so make room for the base of a queen cell, which they proceed to build in the usual manner, and to feed the larva with the usual royal jelly, and in due course of time a developed queen is produced. The subjoined figure shows the appearance of such queen cells built over ordinary cells. The ordinary worker cells, with eggs in them, are

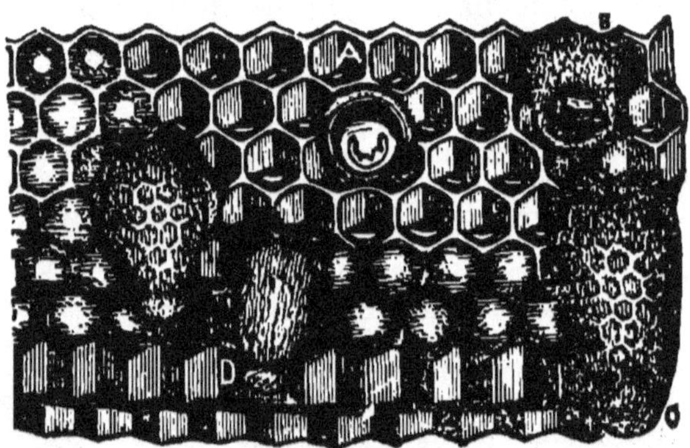

Fig. 23.—QUEEN CELLS BUILT OVER WORKER CELLS.

shown at A; B is a queen cell partly built; and C one completed and closed. D shows a case, which sometimes occurs, of a queen cell built over drone brood. Such cells—which may be known by the absence of indentations on their outer surfaces—are of course useless, as the nature of the drone egg is not altered by the form of the cell or the quality of the food given to the larva.

This phenomenon of queens being reared from worker larvæ caused much astonishment when it was first observed by Schirach and recorded by Huber ; but since the Dzierzon theory has been established it is of course easily understood. If, after a hive becomes queenless, there should be no recently deposited eggs or very young larvæ of workers in the combs, it is of course impossible for the bees to rear a new queen for themselves ; and if they be not provided with a new queen, or with some young larvæ or eggs to rear one from, the colony is certain to dwindle away and perish either from the attacks of robbers or from starvation.

FERTILE WORKERS.

The existence of egg-laying workers in a hive upon certain rare occasions was noted by M. Riem even before Huber's time, and fully confirmed by the latter. The circumstances under which they do occur are given in Propositions 12 and 13 of the Dzierzon theory. They are of course quite useless for keeping up the stock of a hive, as their eggs can only produce drones. Prof. Cook sums up in a few words all that is as yet known about their origin. He says :—

"Huber supposed that these were reared in cells contiguous to royal cells, and thus received royal food by accident. The fact as stated by Mr. Quinby, that these occur in colonies where queen larvæ were never reared, is fatal to the above theory. Langstroth and Berlepsch thought that these bees were fed, though too sparingly, with the royal aliment by bees in need of a queen, and hence the accelerated development. Such may be the true explanation. Yet if, as some apiarists aver, these appear where no (royal ?) brood has been fed, and so must be common workers, changed after leaving the cell, as the result of a felt need, then we must conclude that development and growth, as with the high-holder, spring from desire."

This is evidently one of the matters relating to apiculture about which we have still much to learn. Prof. Cook adds : "Fertile workers seem to appear more quickly and in greater abundance in colonies of Cyprian and Syrian bees after they become hopelessly queenless, than in Italian colonies." If so, it is a point decidedly against the cultivation of those races.

RELATION OF BEES TO FLOWERS.

This subject (the word *flowers* being taken in its widest sense, as applying to all sorts of blossoms of plants, shrubs, and trees) is one of the most important and interesting to which the apiarist can give his attention. The general principle of the relation is very simple, and may be shortly stated here. The detailed investigation and illustration of the subject can only be glanced at in a work of this description. Any one who wishes to follow up the study must apply himself to the works of Darwin, Sir John Lubbock, Herman Müller, and other writers who have done so much for the service of botany, as well as for apiculture, by their labours in this direction. The perusal of such books will be found to be a rich treat, of which we can only give one or two samples which will serve to whet the appetite of the student.

The main principle of the relations between insect-life and plant-life is simply that of mutual advautage, the insects being almost entirely dependent upon the vegetable kingdom for their sustenance, and plants of most sorts being mainly dependent upon the insects for the propagation of their species. A whole host of insects, large and small, but of which the bee is by far the most important, feed chiefly on the saccharine matter secreted in the nectaries of blossoms of all sorts ; and some of them, the honey-bee in particular, require for their own food, or for that of their young, a good deal of farinaceous matter which they find supplied by the fecundating dust of the anthers of the same blossoms, which is called pollen. On the other hand, it is necessary for the fertilization of the plant, that the dust of the anthers should be brought in contact with the pistils of either the same blossom or of some other blossom of the same species. Until a comparatively recent date it was assumed to be the rule, and the intention of nature, that each blossom should be self-fertilizing, except in cases where the stamens and pistils are found in separate flowers ; even then the inter-position of insects would in most cases be necessary for the conveyance of the pollen to the pistil; but it is now made clear to us, by the researches of botanists, especially of Sprengel, Darwin, and H. Müller, that *cross-fertilisation* is one of the chief laws of nature in the vegetable kingdom ; that the pollen of one blossom is intended to be transferred to the pistils of sepa-rate plants of the same species ; and that the means for such transfer are furnished by the visits of insects ; these are at-tracted by the sweets to be found in the nectaries, and in searching for them they become powdered with the dust of the anthers, which is afterwards rubbed off by their contact with

the pistil of the next blossom they visit and that may happen
to have that organ developed. In order fully to understand
the advantages proposed to be obtained by cross-fertilization,
one must read the works of those who have made the subject
their special study; it only concerns us here to note that, as
Mr. Harris says in his book on "The Honey Bee"—

"It must not be supposed, however, that even the hermaphrodite,
or double-sexed, flowers are independent of the visits of bees and other
insects. In all of them, as Darwin has abundantly proved, cross-
fertilization is a most important factor in the continued vitality of any
species, and gives an immense advantage in the 'struggle for existence,' where the conditions for life are not wholly favourable. Indeed,
in many instances, special provision has been made by the Creator
against self-fertilization ; in some cases, by the anthers and pistils coming to maturity, in the same flower, at different times ; in others, by
the placing of the stamens in such a position relatively to the stigma
(or top of the pistil) that it is not possible for the pollen grains of the
one set of organs to fall on the surface of the other."

We shall here give an illustration of one out of the many
interesting cases of special provision, in the structure of the
flower, to secure cross-fertilization by the aid of the bee's
visits.

Fig. 24.—SALVIA OFFICINALIS.

Young Flower visited by a Bee.

In the common sage, *Salvia officinalis*, both the stamens and
the pistil are of a very peculiar form, and the latter is not
fully developed and ready to be fecundated until *after* the
anthers of the same blossom have shed their pollen. The
shape of the flower, and the mode in which the bee enters it
is shown in the above figure, in which the tip of the still
undeveloped pistil is seen just over the back of the bee, which
is forcing its way down to the nectary, through the stamens.
The latter are not visible.

The anthers are shown in the next figure, but on an enlarged scale.

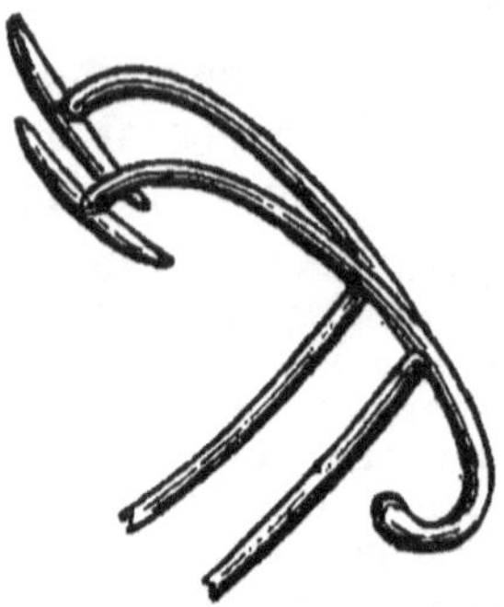

Fig. 25.—STAMENS AND ANTHERS.

The anther cells, instead of being close together, are at the two ends of a long connective, which is attached by a sort of pivot joint at about one-third of its length to the stalk of the stamen. The lower anther cells contain very little pollen, sometimes none at all, while the upper ones are fully developed as shown in the figure. When the bee thrusts its head into the tube, it presses against the lower cells and pushes them back; the connectives revolve on their axis, and the upper anther cells are brought down on the bee's back, the hairs of which brush off the pollen which the bee carries away, and as soon as it meets with an older blossom, in which the pistil is fully developed as seen in the next figure, it is evident that

Fig. 26.—SALVIA OFFICINALIS.
Older flower, with pistil developed.

upon entering the tube of this blossom, the pollen already on the bee's back must be rubbed against the stigma, and the cross-fertilization be thus effected.

Darwin has proved experimentally the great advantages rendered to agriculture by the bee, in the fertilization of clovers, etc.; and Mr. Cheshire and many other writers have demonstrated in a satisfactory manner the services rendered to horticulture by the action of bees upon the fruit blossoms. Finally, the peculiar value of bees, as compared with other insects, in the business of fertilizing plants, is thus described by Herman Müller, in his book on "The Fertilization of Flowers," when speaking of the various fertilizing insects :—

"Bees, which not only feed on the produce of flowers, but nourish their young also thereon, are in such intimate and lifelong relations with flowers, that they show more adaptation to a floral diet, and are more important for the fertilization of our flowers, and have therefore led to more adaptive modifications in these flowers, than all the foregoing orders (of insects) put together. . . . .

"Bees, as the most skilful and diligent visitors, have played the chief part in the evolution of flowers ; we owe to them the most numerous, most varied, and most specialised forms. Flowers adapted to bees probably surpass all others together in variety of colour. The most specialised, and especially the gregarious, bees have produced great differentiations in colour, which enable them, on their journeys, to keep to a single species of flower. While those flowers which are fitted for a miscellaneous lot of short-lipped insects usually exhibit similar colours (especially white or yellow) over a range of several allied species, the most closely allied species growing in the same locality, when adapted for bees, are usually of different colours, and can thereby be recognised at a glance (*e.g.*, Trifolium, Lamium, Tenerium, Pedicularis)."

AXIOM.

"BEES, WHEN FRIGHTENED BY SMOKE, OR BY DRUMMING ON THEIR HIVES, FILL THEMSELVES WITH HONEY, AND LOSE ALL DISPOSITION TO STING, UNLESS THEY ARE HURT." *Langstroth.*

CHAPTER IV.

WHAT BEES COLLECT, AND WHAT THEY PRODUCE.

BEES *collect* three different sorts of raw materials, all of vegetable origin : (1) the sweet liquids secreted by plants in the nectaries of their blossoms, or exuded on any parts of their leafy structure ; (2) the pollen, or fecundating dust of plants ; (3) resinous matter exuded on various parts of some trees and other plants. They *produce*, on the other hand, honey, wax, bee-bread, and propolis. This distinction must be borne in mind if we wish to be precise both in our ideas and our mode of expression. We must look upon bees as manufacturers to a certain extent. The nectar, or vegetable sweet, is not properly called honey until it has gone through the honey-sac of the bee, and been stored and ripened in the cells of the honey-comb ; neither does the pollen of flowers become bee-bread until it has been manufactured by the bees, mixed with a little honey and probably some product of their salivary glands ; and even the resinous matter is not used in the hive in its raw state, but is worked by the bees with some mixture of wax before it becomes what we call propolis. The most important product of the bee is, of course,

HONEY.

The raw material of the honey is entirely a vegetable production ; it is excreted or thrown off by the plant, from the superfluity of its saccharine juices, which, when subjected to chemical analysis, are found to consist of nearly the same constituents as all sugars, starch, gum, and other non-nitrogenous vegetable secretions, namely, of carbon, oxygen, and hydrogen, the two latter in the proportions required to form water. This nectar, therefore, does not contain any of the nitrogenous or of the mineral substances furnished by the soil, and which require to be returned to it, in some degree at least, by the use of manures. Liebig and other chemists have proved that all the elements of the non-nitrogenous vegetable substances are derived from the atmosphere and from rain-water; it is clear, therefore, that no quantity of honey produced in any district can tend to impoverish the soil from which the nectar is collected. While lying in the nectaries of blossoms, and being collected by the bee, or afterwards when being stored in the honey-comb, it may by accident take up some particles of pollen, which will account for the fact that minute grains of that substance are generally discoverable in honey when examined with the microscope. In its passage through the honey-sac of the bee, and in the act of being stored in the cells of the comb, the raw juice goes through a process of ripening, which deprives it of all superfluous watery particles, and while in the honey-sac it is also probably in some way chemically affected by the juices from the salivary glands of the bee. When quite "ripe," it is hermetically sealed in the cells by the worker bees, just as the preserves of a careful housekeeper are closed up so as to save them from the action of the oxygen in the atmosphere. The honey in this ripened state is nearly the same, in point of chemical composition, as ordinary sugar ; but it owes its perfume and flavour apparently to ˜the same volatile oils which attracted the insects to the flowers from which it is derived, and that it is indeed something very different from common sugar is sufficiently clear to every one. Chemical analysis is an invaluable aid in the prosecution of scientific investigations, and it is quite astounding to the layman to observe the nicety of the results which modern chemists can arrive at ; but still there are some subtile peculiarities of matter which seem to evade all analytical examination. The constituent parts of starch and of gum are very nearly the same as those of sugar, and yet how different to our sight, touch, and taste are all those substances ! On this subject Professor Cook remarks as follows :—

"Nectar of flowers and honey are quite different. The former contains more water, is neutral instead of acid, and the sugars taken from the flowers are much modified while in the alimentary canal of the bee in transit from flower to comb. Nectar consists of sucrose, or cane sugar, from twelve to fifteen per cent., and mellose, or uncrystallisable sugar ten per cent. The remainder is mostly water, though there is always a small amount of nitrogenous material. In honey the cane sugar is largely changed to a substance chemically like glucose : the mellose seems also somewhat modified.　There is a little mannite, probably the result of chemical change in the bee's stomach.　The acid condition of honey is plainly recognisable by the taste, as all lovers of honey know."

The small amount of nitrogenous material above referred to is no doubt owing to the minute particles of pollen which get accidentally mixed with the honey in the way already mentioned.

Langstroth, who has left scarcely any point connected with his subject unconsidered, says in reference to this matter :—

"The notion that bees can change *all sweets*, however poor their quality, into *good honey*, on the same principle that cows secrete milk from any acceptable food, is a complete delusion. . . . . That the honey undergoes *no* change during the short time it remains in their sacs cannot positively be affirmed, but that it can undergo only a *very slight* change is evident from the fact that the different kinds of sugar syrup fed to the bees can be almost as readily distinguished after they have sealed them up as before."

This is undoubtedly true ; it is clear that bees do not *secrete* the honey, as they do wax, for instance ; but it is not quite correct to say, in the words of Aristotle, as quoted by Langstroth, that "bees *gather* honey, but do not *secrete* it."　They gather *a substance which is converted into honey*, partly, perhaps, in its passage through their honey-sacs, and partly by a subsequent process of evaporation in the cells of the comb.　The flavour and perfume may be retained, notwithstanding an important change in other respects ; and that some such change does take place is shown clearly by Langstroth himself, though he attributes it chiefly to the evaporation of the watery particles of the honey.　He remarks that "bees are very unwilling to seal it over until it has been brought to such a consistency that it is in no danger of becoming acid in the cells;" and in another place, when speaking of honey gathered from poisonous flowers, he says, "In some of our Southern states, all that is unsealed is rejected.　The noxious properties of honey gathered from poisonous flowers would seem to be mostly evaporated before it is sealed over by the bees."　However this may be, it is evident to our senses that a change does take place, and the conversion of a poisonous or deleterious substance into a wholesome article of food would seem to indicate a decided chemical action in addition to the merely mechanical separation of the watery particles.

The characteristic *structure* of honey is described as follows in an American publication called the *Druggists' Advertiser* :—

" Under the microscope, the solid part of honey is seen to consist of myriads of regularly formed crystals ; these crystals are for the most part exceedingly thin and transparent, and very brittle, so that many of them are broken and imperfect ; but when entire they consist of six-sided prisms, apparently identical in form with those of cane sugar. It is probable, however, that these represent the crystals of dextrose, as they occur in honeys from which cane sugar is nearly or wholly absent. Intermingled with the crystals may also be seen pollen granules of different forms, sizes, and structure, often in such perfect condition that they may be referred to the particular plant from which the juices have been gathered. Crystalline sugar, analogous to grape sugar, may he obtained by treating granular honey with a small quantity of alcohol, which, when expressed, takes along with it the other ingredients, leaving the crystal nearly untouched. The same end may be attained by melting the honey, saturating its acid with carbonate of calcium, filtering the liquid, then setting it aside to crystallise, and washing the crystals with alcohol. Inferior honey usually contains a large proportion of uncrystallisable sugar and vegetable acid. When diluted with water honey undergoes the various fermentations, and in very warm weather an inferior grade of honey will sometimes undergo a change, acquiring a pungent taste and a deeper colour. The usual adulterations of honey are with various forms ot starch, as those of the potato and wheat, and with starch and cane sugars. The starch is added to whiten dark honey, and to correct the acidulous taste which old honey is apt to acquire, as well as for the sake of increased weight. The presence of starch may be readily detected by the usual iodine test."

ADULTERATION OF HONEY.

The honey industry is so injuriously affected by any attempts to impose adulterated stuff upon the public in place of the pure article, that it becomes a matter of importance to the bee-keeper to know how to detect and expose any such attempts at imposition. In Quinby's "New Bee-keeping" we find the following passage, which is worthy of particular attention :—

" The first fact to be understood is that all granulated or candied honey is presumably pure. The natural inference is that such is the best to buy. It is also well established that all pure honey will, as a general rule, granulate if exposed to a sufficiently low temperature. To this rule exceptions have been reported, and such have occurred under my own observations, as will soon be noticed. Thus, ordinary honey remaining liquid in cold weather, when exposed to the air, should be regarded as suspicious, and put to a test. The presence of

glucose in such honey may be ascertained as follows :—Place a small
quantity in a cup, and add to it some strong tea ; if the poorer grades
of glucose are present it will turn dark like ink ; if it is combined with
the better quality of glucose, the fact may be determined by the use of
a little alcohol ; pure honey will unite with alcohol, but glucose has no
affinity for it, and they will separate, like oil and water.　A common
method of adulteration has been practised by placing a piece of fine
comb-honey in a jelly cup and filling it up with glucose.　If this were
pure honey it would become candied and conceal the comb.　Yet these
are found unchanged upon our grocers' shelves the year round.　If
honey is put in a can, and heated and sealed, the same as fruit is
canned, it will remain liquid until opened.　The specimens of comb
mentioned above could not have been thus treated, as the process
would have melted the comb."

The only thing which appears to call for remark in the foregoing quotation is the expression that the honey will granulate
"if exposed to a sufficiently low temperature," which might
lead to the idea (which is sometimes erroneously entertained)
that the granulation or solidifying of the honey is owing to a
freezing process ; it is, indeed, sometimes rather loosely termed
congealed honey.　In this climate the honey very often granulates most rapidly in the warmest summer weather, and in the
height of the honey season it will sometimes become quite
solidified within a week or two after being extracted.　This
appears to be especially the case with white clover honey.
Here at Matamata I have known it to become quite solid in
the tank in forty-eight hours from the time it was extracted,
and this in the middle of summer.　Perhaps the most notable
exception to the rule of pure honey granulating easily may be
found in some of the sage honey of California.　I have kept
some, obtained from Mr. Wilkin, of San Buena Ventura, all
through the winter months, without its showing the slightest
sign of change from its liquid state.

In the process of ripening, most honeys become darker as
well as thicker ; so that clear, transparent honey is not *always*
the best.

Mr. Otto Hehner, analyst to the British Bee-keepers' Association, has reported pretty fully on the subject of adulterated
honey, in a paper read before the conference of bee-keepers at
the International Health Exhibition, in the month of July
1883.　In that paper he informs us—

"There are three classes of manufactured honey : first, honey made
from ordinary sugar, and essentially consisting of cane-sugar syrup ;

second, that obtained by the action of an acid upon cane sugar, and consisting, as does genuine honey, of water, dextro and levogluoose; and third, the product of the action of acid upon starch, called corn syrup. I have never met with any samples of the first of these three classes, and I doubt whether any such article can now-a-days be found, although in older works on adulteration their occurrence is asserted. The second kind is also very rare, but yet it exists; but the third, starch syrup, is the main substitute and adulterant used at the present time."

After describing the characteristics of the first two sorts of adulteration, and the mode of detecting them, he goes on to the third sort, which is the most important, as being the most likely to be met with :—

"Corn or starch-syrup, lastly, differs in almost every respect from the genuine product. It throws down abundant precipitates with lead or barium solutions, often with alcohol; it does not ferment completely, but leaves about one-fifth or one-sixth of its weight as unfermentable, gummy residue, and, examined by the polariscope, turns the ray of light powerfully to the right.

"These few simple tests readily enable us to distinguish these products from each other, and from honey. Examined with the microscope they all are found to be devoid of pollen; and, in consequence, are without the delicate aroma, the bouquet, which is inseparable from the product of the flower and the bee.

"From the very variable amount of pollen granules met with in different honeys—some samples which I have examined containing enormous numbers, others but very few—there appears to be a considerable difference in the degree of cleanliness with which bees store the honey. Some flowers yield an infinitely larger number of pollen granules than others, but the importation of the latter to a greater or less extent into the honey itself appears to me to depend mainly upon the bee itself."

The inference to be drawn from the latter part of this quotation, that honey owes its flavour and bouquet to the pollen grains which may chance to be mixed with it, is one, the correctness of which appears to us to be more than doubtful.

HONEY DEW.

Bees sometimes gather a quantity of saccharine liquids found on the leaves or branches of trees and other plants, or dropping from them on to the ground. This liquid is generally called honey-dew; but there appears to be a great diversity of opinion amongst otherwise well-informed writers, as to its nature and origin, some even describing it as " a dew that falls at night, and is sweet like honey;" while some assert that it is the secretion of aphides, or plant lice; and others, that it is simply an exudation from the plants or trees on which it is found. The first-mentioned idea is now quite exploded; the two latter, between them, contain the truth. German apiarists make a distinction between *Honigthau* (honey-dew), which they look on as an excretion of aphides, and *Nebenblatthonig*, which they consider to be exuded by the leaf-buds of certain plants or trees. Liebig asserts that when

"the quantity of sugar present (in the sap of a tree) is greater than can be exhausted by the leaves and buds, it is execeted from the surface of the leaves or bark. Certain diseases of trees, for example that called honey-dew, evidently depend on the want of the due proportion between the quantity of azotised and that of the unazotised substances which are applied to them as nutriment. . . . . When a sufficient quantity of nitrogen is not present to aid in the assimilation of the substances destitute of it, these substances will be separated as excrements from the bark, roots, leaves, and branches. The exudations of gum, mannite, and sugar, in strong and healthy plants, cannot be ascribed to any other cause."

He instances several cases of copious exudation of a thick sweet liquid from the leaves of trees, more especially from those of the linden, during a hot and dry summer. Ludwig Huber, in his *Bienenzucht* (1880), mentions three sorts of *Honigthau*, one which is voided by the plant lice, which is the worst; "it is tough, tasteless, and unhealthy for the bees in winter;" another is obtained from the leaves of the oak, which, when bitten by a certain kind of beetle, give out, " by favourable temperature, a sweet liquid, which the bees eagerly lick up;" and the third sort is exuded from the leaves of the linden, oak, plum, and other trees, generally in the early mornings, by a change of temperature after very close, hot weather. Most of the American writers on bee-keeping appear to have had little or no personal experience of honey-dew. Professor Cook mentions it as occurring in California, in a case where it was unmistakably exuded from the tree leaves; and Langstroth quotes the English entomologists, Spence and Kirby, as to clear cases of plant-louse origin; also Bevan, who, however, does not suggest anything as to the cause of the phenomenon, but merely says :—

"Honey-dew usually appears upon the leaves as a viscid transparent substance, as sweet as honey itself, sometimes in the form of globules, at others resembling syrup. It is generally most abundant from the middle of June to the middle of July—sometimes as late as September. It is found chiefly upon the *oak*, the *elm*, the *maple*, the *plane*, the *sycamore*, the *lime*, the *hazel*, and the *blackberry* ; occasionally also on the *cherry*, *currant*, and other fruit trees. Sometimes only one species of trees is affected at a time. The oak generally affords the largest quantity. At the season of its greatest abundance the happy humming noise of the bees may be heard at a considerable distance, sometimes nearly equalling in loudness the united hum of swarming." Langstroth adds, " that in some seasons the bees gather large supplies from these honey-dews, but it is usually abundant only once in three or four years. The honey obtained from it, though seldom light coloured, is generally of a good quality."

Some writers in the American Bee Journals at present, however, speak of the honey gathered from honey-dew in some of the States as being wretched stuff, neither fit to be sent to market, nor to be fed to the bees in winter. No doubt there are many different qualities of honey-dew, as well as of blossom honey, and bee-keepers in New Zealand and Australia should be ou their guard to ascertain all the sources from which their bees gather their stores, and to test the qualities of the different products.

WAX.

Previous to Huber's time it was generally believed, and asserted by writers upon apiculture, that wax was *collected* by the bees, or formed by them from bee-bread, either in its crude state or after undergoing a process of digestion. The accurate observations of Huber, however, led him to doubt the correctness of that theory, and he ultimately proved its utter fallacy by careful experiments made in the following manner. He confined bees to their hives, without a particle of pollen, and fed them with sugar syrup, and at the end of a few days they had built several beautifully white combs. They were then deprived of these, and supplied with honey and water, when combs were again constructed. This was repeated seven times ; all the time the bees were prevented from flying, thereby proving that wax is secreted, and not gathered, by them.

Langstroth remarks, with his usual sagacity and caution, that although Huber has clearly proved

" that bees can construct comb from honey or sugar, without the aid of bee-bread, and that they cannot make it from bee-bread, without honey or sugar, he did not prove that they can continue to work in wax when *permanently* deprived of bee-bread. . . . . Some bee-bread is always found in the stomach of wax-producing workers, and they never build comb so rapidly as when they have free access to that article.   It must therefore either furnish some of the elements of wax, or in some way assist the bee in producing it.   Further investigations are necessary, before we can arrive at perfectly accurate results."

He further points out the fact that, while honey and sugar contain by weight about eight pounds of oxygen to one of carbon and hydrogen, the wax contains only one pound of the first to more than sixteen of the two latter; and that, as the combustion of oxygen is the great source of animal heat, the great quantity consumed in the conversion of honey into wax

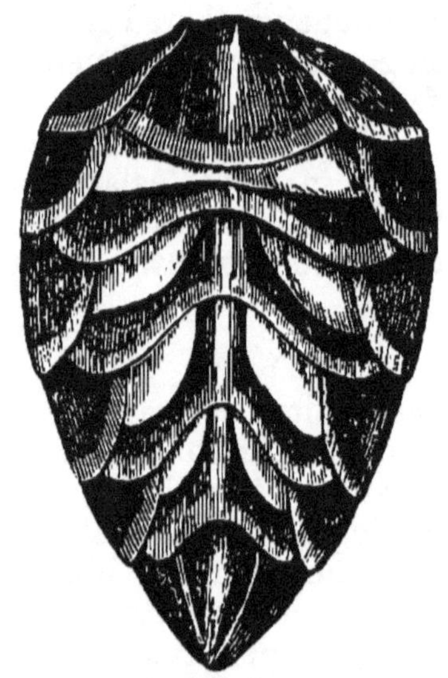

Fig. 27.—UNDER SIDE OF ABDOMEN OF WORKER BEE, SHOWING WAX POCKETS AND WAX SCALES.

"must aid in generating the extraordinary heat which enables the bees to mould the softened wax into such exquisitely delicate and beautiful forms."   The force of this observation will be seen when we recollect that wax requires a temperature of about 145° to melt it, though it may be moulded, by pressure, at 100° or less.   Is it not probable that the way in which "bee-bread assists the bee in producing the wax," as Langstroth expresses it, is that its nitrogenous qualities serve to keep up the bodily strength of the insect during the exhausting work of secreting the wax and building the comb ?   This appears to be Professor Cook's view ; he says,

"That nitrogenous food is necessary, as claimed by Langstroth and Neighbours, is not true. Yet, in the active season, when muscular exertion is great, nitrogenous food must be imperatively necessary to supply the waste and give tone to the body. Secretion of wax demands a healthy condition of the bee, and so indirectly requires some nitrogenous food."

At all events, it is now well known that the wax is exuded from the body of the worker bee, and formed in thin flakes in what are termed the wax pockets, of which four may be observed in the foregoing engraving, on each side of the centre line on the under part of the abdomen, and which are, in fact, the folds of the shell-like plates covering the abdominal rings.

The wax can only be secreted when the temperature of the hive is above a certain point, and during the time of secretion the bees appear to hang in clusters or festoons, in a state of absolute repose.   In the height of the honey season, or so long as new comb is required, this secretion goes on night and day. The constituents of wax, according to the analysis of Hess, are—

Oxygen ...	... 7.50
Carbon ...	... 79.30
Hydrogen	... 13.20
	100·00

Langstroth says that "careful experiments prove that from thirteen to twenty pounds of honey are required to make a single pound of wax." This has been until lately accepted as a well-ascertained fact ; but within the last few years some American apiarists have begun to doubt if quite so much honey was consumed, and lately it has been stated, on the strength of some isolated experiments, that the bees do not consume more than eight pounds of honey in order to secrete one pound of wax. Many more careful experiments will be requisite before this can be satisfactorily proved or disproved. In the meantime it may be asserted that something between *eight* and *twenty* pounds are required, and we cannot be far astray if we assume the mean of these figures, or fourteen pounds, to be correct for any purposes of calculation.

BEES WASTING WAX.

It has been generally asserted and believed that wax secretion only takes place as required; but recently two or three leading bee-keepers—Mr. Doolittle amongst the number—have expressed the opinion that during a heavy flow of honey the secretion goes on, whether wax may be required for comb-building or not; and that at such times, if not needed, the wax goes to waste. From my own observations I think these cases are very rare. A certain amount of wax is wasted while comb-building *is* going on, as may be seen by examining the bottom board of a hive shortly after a swarm has been placed in it, when small scales of wax will be found scattered under the place where the bees are working. In accounting for this waste, I am of the same opinion as Mr. Dadant; he says:

"The cause can be told in two words, *cold nights*. When bees harvest honey in large quantities, the weather is often cool in the nights, and very hot in the day. The wax producing bees hang in clusters; but in a cool night, those which are on the outside of the cluster feel the change of temperature, and when the scales of wax come out of the rings of the abdomen, if they are not at once taken by other bees and fastened to their places, they become too hard for easy manipulation, and are allowed to drop to the floor." He quotes an instance in support of this view where he found about three dozen bees of a small swarm, upon which the wax had cooled so promptly that the little white scales were still fastened in the rings of their abdomen.

EXTRAVAGANT WASTE OF WAX.

There is a much more serious waste of wax going on in these colonies, that might easily be avoided. The number of bush hives taken by bushmen and country settlers every season must be enormous, yet probably in not one case out of a hundred is the wax saved. A country settler once informed me that he and others had taken about fifty bush hives during the season, and when asked what he had done with the wax, he replied, "Oh, we threw it away." Now each bush hive will, on an average, yield four pounds, or more, of clean wax, worth at the present time from tenpence to one shilling per pound, with a market demand for it far beyond the supply. Careless bee-keepers also waste much wax by not taking care of the pieces of comb that will accumulate about an apiary. If all the pieces, however small, were saved and put away in some moth-proof place till the end of the season, and then cleaned, the proceeds from this source would make a sensible addition to the profits of the apiary.

It may not be out of place here to show the importance of the trade in beeswax, in the United Kingdom alone, during the years 1882 and 1883. The imports in the former year amounted to 1,776 tons 18 cwt., valued at £126,926, and in the latter to 1,409 tons 12 cwt., valued at £87,146. The exports of wax from Chili to all parts in 1883 were 93 tons 1 cwt., valued at 85,617 dollars. These figures are taken from official records, and will serve to give some idea of the importance of wax as an article of commerce.

METHOD OF RENDERING WAX.

No doubt one great cause for the waste of wax in this part of the world has been the want of means and knowledge to clean it in a quick and inexpensive manner. This want is now

Fig. 28.—THE GERSTER WAX EXTRACTOR.

supplied by what is called the "wax extractor," a figure of which is placed above, and the following is the method by which it is accomplished :—

The basket B is made of perforated tin, and it is into this that the pieces of comb, etc., are to be put. When the basket is filled, the cover of the vessel A is to be removed, and the basket placed inside, resting on the fixed shallow pan, shown where the side is cut away. This pan has three pieces of tin fixed near its inside rim (only two of which are visible), to support the basket a little distance from the bottom. The spout D is fixed to the pan inside, so as to take all the wax as it falls from the perforated basket above. Now, to set the machine working, we have only to supply steam around the basket. This is done by setting A over a pan or kettle of boiling water. The steam then melts the wax, which will run out at D, while the refuse is retained in the can.

Fig. 29.—JONES' WAX EXTRACTOR.

Another wax extractor, an improvement on the "Gerster," is shown above. It was invented by Mr. D. A. Jones, of Canada. The improvement consists in carrying a perforated cone up through the centre of the basket to near the top, thus allowing the steam to penetrate more easily through the mass, saving both time and fuel.

The extractor which I have in use, and which may be considered a rather large one, measures as follows : outside can, diam. $13\frac{1}{2}$ inches, depth 13 inches ; inside basket, diam. $11\frac{1}{2}$ inches, depth $11\frac{1}{2}$ inches ; boiler, 11 inches deep. The can and basket are made of stout tin, the boiler of galvanised iron ; cost 30s.

There is also another very simple way of cleaning wax, in places where the apiarist may not find it convenient to use the " wax extractor ; " i.e., by taking small bags made of coarse scrim, into which the combs, after being squeezed into balls, should be placed. These bags are then put into a kerosene tin partly filled with water, and boiled ; this melts the wax, which floats on the top of the water, while the refuse remains in the bags. The wax is skimmed off and put into vessels containing some *hot water*, and allowed to cool gradually. By this method any remaining dirt or foreign matter will be precipitated, and can be scraped off the bottoms of the cakes when cool. To get all the wax out of the bags it is necessary to press them ; this can be done by having a piece of inch board cut rather less than the inside dimensions of the tin, with a few holes bored in it. This should be pressed down on the tops of the bags, and held in that position while the wax is being skimmed. By adding a little vinegar to the water in which the wax is melted, it may be separated from the refuse much more readily.

COMB.

Wax, after being produced by the bees, is formed by the workers into comb, which consists of hexagonal-shaped cells of two sizes—one for the deposit by the queen of the worker eggs, the other for the same purpose, for drone eggs ; and these are known to apiarists by the names of "worker" and "drone" comb (Fig. 30).

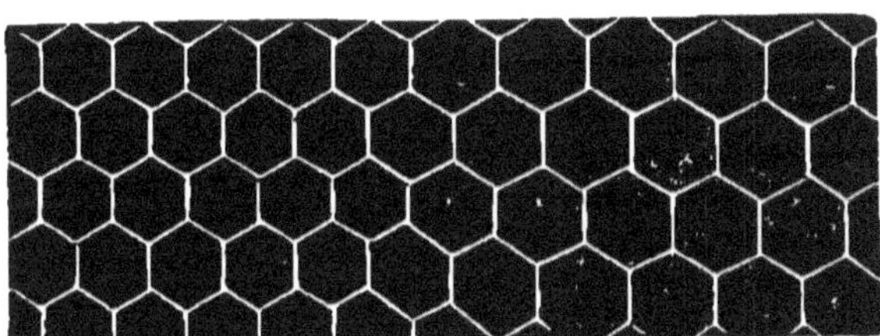

Fig. 30.

WORKER CELLS. DRONE CELLS.

HOW CONSTRUCTED.

The wonderful instinct of these little workers is amply shown in the construction of the comb; for there is no other form known to mathematicians in which the cells could be constructed—1st, to occupy the least possible space ; 2nd, with a view to consume the least material ; 3rd, for the comfort and health of the young bees. The cells are constructed on both sides of the foundation, in a horizontal plane to it, which is

perpendicular. Some of our greatest naturalists have made
the process of building up honey-comb their special study.

" The expedients tried by Huber unfolded the whole process. He
was enabled to bring each bee so completely under view that it could
be seen to extract with its hind feet one of the plates of wax from
under the scales where they were lodged, and, carrying it to the
mouth, in a vertical position, turn it round ; so that every part of its
border was made to pass in succession under the cutting edge of the
jaws. It was thus soon divided into small fragments ; and a frothy
liquor was poured upon it from the tongue, so as to form a perfectly
plastic mass. This liquor gave the wax a whiteness and opacity
which it did not possess originally, and at the same time rendered it
tenacious and ductile. These materials, thus blended, having been
accumulated in the hollow of the teeth, issued forth like a very narrow
ribbon. The tongue, during this process, assumed the most varied
shapes, and executed the most complicated operations; and after
drawing out the whole substance of the ribbon in one direction drew
it forth a second time in an opposite one. It was, doubtless, the
issuing of this masticated mass from the mouth that mislead Reaumur
and caused him to regard wax as nothing more than digested pollen."
—*Bevan*.

ADVANTAGES OF THE HEXAGONAL FORM OF CELLS.

There are only three geometrical figures into which a given
plane surface can be divided into perfectly equal parts—the
square, the triangle and the hexagon; and of these three the
form which most nearly approaches that of a circle, and there-

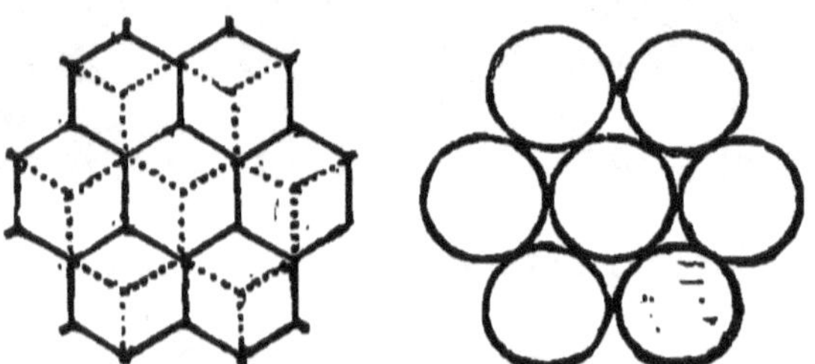

Fig. 31.—HEXAGONAL CELLS. Fig. 32.—CIRCULAR CELLS.

fore the most suited for the development of the larva and
nymph forms of the young bees, is the hexagon. The above
figures will exhibit at a glance the loss of space and the
waste of material that must result if the cells of the comb were
built of circular, as compared with the hexagonal form.

In the *American Apiculturist* for March, 1885, there is an interesting article on the "Origin of the Cells of the Hive Bee," by A. Todd, and also a description of "Holman's new illustration of cell formation" in vegetable structure. The latter consists of an apparatus for forcing soap bubbles between two parallel plates of glass, when it is seen that the spherical bubbles, when crowded together, are changed into "polygonal forms analogous to those that we see in cross sections of wood fibre." These new forms are nearly all hexagonal, though by no means regular hexagons. Founded on this and upon other considerations—such as the natural tendency of bees to build cylindrical cells when working with an excess of material—Mr. Todd informs us that "Herr K. Mullenhoff has found a quite simple and satisfactory solution of the question" (the form of the hive bee's cells), "which neither admits of any mysterious instinct, nor, on the other hand, credits the bee with the knowledge of the differential calculus." He argues that the bees are in some way forced to build the foundation of the cells and the side walls of the hexagons, in consequence of the position they assume when beginning to build, and in consequence of the natural tendency of the soft wax to take form in accordance "with the laws which Plateau has discovered for his equilibrium figures." Mr. Todd considers this solution very satisfactory, as "not a few writers, even to the present day, maintain that we have here a typical case of 'instinct,' in the old acceptation of the word; that is, of blind, unconscious, untaught action, producing results which men can only reach by dint of highly cultivated reason." The line of reasoning intended to confound those old-fashioned writers is scarcely a convincing one. Somehow, between the bee and the wax, the comb receives the form we admire and find to be the best. If the result can be proved to be partly owing to the wax being affected by a law of nature, which Plateau has been so fortunate as to discover, then it only demonstrates the truth that inert matter, as well as the living organism, exhibits a "blind, unconscious, untaught action," leading to some end intended by the Giver of that law. It would be equally silly to attribute to the bee, as to the wax, a knowledge of the differential calculus, and therefore we are thrown back upon the quality of instinct, which is admirably defined by the words quoted above.

The bees, if left to themselves, usually build a large amount of drone comb for storage of honey; and thus in another season a great and what appears to be an unnecessary quantity of drones appear on the stage. This is now obviated by the introduction of artificial comb foundation; for, as will be seen, the apiarist now can regulate the work, and so force the bees to produce either drones or workers, as he may most require.

POLLEN AND BEE-BREAD.

Pollen is the name given to the dust-like particles of farinaceous matter which constitutes the fecundating principle of the stamens of flowers and blossoms of all sorts. The manner in which it is collected by bees, and the part it plays in the "relation between insects and flowers," have been already described in Chapter III. It is of great importance in the economy

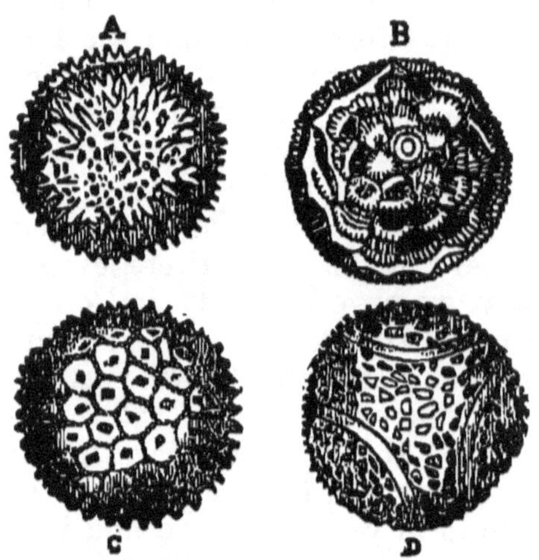

Fig. 33.—POLLEN GRAINS UNDER THE MICROSCOPE.

of the hive, as, after being mixed with a little honey, and packed in the cells of the brood combs, it forms the bee-bread, which is indispensable to the nourishment of the young bees, and without which, as has been proved, no brood can be raised. It is very rich in nitrogenous substances, which are necessary for the formation and maintenance of muscular tissue, and therefore to the development of the young bees.

The grains of pollen, although so minute as to form an almost impalpable dust, exhibit when viewed through a powerful microscope very beautiful and distinctive markings, according to the plants from which they are obtained. The engraving (Fig. 33) shows four different sorts of grains; that marked A being from the hollyhock; the sources of the other grains are not given in Root's "A B C of Bee Culture," from which the illustration is taken.

Previous to Huber's experiments, bee-bread was supposed to be used in comb-building. He, however, proved, as we have seen, that comb could be built by the bees in confinement, by being fed with honey or sugar syrup alone. He was not long in discovering that pollen was used for the nourishment of the young bees. Confining some bees to their hives without pollen, he supplied them with larvæ, honey, and eggs. In a short time the young all died. A fresh supply of brood being given them, with plenty of pollen, the development of the larvæ proceeded in the natural way. The following analysis of bee-bread is taken from the *Journal of Horticulture :*—

Artificial nitrogenous organic substance	36·59
Water	12·74
Ash	2·72
Albuminous	21·75
Sugar	26·20
	100·00

Here we find nitrogen and albumen predominating.

The mode in which the bee-bread is deposited in the comb-cells is graphically described in the *British Bee Journal* for May, 1876, as follows :—

" The pollen laden bee, upon entering the hive, makes directly for the brood nest, and where its load is required, it quickly disencumbers itself. Sometimes the nurse bees are in want of the all-necessary pollen, and nibble it from the legs of the worker without ceremony ; but more often the bee goes to a cell devoted to pollen-storing and hangs by its first pair of legs to another cell immediately above, and (as it were) kicks the balls of pollen into the proper receptacles. Here they are mixed with a little honey and kneaded into a stiff paste, which is then rammed hard against the bottom of the cell for future use, the bee using her head as a battering ram ; these operations are repeated until the cell is almost filled with the kneaded dough, when a little clear honey is placed on the top, and it is sealed over and preserved as bee-bread. If a cell full of pollen be cut in two longitu-

H

dinally its contents will, as a rule, be found of many colours, stratified, the strata of varied thickness standing on edge, as if the bees, instead of storing bread had stored pancakes."

ARTIFICIAL POLLEN.

In cold climates, to induce brood-rearing in early spring, before there is a supply of natural pollen, recourse is had to a substitute. Artificial pollen, as the substitute is termed, generally consists either of pea-flour, wheaten flour, rye-flour, or a mixture of finely ground oats and wheat. This is placed in a sheltered position in the apiary, with short straws or shavings mixed with it, for the bees to alight upon, and a little honey or syrup near to attract the bees and get them started, when there will be no further trouble, save the replenishing the material, if required. This is unnecessary in most, if not all, parts of New Zealand and Australia, where bees can work freely all through the winter, and gather abundance of pollen all the year round.

PROPOLIS.

This is a substance used by the bees for glueing things together, and for stopping up all crevices in their hives. In order to make it they gather the resinous matter which exudes from some trees ; or when this is scarce, they will take varnish, or even tar, whenever they can find it. They carry this substance home in their pollen baskets, and use it, mixed with wax, wherever they want to fasten any loose parts, or to fill up joints to exclude enemies or air. They make a very liberal use of it at the end of the honey season. It is also used for other purposes by them, as the following anecdote will testify:—

"A snail having crept into one of M. Reaumer's hives early one morning, after crawling about for some time, adhered, by means of its own slime, to one of the glass panes. The bees having discovered the snail, surrounded it, and formed a border of propolis round the verge of its shell, and fastened it so securely to the glass that it became immovable."—*Bevan*.

I have seen the same thing occur at the Thames, when transferring bees from an old box to a bar frame hive.

CHAPTER V.

THE APIARY.

LOCATION.

TAKING into consideration the climate, the native flora, and the results which have heretofore followed the introduction of bee culture, according to the reports furnished to the *New Zealand and Australian Bee Journal*, in all parts of Australasia, I feel safe in saying that there is no part of these colonies which is at all fitted for European settlement, where the culture of bees may not be carried on to a greater or less extent with advantage. But it does not therefore follow that every district is adapted for the working of extensive apiaries. No person should attempt the establishment of a large apiary without first making himself acquainted with the resources of the neighbourhood, and to do this effectually he must first have a knowledge of the flora which is best suited to his purpose. A careful perusal of the chapter on "Bee Forage" will aid him in that respect, so that it is here only necessary to draw attention to the subject.

GENERAL ARRANGEMENT.

The first consideration should be to have the apiary as convenient to the dwelling of the person who is to work it as may be compatible with a due regard to aspect and shelter. If it can be laid out so as to be within view of the dwelling-house, it will be all the better. It should not, however, be so near to a public road, or to a railway, that the ground could possibly be shaken by the passing traffic. Shelter, to protect the hives from the prevailing high winds, is absolutely necessary for the welfare of the inmates. Bees do not thrive nearly so well when their hives are exposed to cold and stormy winds, and especially in early spring, when it is so essential to the bee-keeper to keep them in the best possible condition. Much time is saved in a well-sheltered apiary, as all the necessary manipulations can be got through more rapidly and securely, and with greater satisfaction to the apiarist. At the same time there must be free access to the hives on every side ; and there can be no greater mistake made than that of placing hives with their backs close up to a hedge, wall, or paling ; first, because the hives have to be manipulated from behind, and from the side, but never from the front ; and secondly, because such a position exposes them to the ravages of spiders and other insects, and favours dampness. If the shelter is to be secured by planting, I would advise the selection of trees or shrubs which will not grow very high ; ten to twelve feet is high enough for shelter, and if there be no higher trees in the immediate vicinity of the apiary, it may save much trouble in climbing after swarms. There are many kinds of quick-growing evergreen shrubs suitable for shelter hedges, from which a choice may be made ; and the bee-keeper should, of course, make it a point, when planting, to get something that will also be ornamental and afford forage for bees.

The apiary must be well fenced in, so as to be secure from cattle or poultry ; the ground should be dry, level, or gently sloping to the front, so that each row of hives may be on a slightly higher level than that in front of it, and clear of everything that would tend to impede a free movement about the hives. Swampy or badly drained places must be avoided, as excessive moisture is very injurious to bees. Some writers recommend spreading a layer of sand, or even sawdust, over the ground ; but having tried both sand and grass, I prefer the latter. Grass, if kept trimmed, looks very neat and tidy, keeps the ground cool in hot weather, as compared with sand, and as a contrast to the white hives is a great relief to the eye in bright sunshine. My method is to sow a mixture of perennial rye-grass and white clover seeds over the ground with a little bone dust in spring or autumn, and roughly rake them in, and by the end of the season there is a good permanent grass-plat. A small lawn mower is useful to keep it nicely trimmed.

SHADE.

There can be no doubt that shade for the hives from the hot mid-day sun of our summers is a desirable thing, but there seems to be no convenient way of obtaining it. If they are placed under fruit trees, which, according to my idea, are the most suitable, these must be kept trimmed very high, or else the branches will be in the way. The same difficulty occurs with evergreens, besides being very likely to keep the hives continually damp in winter. Where only a few hives are kept, shady spots may be selected without inconvenience in other respects, but in a large apiary it is rather a difficult matter to supply shade to each hive without involving disadvantages which are more than equivalent. I have recommended and have tried fruit trees, but I find they are a great hindrance to rapid work where many hives have to be gone through. A. I. Root recommends planting a grape-vine on the sunny side of each hive, and training it on a trellis high enough to afford shade ; but in these colonies, where the sun is so nearly vertical in midsummer, we require the shade directly over the hives, and the vines could not be trained to effect that without great inconvenience in the working of the hives. Some American apiarists use and recommend loose shade-boards laid on top of the hives, and kept in place by heavy stones. This is a contrivance so clumsy and unsightly, and attended with so much inconvenience in working, that it could scarcely be justified unless the necessity was very urgent. Here, however, with the hive recommended, which has not a flat cover, but one sloping to each side, with a considerable space between the mat and the inside of the roof, and ventilating holes in front and rear, I find that when the covers are painted white, and careful attention is paid to the ventilation, very little inconvenience is felt from want of shade, and I have therefore of late dispensed with all contrivances for obtaining it, except such temporary expedients as throwing some branches of ti-tree over newly hived swarms for a while. If, however, trees are considered to be desirable, I believe fruit trees are the best to plant, as they afford shade just when it is required, and do not obstruct the sun's rays or occasion dampness in winter.

WATER.

A supply of water for the bees is indispensable, and the nearer it is to the apiary the better. A large number of bees belonging to each hive will be occupied in carrying water all through the breeding season, and if they have to go a distance to procure it so much labour will be unprofitably expended. A shallow stream affords about the best drinking place ; if the only water near the apiary happens to be deep and not accessible by means of flat sloping edges, some contrivances of wicker work, or thin boards with a number of holes bored in them, should be provided to float on the surface and enable the bees to rest on them and sip up the water without risk of being drowned.   In the absence of any natural source it would be advisable to supply the water artificially in shallow troughs placed near at hand, under the shade of a tree or hedge.   For two or three hives an arrangement like that shown in Fig. 34 will be found to answer very well.   The neck of a glass bottle

Fig. 34.—WATER-BOTTLE.

is let down a short distance into a hole in a block of wood, on the upper surface of which are cut shallow grooves from the centre to the outer edges.

The bottle is filled with water and turned bottom upwards, with its neck in the central hole in the block.   The atmospheric pressure prevents the water from running out faster than it is taken away or evaporated from the surface of the block.

AREA OF GROUND.

If the space available for placing the hives happens to be very limited, one hive may be allowed for every 50 square feet, so that a space of 100 feet square would be sufficient for two hundred hives. This supposes the hives to be placed in rows six feet apart from centre to centre in each row, and eight feet from the centre of one row to that of the next behind it. If space permits, however, it will be more desirable to have them eight or even ten feet asunder in each row; there will then be less risk of loss of young queens when returning from their first flight, or even of mistakes being made by worker bees returning loaded to their hives, and also less chance of inconvenience from "robber bees" when extracting honey, or in any other way working at the hives. Allowance has also to be made for the apiary buildings, consisting of an extracting house and honey store, a workshop and store for hives and implements, and a fumigating room. These buildings may be combined in one plan or separate according to circumstances, but in the former case they must be central in the apiary, and in any case the extracting house should be as near the centre point of the whole number of hives as may be possible, and the other buildings not far from the apiary.

ARRANGEMENT OF HIVES.

For the convenience of the bees, in giving them a free flight to the entrance of their hive, and for that of their master,

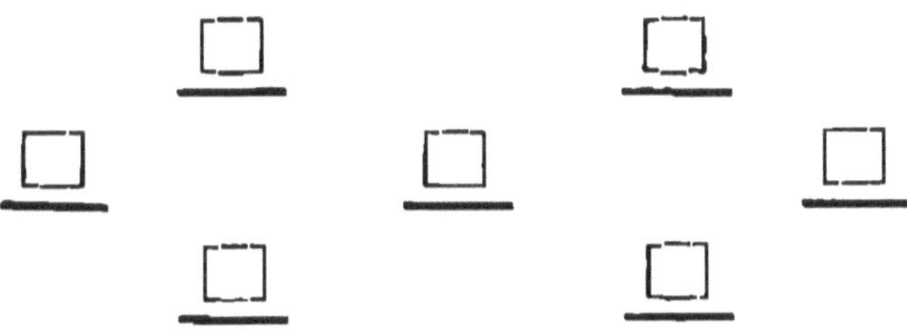

Fig. 35.—ARRANGEMENT OF HIVES.

that he may, when working at any one hive, not be in the line of flight to that just behind him, the rows must be so arranged that the hives in each shall be opposite the centre of the spaces betweeu those in the row either before or behind it, as shown in Fig. 35.

For the same reasons which recommend this arrangement as to plan, it is very convenient, if the ground slopes gently—but very gently, so that a barrow or light carriage can be easily wheeled to any part—from rear to front, in order that each successive row may be a little above the level of that in front of it, something like the seats in a theatre. If space and the natural formation of the ground admit it may also be found a very convenient plan to arrange the hives in three or four long and curved lines, so as to form something of a horse-shoe shape, having the extracting house, etc., in the centre, with a wing of the apiary on each side, instead of being placed in a compact square with so many rows one behind the other as that arrangement requires. A great number of hives placed in this latter form, with the geometrical precision of the squares on a chess-board, present probably the greatest difficulties to both bees and bee-keeper in finding the position of any one particular hive, while the longer lines, with not more than three or four rows to choose between, and perhaps with occasional breaks or gaps in the line, can be made to offer a variety of land-marks which will easily impress themselves upon that sense, whatever it may be, of the bee which guides it in seeking its hive; and also on the mind of the bee-keeper, who if he devotes full attention to his business, will generally have something to remember about most of the hives individually, and should have a pretty clear idea in his head as to where each is to be found.

When placing the hives in position, the bottom boards have first to be put down firmly, perfectly level from side to side, but with one inch of dip from the back to the front. The stands, as fully described in the next chapter, are nailed to the bottom boards and keep them a sufficient height off the ground to avoid dampness. As the bottom boards are laid, the hives can be placed upon them ready for use, the entrances being closed until they may be required.

EXTRACTING HOUSE AND HONEY STORE.

The extracting house should, as I have already said, be nearly in the centre of the apiary, so as to make the average distance from the hives to it the least possible. If the same building is intended to serve as a honey store it must be more carefully built, and, of course, much larger than would be necessary for an extracting house only. It should be well floored, lined and ceiled, well ventilated and lighted, but all openings for ventilation, and such windows as are intended to open, should have an outer guard of wire gauze fine enough to keep out bees. The portion of the building required for storage should be partitioned off from the extracting part and have a northern or north-eastern aspect, and will be all the better by being painted a dark colour outside in order to absorb the heat of the sun. If honey storage be provided in another building, as in the case described below, a much smaller house will answer the requirements of the extracting and ripening process (to be described in Chapter VII.). It need not be close lined or ceiled, but must still be made bee-proof, and well ventilated.

WORKSHOP AND HIVE STORE.

This building should be situated either within the apiary inclosure or very convenient to it, and should be large enough to contain all the spare hives and supers in winter, and fitted up with the necessary appliances for making or putting together hives, frames, packing-cases, etc. A small lean-to at the back or end of the workshop may be made to serve as a

FUMIGATING HOUSE.

The use of this building will be explained in Chapter XVII. The one I have in use at Matamata is very convenient, and a description of it will best illustrate the mode of arrangement I would recommend. Our workshop is 34ft. long, with 10ft. studs ; against the back of this is built a lean-to 10ft. wide and the length of the shop. There is a drop from the floor of the workshop to the floor of the lean-to of about 10in., which allows of that much more pitch in the roof, the back studs being 8ft. At one end of the lean-to is the office, 8ft. x 10ft., at the other end a comb-honey room, 12ft. x 10ft., leaving the centre compartment 14ft. x 10ft. This is the fumigating room. The whole of the lean-to, including partitions and roof, is close lined with tongued and grooved lining, making the different compartments as nearly smoke-tight as possible. In the centre of the partition, between the honey and fumigating rooms, is a door leading from one room to the other, and a window in the centre of the back of the fumigating room. A passage is left down the centre of the room, from the door, 3ft. 6in. wide, and on each side of this passage 2in. x 3in. scantling are nailed in an upright position from floor to ceiling, 20in. apart, the narrow edge of the scantling towards the passage. On each side of the scantlings 3in. x 1in. battens are nailed in a horizontal position to carry the frames. These are nailed a sufficient distance above each other to allow a space of about 1½in. between each tier of frames. The space on the side of the window and immediately opposite it is unoccupied, so as not to block out the light. When we are stowing away our spare combs they are carried into this room and hung on the battens exactly as they hang in the hives, the battens answering as rabbets. The combs are kept about an inch or so apart. As soon as we detect the slightest sign of the bee moth we fumigate the room in the usual way. This sized room has a capacity for storing about 2,500 combs.

STOCKING THE APIARY.

The best time of the year to start an apiary is the spring, although there can be no objection to procuring bees at any time during the summer months. It is not advisable for a beginner to purchase his first bees in the autumn, unless he can get established colonies in movable frame hives from some reliable person, and can also get advice occasionally, if it should be necessary, from some experienced bee-keeper. For various reasons I would recommend the novice to deal with an advanced bee-keeper, one who knows exactly what he is selling, in preference to picking up apparent bargains from box-hive men. When sufficient experience has been gained, so that the purchaser can judge of what he is buying, there can of course be no objection to his dealing wherever he finds it most advantageous.

The value of a swarm or stock of common bees, like that of everything else, depends so much upon the circumstances of demand and supply that it may be best described in the ordinary phraseology to be " that which it will fetch in the market." There is and can be no fixed price. One person may sell a swarm for five shillings, for which another would, and with reason, ask a pound. A good strong swarm (over five or six pounds weight of bees), with a fertile queen not more than two years old, if obtained near at hand in the spring, I consider worth fifteen to twenty shillings; a similarly good established colony in a Langstroth hive, just at the commencement of the swarming season, should be worth half as much more, besides the value of the hive and combs. For the sake of economy it is as well to start with common (black) bees, and, if considered desirable, purchase an Italian nucleus colony or queen later on and Italianise the common stock, according to the instructions given in Chapter XII.

MOVING HIVES.

Beginners are apt to think that bees may be moved about their grounds from one place to another without making the slightest difference to the colony. A little knowledge of the habits of bees would at once convince them of their error. The usual range of a bee's flight is from one and a half to two miles. For the first day or two after a newly-hived swarm has been placed in position the bees, on taking flight, will fly and dodge about the front of the hive for some time, with their heads towards it, marking the surroundings; presently they begin to fly in circles, enlarging each one till out of sight. As soon as they are used to the locality they do not need to take these precautions, and consequently fly direct from the hive. Now, if it should be moved to a new position within the range of their flight, the bees when returning, recognising some of the old landmarks, will make direct for the old locality. On arriving there, and finding their hive gone, they will hover about the place till they are starved to death or die with cold; or possibly, if there is another hive near, will try to enter it, and so meet death that way. This will happen even if the hive has only been shifted a few rods.

Moving established colonies short distances should always be avoided if possible, but if it is absolutely necessary, they should only be shifted a few feet every day, till in their new position. If there is more than one to move, the first must be allowed to get a considerable distance away before the next is moved, and so on.

Natural swarms may be located anywhere as soon as hived, but if an established colony is procured near at hand it would be better to move it, say, two or three miles away, and let it remain there about three weeks before bringing it back. By doing this very few bees would be lost.

Where bees are to be moved in a cart or wagon a good thick layer of straw to stand the hives on will prevent jarring to a considerable extent. If movable comb hives, the frames should be secured so that they cannot move. I am frequently obliged to shift bees at Matamata, and I secure the frames by putting two end bars between each pair and wedging the last frame tight from the side of the hive. The bottom board is secured to the hive body by screwing two thin battens on each side to both parts, a strip of wire cloth is fastened over the entrance, and the cover is also made secure. The hives are prepared in this way over night and shifted as early as possible the next day.

When moving bees in box-hives I have always practised the method recommended below to a correspondent of the *New Zealand and Australian Bee Journal*, who applied for advice in the following manner : " I have purchased 10 boxes of bees at a farm eight miles off. I want to get them over to my own apiary, and to put them into frame hives. The road is a rough one. I propose about the end of September to drive the bees from each box, and to bring them over here in empty boxes covered with cheese cloth ; then to hive them on full sheets of foundation, and to utilise the old combs and honey as best I can. They will most probably be a good deal broken up on the journey over. Can you advise any improvement on this plan ?"

To this I replied as follows :—The present is a very good time to move bees in box hives, but we do not think the above is the best method that could be adopted, for several reasons. First, a good deal of brood would be destroyed, which at this time of the year is of greater value than later on. Second, you would need to feed liberally to get your foundation drawn out, and unless the colonies are very strong and the bees crowded together by division boards, very little comb building would go on for the next two or three weeks. If the bees are in common boxes we would advise you to adopt the following plan in preference to the one you suggest :—Take some scrim or cheese cloth, tacks, paper, and your smoker with you to the boxes the afternoon or evening before you intend to move them. Cut the scrim into pieces that will cover the bottoms of the boxes. Now light your smoker and blow a few puffs into the entrance of No. 1, turn it bottom up and give a little more smoke. You can now fit in some paper wads between the combs to steady them. This done, the scrim can be tacked on if all the bees are in, if not turn the hive up again, and proceed in the same way with the others. Tack on the scrim when all the bees are in and keep the boxes bottom up till you have them in your own apiary. Have your bottom boards laid permanently, place a box on each, and liberate the bees ; in a week or so transfer the bees and brood to your Langstroth hives in the usual manner. By this method you will save yourself the trouble of getting extra boxes and driving the bees, save the brood, and your colonies will be in much better condition when transferred. We moved 58 colonies in two waggons a distance of 40 miles over a very rough road and a high range of hills— two days on the road—and did not have one mishap by the above plan.

My correspondent informed me some short time afterwards that he had followed this advice and had been completely successful.

HOUSE APIARY.

A few years ago bee-sheds and bee-houses were considered quite indispensable, at least by British bee-keepers. In most of the English works on bee-culture, and in bee-catalogues published seven or eight years ago, or even at a later date, these things are figured in all designs. Very few however seem to advocate their use now. They are in the first place expensive, and in the second inconvenient. Possibly a house apiary might be found of some use in rearing queens very early in the season, as the work could be carried on at any time independent of the weather ; but for raising honey I much prefer to have the hives outside. I have had several years' experience with bee-houses, and desire to have no more. Should any of my readers however wish to erect an ornamental house apiary, the following illustration will give an idea of one suitable for twenty hives.

In this house, it will be seen, there are places for six hives on each side, and four at each end ; one row standing on the floor, and the other on a shelf sufficiently high up to allow of working the lower hives without inconvenience.

Fig. 36.—HOUSE APIARY FOR TWENTY HIVES.

Axiom.

"Bees dislike the offensive odour of sweating animals and will not endure impure air from human lungs."

Langstroth.

CHAPTER VI.

HIVES, FRAMES, AND SECTION BOXES.

AMONG all the aids to improved apiculture which have been invented within the last thirty-five years, the movable comb-hive stands first in importance. The most advanced improvements in our system of working are traceable back to, and are the result of, this invention. Although the honey-extractor and comb-foundation are invaluable helps which no one can now afford to dispense with, they could not be made use of without the movable comb-hive. Thirty-five years ago the

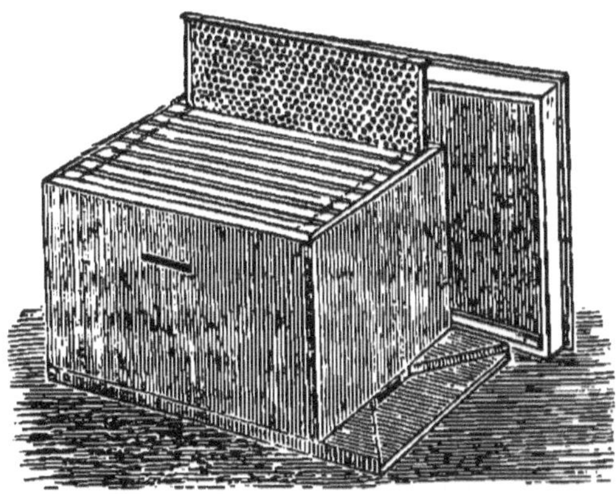

Fig. 37.—MOVABLE COMB-HIVE.

average bee-keeper knew little or nothing of the internal economy of the hive, as compared with the degree of knowledge which the merest tyro can now acquire before even commencing to practise. Although Huber had shown, nearly a hundred years ago, how to obtain control of the combs to a certain extent, and especially for purposes of observation, yet his 'leaf or book-hive" was impracticable in the hands of an ordinary person, and for the ordinary purposes of bee keeping. After Huber's time the attention of apiarists was for many years directed chiefly to improvements in straw skeps or in wooden box-hives, in order to obtain the surplus honey in good condition, and without the destruction of the bees; and the first great step in advance was made some sixty years after the publication of Huber's discoveries, when the Rev. L. L. Langstroth, of America, gave to the world the present simple form of movable comb-hive. No doubt Langstroth was greatly assisted, as he has himself informed us, by his knowledge of what Huber had done; and it is a remarkable coincidence, that at the very same time he and another enthusiastic apiarist, Dzierzon, in Germany—already noticed in Chapter I.—were, unknown to each other, pursuing the same object, and that the latter produced also in Germany, nearly simultaneously with the former in America, a movable comb-hive. The two inventions were, however, quite independent of each other; and although the grand principle of having the combs in movable frames is common to both, still there is a very marked superiority in the practical working out of the American one, and it is quite certain that nothing like a simple and practicable form of movable comb-hive had been invented or was known anywhere outside of Germany, until that one was introduced. Mr. Langstroth not only gave us a hive which, after the lapse of so many years, stands pre-eminent at the present day, but he also gave us his book, " The Hive and Honey Bee," which, although now necessarily somewhat behind the times in the practical work of the apiary, must always be of the greatest value to the advanced bee-keeper, containing, as it does, a full and interesting account of the writer's able researches in apiculture. I shall always have a grateful remembrance of the name of Langstroth, as I feel indebted to his work for my first insight into the advanced system of bee culture, and for the foundation of my present knowledge of the art. He has held up Huber as the " Prince of Apiarians," and I think he may himself be justly called the Huber of America.

CHOICE OF A HIVE.

No sensible person who intends to devote himself to apiculture would think nowadays of beginning without first getting some advice from a practical bee-keeper or studying some of the bee literature of the day; there is, therefore, no likelihood of his being led through ignorance into the adoption of any form of box hives, so that any discussion of their shortcomings or warning against their use is here unnecessary.

Before proceeding to speak of the varieties of hives now in use I will mention a few of the most important requisites of a complete or ideal hive, and the reader will then more easily understand much that is to follow.

AN IDEAL HIVE.

A complete working hive should be so constructed as to allow of any and every portion of the interior being inspected at pleasure with little trouble or loss of time. Its construction should be as simple as possible, consistent with strength, good workmanship, completeness, and durability. It should permit of all necessary operations (such as removal of combs, bees, brood, and surplus honey) being performed without necessarily killing a single bee. While affording ample protection from the weather, it should permit of increased or diminished ventilation at a moment's notice. It should be capable of being contracted, as regards working room, to the smallest space required at any time for the stock, and enlarged to any size that may be found necessary. The entrance should be so arranged as to be easily enlarged or contracted whenever required; the hive should permit of the surplus honey being stored in the best and most convenient form for depriving; and last, though not least, it should have as few loose parts belonging to it as possible.

VARIOUS FORMS NOW IN USE.

There are various modifications of the original Langstroth hive in use at the present time, the main difference being in the size. The principle is the same in all, namely, that the combs are built within light wooden frames, suspended in a case, which forms the body of the hive. These frames can be inserted in or removed from the hive, or made to change places with each other, at pleasure; hence the name of movable comb or movable frame hives. As the size of the body is guided by the dimensions of the frames, it is usual to speak of the latter only when comparing different hives. Now the principal frames—modifications of the "Langstroth," which is 17⅝in. by 9⅛in.—are known by the following names :—The Quinby, 18½in. by 11¼in.; the American, 12in. by 12in.; the British Standard, 14in. by 8½in.; and the Gallup, 11¼in. by 11¼in. All these are the outside measurements of the frames. There are a few other sizes used, but the foregoing are the principal ones. All these frames have their advocates, but it has been frequently asserted, and I believe with truth, that there are more of the original Langstroth size in use than of all the others put together. The general opinion of late years has been in favour of a shallow frame, of medium length, as being best for the storage of surplus honey and most convenient and handy for manipulation. Of the frames mentioned above, two answer this description, viz., the "British Standard" and the "Langstroth." For several reasons I consider the latter preferable.

To give a detailed description of each of the hives enumerated above would occupy too much space, and I feel convinced that it could only serve to confuse beginners, who require at the start not so much to know the particulars of all hives as to be guided in the selection of the best. It would be impossible in a work of this kind to enter into the details of all the different hives and systems of management without extending it to an unreasonable length, nor do I consider it either necessary or desirable for the reason already given. I shall therefore, with only occasional reference to other appliances and systems, and without dogmatically condemning any, confine myself as much as possible to describing and recommending such as I have found by experience to be the best.

THE LANGSTROTH HIVE.

This hive, which has now stood the test of thirty years' trial, is more in favour to-day than at any previous time. In 1851 Mr. Langstroth gave it to the world, and, as Professor Cook says, " it left the hands of the great master in so perfect a form that even the details remain unchanged by many of our first bee-keepers."

Early in 1878 it was my good fortune to become acquainted with this particular hive. For years previous to that time I had been trying different forms of frame hives, but none pleased me so well as the Langstroth. I finally adopted it in its most simplified form as a ten-framed hive, and now, after seven years' experience, I have the fullest confidence in recommending it to all who require a good hive ; and this confidence is supported by the knowledge that it has given entire satisfaction to all who have heretofore adopted it upon my recommendation. Many thousands of these hives are now in use throughout Australasia. It has practically become the standard for these colonies, and that is another and a very urgent reason why it should be adopted by beginners. I would not, however, wish it to be supposed that I lay too much stress upon the form of the frame or hive, or that the novice should entertain any idea that success depends on the precise form of either. Without intelligent management the best hive may be of no more value to its owner than a common box. Bees may be *kept* in either ; but it has been well observed that "everyone who keeps bees is not necessarily a bee-keeper," the latter term as now used being synonymous with "bee-master."

There is one rule, in the correctness and importance of which the advocates of every variety of hive will agree. Whatever sort may be adopted, *all* the hives in the apiary should be the same, and none other should be tolerated amongst them, except of course for experimental purposes.

GENERAL DESCRIPTION.

Fig. 38 gives a view of a complete Langstroth hive, as now generally made and used in New Zealand. It consists of four principal parts—the floor-board, the hive, or brood chamber, the super, and the cover. The first two and the last are indispensable parts of any hive ; the super may consist of a whole story (same as the brood chamber), or of a half-story or there may be two or even more supers between the brood chamber and the cover. The outside dimensions of the brood chamber and of the one-story super are $20\frac{1}{4}$ inches by 16 inches, by 10 inches in height, including the rabbet, or $9\frac{3}{8}$ inches from joint to joint. The half-story supers are, of course, the same in length and breadth, but only $5\frac{3}{4}$ inches in height, including the rabbet. The bottom, or floor-board, is the same width as the brood chamber, but four inches longer ;

it is raised about four inches from the ground by the two
stands which are nailed to the bottom, and has a sloping
alighting board in front (which may be nailed to the front
stand, or used as a separate movable part), to facilitate the
entrance of any bees that may fall short of the flat floor-board
when returning laden to the hive. A convenient entrance,
which can be enlarged or contracted at pleasure, is provided
by cutting a **V** shaped piece out of the surface at the front end
of the board, to a depth of $\frac{3}{8}$ inch. The wide part of the
depression so formed is at the outer edge, joining the sloping
alighting board ; the angular end is five inches back towards

Fig. 38.—TWO-STORY LANGSTROTH HIVE.

the centre of the board. When the back of the brood chamber
is flush with the back of the floor-board, the entrance for the
bees is quite closed ; according as the hive is shoved forward,
the length of the opening is of course increased, and can be
easily fixed to suit all requirements. The roof fits down upon
either the brood chamber or any of the supers, in the same
way as these are made to fit one on the other. The ends are a
couple of inches higher in the centre than at each side, so that
the covering boards, which are made to project two inches in
front and rear, and an inch and a half at the sides, slope to
each side, and cast off the drops of rain. Ventilation is provided for by two round holes, one in each end of the cover,
which are protected by pieces of wire gauze or perforated zinc.

The interior of the hive is shown in Fig. 39. The brood chamber is intended to contain ten narrow frames of comb; the super, if worked for extracting, has generally nine of the same frames, the intermediate spaces being left a little wider than in the brood chamber. If worked for comb honey, the super contains only seven broad frames, fitting close together, each frame containing four or eight section boxes of the sort to be described further on. The half-story supers are made to contain either shallow frames, with section boxes, or a section rack. An inside covering mat is placed on top of the frames in the hive or super just under the cover. The stands and the bottom of the floor-board may be painted a dark colour; the hive itself is better if either white or a light tint; the top of the roof should in any case be of a pure white, in order the better to cast off the hot rays of the sun in summer.

These hives can be procured at very moderate prices, and very complete in every respect, from the manufacturers. For the convenience of those who wish to put their own hives together and to save expense in the transport, they can be had *in the flat*, that is, all the separate parts complete and ready to be nailed together, and packed as close together as possible in crates or packages containing generally four one-story hives, or three of two stories, or of one and a half. If the beginner decides upon getting his hives in this way, and if there be no hives already in use in his neighbourhood, or no one to show him how to set to work, it will be advisable for him to procure one hive complete and fitted together, to serve as a pattern, and the rest in the flat.

INSTRUCTIONS FOR MAKING.

To those not skilled in the use of tools, or who have other occupation, it may be found the most profitable to purchase all their hives as they require them. On the other hand, there will always be some having plenty of spare time, who would prefer to occupy a part of it in making their own hives, had they clear instructions how to proceed. I shall now endeavour, with the aid of illustrations, to give plain instructions for making the Langstroth hive, the one I have already advised my readers to adopt.

The first, and it will be found the principal point to be observed, is to use none but *thoroughly seasoned* wood. Of course all know that unseasoned wood will shrink and twist after it is made up; but some inexperienced person may say, "Well, what does it matter in a beehive, supposing it is a little twisted or a quarter of an inch larger or smaller than the prescribed size?" Let me remind those inclined to look at the matter in this light, that a complete hive is composed of a number of movable parts, each part of which ought to fit in its place like a piece of cabinet work, or the hive, as a whole, cannot give satisfaction. It is absolutely necessary that every movable part of a complete hive should be interchangeable with like parts in all other hives throughout the apiary, let them number two or two thousand; they must therefore be exactly alike, and not liable in the slightest degree to alteration by twisting or shrinking. Even to be within an eighth of an inch is not near enough—they should be exact. It is not always possible to purchase seasoned wood when required, and as an extra price is charged for it when it *can* be got, it is just as well for the bee-keeper, if he has the convenience, to get the driest timber available, and season it himself. Some place under cover can nearly always be found to stow a few boards to season. So that the rain is kept out, the more open the place is to the wind the better. The autumn is a good time of the year to purchase timber for the purpose, as dry boards can be got, and there is enough time for seasoning purposes before it will be necessary to make them up. Care must be taken, when stacking boards, to put thin strips of wood between each two to allow the air to circulate freely around them. I have dwelt on this matter at length, knowing by experience how important it is.

The thickness of the timber used principally throughout the hive is seven-eighths of an inch; and as one-inch boards—the nearest size most easily obtained—when well seasoned are a sixteenth less, there is just sufficient substance left to allow of a smooth surface being put on one side with a plane. The body of the hive, with which I will start, is 10 in. deep, and takes exactly 5 ft. 11 in. of board to form the two sides and two ends; so that boards 12 ft. x 1 ft. will cut two bodies, allowing two inches for saw cuts and waste. I would advise getting 1 ft. boards, as the exact 10 in. can be cut after they are seasoned, while the 2 in. strips will come in for making frames. Plane the 12 x 1 ft. board on one side, reducing the thickness to seven-eighths of au inch, and run a trying plane along one edge till the edge is perfectly straight. Mark the depth (10 in.) from the straight edge, and rip off the strip; now cut from your 12 ft. board four pieces 16 in. long, for end pieces, and four 19½ in. long, for sides; then set your gauge to mark three-eighths, and take each of your end pieces, lay them on the bench, planed side up, and run your gauge along the rough edges, marking for the rabbet D shown in Fig. 39.

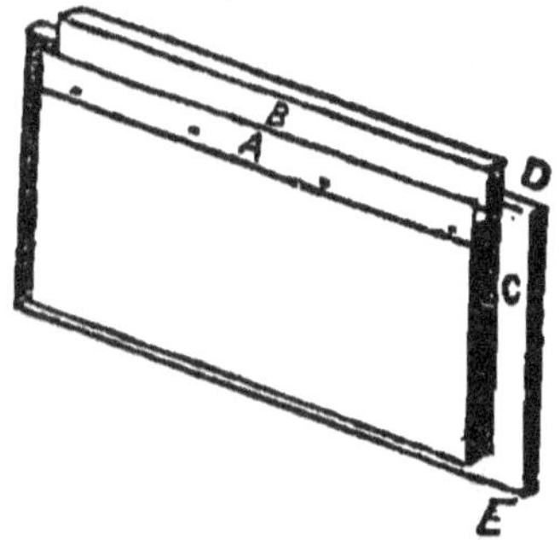

Fig. 39.—END OF HIVE (INSIDE VIEW)·

Next hold the pieces on their edges, and mark with the same gauge *in* from the planed side. This will show the piece to be taken out to form the rabbet D. The pieces should now be

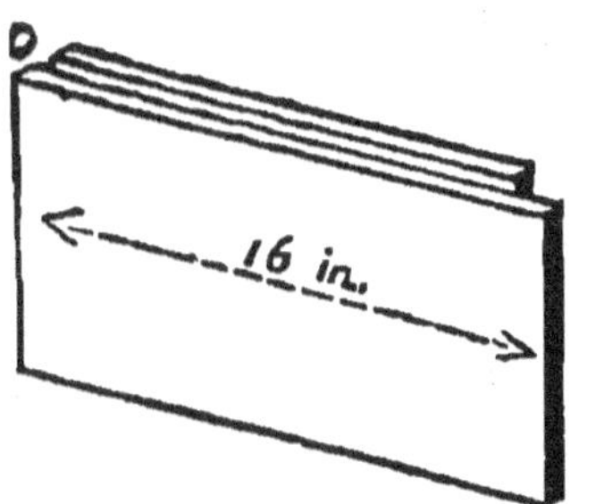

Fig. 40.—END OF HIVE (OUTSIDE VIEW).

turned the planed edges up, and the same gauge run along the edges *from* the planed side of the board to mark for the rabbet E, shown in Fig. 39. Now lay the ends flat (rough side up),

and mark with same gauge *in* from the edges for rabbet E, and
also run the gauge down the ends of the boards *from* the planed
sides to mark for rabbet C (Fig. 39).　Before shifting the gauge,
the rabbets on the side pieces can be marked.　The rabbets D
and E (Fig. 41) are marked exactly the same as the rabbets D
and E in Fig. 39.　We have now the rabbets D, E, and the edge
of C marked.　The gauge will then require to be set at seven-
eighths of an inch to mark *in* from the ends of the end pieces
on the inside for rabbet C, Fig. 39, and also down from the top
edges for rabbet B, Fig. 39.　All that is wanted now is to reset
the gauge to a quarter of an inch, and mark on top edges *from*
the inside for rabbet B, Fig. 39.

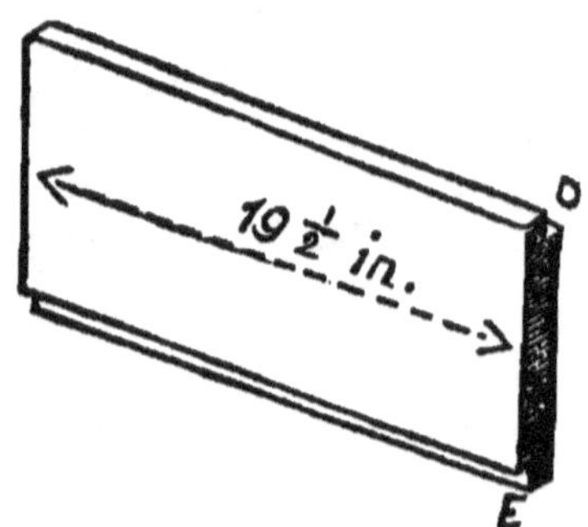

Fig. 41.—SIDE OF HIVE (INSIDE VIEW).

The rabbets, being all marked, will require cutting out.
The best tool for this purpose I have ever used was an iron
plough (American).　With this tool, fitted with a ⅜-iron and
set to a ⅜-gauge, it will scarcely require the marking gauge to
be used.　By cutting out the rabbets D and E (Fig. 39) first,
some little labour will be saved when cutting C.　If a saw cut
is put in across the latter it will expedite the cutting.　For
C and B a ⅞-iron will be required, with the gauge of the plough
set to the proper depth.　After the rabbets are cut, strips of
tin, 1½in. wide by 14in. long, should be folded in the centre to
form the metal supports A (Fig. 39).　These are tacked on, as
shown, so as to allow the upper edges to project above the
lower part of rabbets about one-eighth of an inch.　Metal
supports, or, as they are commonly but incorrectly termed, "tin
rabbets," are for supporting the frames, the projecting ends of

which rest on them ; but I shall have more to say respecting these in another place.

Tho ends and sides being properly formed will have the appearance of the figures and will themselves suggest how they should be put together. Fig. 42 represents the two ends and one side nearly in place, the ends of the side pieces dropping into the rabbets C (Fig. 39) should fit nicely, and be firmly nailed with three $2\frac{1}{2}$ in. wire nails at each end. These should not be driven through the end pieces into the sides, but through the sides into the ends, dovetail-fashion.

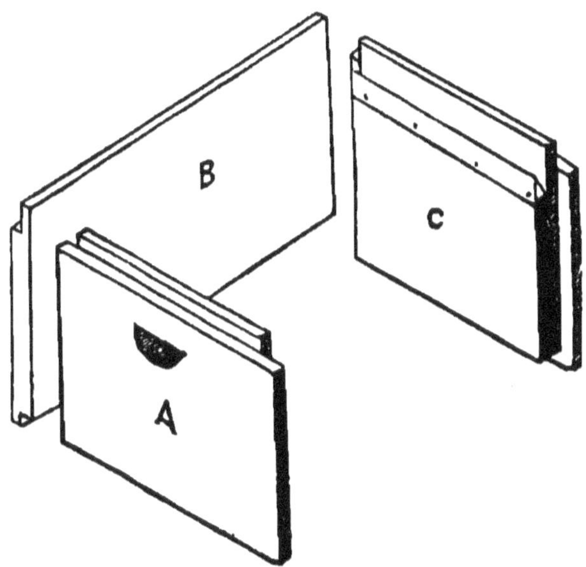

Fig. 42.—SHOWING HOW THE DIFFERENT PARTS OF THE HIVE GO TOGETHER.

BOTTOM BOARD.

For the platform of the bottom board a piece of board 2ft. long, 16in. wide, by 1in. thick, is required. Although this may be made out of two or more pieces, it is much better to have it in one, as the joints give facilities for moths and other insects to deposit their eggs where it is difficult for the bees to get at them. The entrance A (Fig. 43, next page) is cut out of one end $\frac{3}{8}$ths of an inch deep, starting $1\frac{1}{2}$in. from each side and running back 5in. to a point, as shown. After marking it out, a saw-cut can be run on each side to save labour in chiselling. The stands

B B are 4in. wide, 1½in. thick, and 16in. long ; nailed on edge, 3in. back from each end.   These pieces keep the hive a sufficient height off the ground and prevent the bottom board twisting.

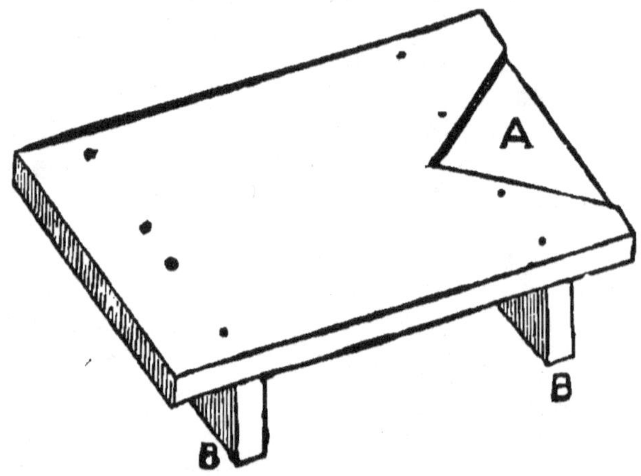

Fig. 43.—BOTTOM BOARD.

ALIGHTING BOARD.

This is a very necessary part of the hive.   Placed in front of the entrance, it makes a capital landing stage for the bees, and thus saves many from falling to the ground when heavily

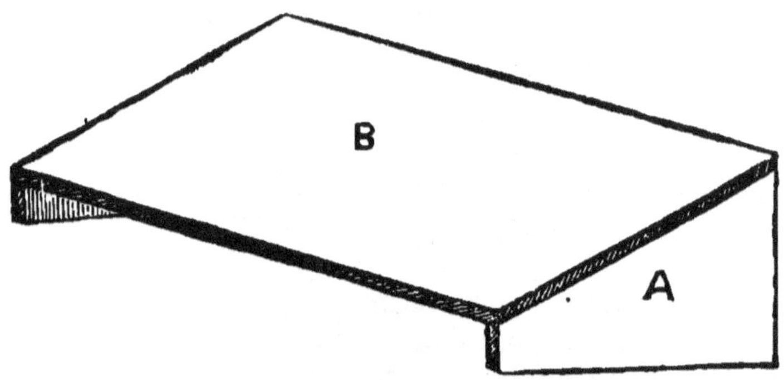

Fig. 44.—ALIGHTING BOARD.

laden.   The platform B (Fig. 44) is 16in. long, 9in. wide, and ⅝in. thick.   The upper edge should be slightly bevelled, to fit snug against the bottom board.   The pieces A are 8in. long and 4in. wide at their widest part, tapering down to 1½in. at the outer end.   The handiest way of making these is to cut them out of a board 5½in. wide and 1in. thick. Every eight inches of the board will make two without any waste and save a deal of cutting.

COVER.

This is a part of the hive that requires to be very carefully

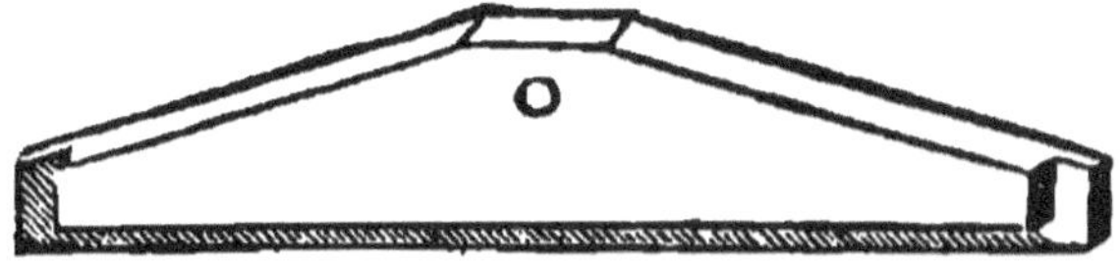

Fig. 45.—END OF COVER (INSIDE VIEW).

made. Leaky covers are an abomination. The ends (Fig. 45) are made of ⅞in. timber, 16in. long, 4in. wide for 1¼in. in the centre, then tapering down to 1¾in. wide at each end. A rabbet, ⅜in. wide by ⅜in. deep, is taken out of the lower edges on the inner or rough side, to allow it to fit over the rabbet on upper edge of the body of hive, and another rabbet is cut in the ends, as shown, ⅞in. by ½in. deep, for the sides to house into. An inch hole for ventilation (shown in figure) should be bored in the centre, and have a piece of perforated zinc tacked

Fig. 46.—SIDE OF COVER (INSIDE VIEW).

over it. The side pieces (Fig. 46) are the same thickness as the ends—19½in. long, 2in. wide on the insides, and 1¾in. wide on the outsides, the upper edges being bevelled ¼in. to give them a similar slope to the end pieces. The lower inside edges of these are rabbeted the same as the ends. The ridge board

Fig. 47.—RIDGE BOARD OF COVER.

of these are rabbeted the same as the ends. The ridge board (Fig. 47) is 2ft. long, 4in. wide, and ⅞in. thick. This should be rabbeted on the under side in a sloping manner, similar to the rabbets shown in the figure, tapering off from nothing at the edge to ⅝in. at the deepest part.   The width of each rabbet from the edge is 1⅛in., leaving 1¾in. of the full thickness in the centre, corresponding with the top centre of end pieces. When making my hives by hand I had an iron fitted to my plough made the shape of these rabbets, which was the means of saving much time and labour.

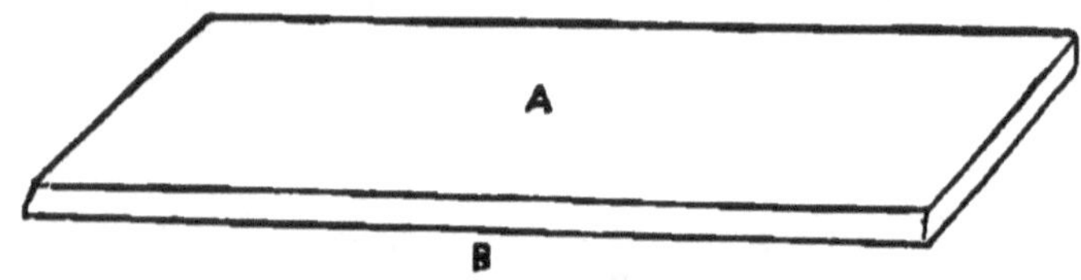

Fig. 48.—ROOF BOARD OF COVER.

The roof boards (Fig. 48) are made of ⅝in. timber 2ft. long by 8in. wide, the lower edge being slightly bevelled to suit the slope of the cover.

To put the cover together, the sides and ends are nailed first ; then place the ridge piece on, allowing it to project an equal distance at each end, but before nailing it, put on one of the roof boards in its place—the upper edge under the ridge, and nail through both ridge and board to the end pieces. Now nail the other board on in the same manner, and fasten both boards securely round the sides. The engraving (Fig. 49) shows the cover finished.

We have now gone through the whole hive, with the exception of the frames, and if every part is made according

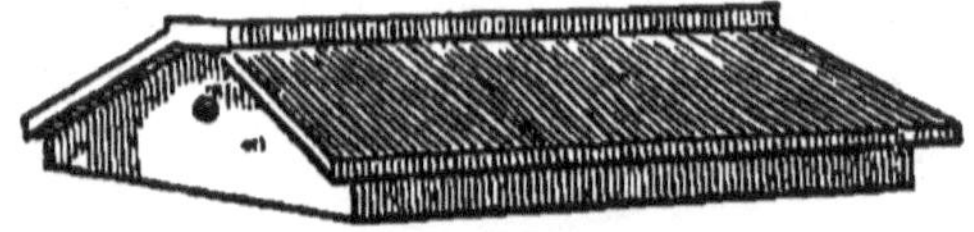

Fig. 49.—COVER COMPLETE.

to the foregoing instruction, they will fit each other like a glove, and when two or three story hives are required, it is only a question of having extra bodies similar to the one already described.

HALF-STORY HIVES.

These are always used for the purpose of raising comb-honey in, and are made exactly similar to the full hive described, but only 5½in. deep.

HIVE CRAMP.

To secure the best results in utility and appearance, it is just as necessary to be as careful when putting the body of the hive together as when making the different parts. Although the ends and sides will, if properly made according to the instructions given, go together nicely, something more than hand pressure is required to hold them in place while nailing them. As each body may at any time be required to take the place of a top or lower box in any hive throughout the apiary, all of them should be so firmly fastened as to stand ordinary knocking about without getting out of the square.

The following engraving shows a very useful hive cramp atht I have had in use for some years :—

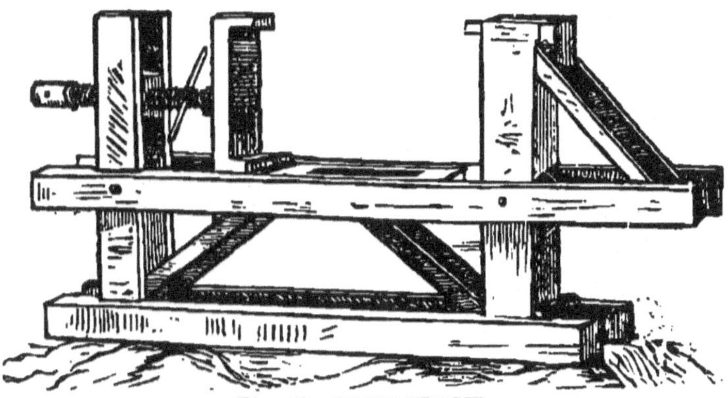

Fig. 50.—HIVE CRAMP.

It is made by taking four pieces of timber 6in. x 2in. three feet long for uprights, and two 4in. x 3in. five feet six inches long for horizontal pieces to form the platform and to bolt the uprights to. Two uprights are halved into each horizontal piece two feet ten inches apart, and sixteen inches from the upper ends of uprights. The uprights are let in to the horizontal pieces on the inside, as shown. A scarf four inches wide by three-quarters of an inch deep is made in each upright. Another scarf six inches wide by one and a-quarter inch deep is made in each horizontal piece ; so that when the pieces are put in position they are flush with each other. All parts go together on their flat. Two 4in. x 3in. four feet six inches long are fastened in a similar manner to the lower ends of uprights. The two parts of the framework are held parallel to each other six inches apart by two spreaders eight inches long of 6in. by 1in. stuff let in flush in the upper horizontal pieces between the uprights, as shown in engraving; these spreaders form part of the platform. Similar spreaders can be nailed across the lower horizontal pieces. Two $\frac{5}{8}$in. iron bolts run through the double part of horizontal and upright pieces with a nut and screw hold all firmly together. The uprights and horizontal pieces should be exactly at right angles to each other. If made correctly, when the frame is standing erect, the platform formed by the 4in. by 3in. pieces should be exactly the width of a hive (16in.) below the upper end of uprights.

The next things required are the two jaws, one stationary and the other movable. To make the stationary jaw, one pair of the uprights is made use of. To these are screwed two pieces of 1$\frac{1}{2}$in. square, one at the top and one close down to the platform. If thought necessary a stay may extend from the back of each upright to the ends of the horizontal pieces. For the movable jaw a piece of 9in. by 2in. sixteen inches long is required. A carpenter's wooden bench screw is made to work between the other pair of uprights by fastening the screw-block between them. The end of the screw is now fastened to the movable jaw in the usual manner. To steady the jaw when screwing up, a short piece of inch board is nailed on the bottom of it, which slides between fillets nailed on each horizontal piece; this forms the cramp. When working it, one side of the hive is placed on the platform between the jaws; the two ends are then put in their proper position, and the remaining side; we have now only to give the screw a turn and all the parts of the hive are cramped together close and held firm and square while nailing them. When one side is nailed the screw is loosened, the hive turned over, and the other side is nailed in the same manner. The cramp is also useful for putting the ends and sides of the covers together, and for many other purposes; in fact it is one of the most convenient appliances a bee-keeper can have in his workshop. It is easily constructed, and if made by a carpenter should not cost more than 30s.

A simple form of cramp may be made of timber 2in. square. Cut two pieces 24$\frac{1}{2}$in. long and two pieces 20$\frac{1}{4}$in.

Halve them together at the ends, and nail firmly, taking care that the pieces are nailed squarely to each other. To make the frame more secure, a broad piece of hoop iron should be nailed across each corner. Two such frames are required, and a few thin hardwood wedges. The inside measurement of the frames will be $\frac{1}{4}$in. longer and $\frac{1}{4}$iu. wider than the outside dimensions of the hive. This extra space allows them to slip over the parts easily, and gives room for wedging.

NUCLEUS HIVE.

This hive is used both for the purpose of rearing queens and keeping spare ones in till required.

Fig. 51.—NUCLEUS HIVE, WITH COVER AND MAT.

It is the same length and depth as the ordinary hive, but is usually made to contain three frames only. It may be made out of light material and in a more simple manner than the ordinary hive. The bottom board can be nailed on, and the entrance cut out of one end of hive; the figure does not show the bottom board. The width inside should be $4\frac{3}{4}$in. or 5in. to take three frames. A full explanation of the use of it is given in the chapter on queen rearing.

OBSERVATORY HIVES.

There is no branch of entomology so interesting to the ordinary individual as the natural history and habits of bees. To the majority of people there is not a more pleasing sight than the interior of a beehive during the busy season, if the working of the bees can be observed leisurely without danger of receiving a sting. To watch the queen surrounded by some of her subjects as she moves in a stately manner from cell to cell depositing her eggs; to view the worker bees building comb, bringing in and storing honey and pollen; to see the young bees gnawing their way out of the cells and first catching sight of their future surroundings, is indeed interesting. Hives for observatory purposes can be so constructed as to allow of the interior being examined at pleasure by the most timid person without disturbing or exciting the bees.

The most simple form of this hive may be made to hold one frame only, but for several reasons I prefer one that will take

Fig. 52.—OBSERVATORY HIVES.

three.   The engraving represents the one I have in use.   This, as will be seen, has three leaves (cases), each of which holds an ordinary Langstroth frame.   The centre one is a fixture, while the two outside leaves may be opened to allow of both sides of each comb being inspected.   The woodwork of the cases is made of battens 2in. wide by 1in. thick ; a narrow groove to take the glass sides is run along each inner edge, so that the sides may be just 1$\frac{5}{8}$in. apart, the length and height of each case inside being the same as a Langstroth hive.   A one-inch hole should be bored out of the bottom of the front end of each face to form an entrance for the bees.   Next cut an entrance $\frac{3}{8}$in. deep in the bottom board, somewhat like an ordinary entrance ; then over the back part of this tack a flat piece on tin having two tubes, 1in. in diameter and 1in. high, soldered on in the right position, so that when the tin is on, the outside cases will fit over these tubes, that is, the tin tubes will fit in the holes bored in the bottom of each case. Now if the centre case is nailed to the bottom board, and the side cases fitted over the tubes and hinged on top to the centre one, it will be pretty near complete. It must, of course, be seen that the entrances to each case through the tubes are clear, and that there is no place the bees can get out but through the entrance in the bottom board. We have now only to make a cap to fit over the cases when closed, to darken the interior when not in use, and the hive is complete.

Every bee-keeper should possess an observatory hive, as it will not only be interesting aud instructive to himself, but a source of amusement to his family and friends. To stock it, take an ordinary frame of brood, with the adhering bees *and the queen*, from a hive and place it in the centre compartment; a frame of honey and an empty comb or frame of foundation can be put in the outside cases ; close the entrance with perforated zinc or wire cloth ; put on the cap, keep the hive in a cool place, and let it remain closed for 48 hours, when it may be put on its stand and the entrance opened at dusk in the evening. In a week the queen may be removed if it be desired, aud the process of queen-rearing be observed.

TIMBER FOR HIVE-MAKING.

Soft porous timber is usually recommended as being the best for hives, though it should at the same time be of a tough lasting nature. Wood that is soft and porous is a better non-conductor of heat than hard close-grained wood ; hence hives made of the former give better protection to the bees both in winter and summer than would hives made of the latter kind of wood. It is also preferable to use straight grained timber. more especially for the frames, as being most easily worked and not so likely to twist. The bottom boards and stands may be made of a harder timber in order the better to withstand any dampness that may arise from the ground, as these would be the parts of the hive that would be most affected by it.

PAINTING HIVES.

Hives should never be used until they have been painted. They ought to receive three coats to start with. The first coat should be very thin, and may consist principally of red lead and linseed oil—I believe equal parts of raw and boiled oil to be the best. The heads of the nails should be punched in before applying the first coat, and the holes be puttied as well as the joints before giving the second. The colour of the second and third coats ought to be white or nearly so ; on no account should hives that are to be exposed to the sun's rays be of a dark colour. The whiter the hives are the cooler will the interior be in hot weather. White paint, made even of the very best lead, does not last long when exposed to the sun ; but if a little black be mixed with it, just sufficient to give it a very slight slate tint, it will last much longer. Rubber paint for the last coat answers very well indeed, and is more lasting than white lead.

The under side of the bottom boards and stands may be painted with a cheaper paint ; I find hematite paint to answer the purpose, but the part exposed on the upper side and the alighting board should be white.

FRAMES.

From what has already been stated at the beginning of this chapter, the reader will understand that the frames of a hive are those movable structures, suspended within the body, in which the combs are built. There are two kinds generally used, each adapted for a different purpose, viz., the narrow or brood frame and the broad or section frame. The narrow frame is invariably used in the lower or breeding part of the hive as being most suitable for the brood combs, and is also used in the super when working for extracted honey. The broad frame is specially adapted for holding section boxes when raising comb-honey, and is rarely used for any other purpose.

NARROW OR BROOD FRAMES.

It is of great importance that the timber used for making frames as well as the hives should be thoroughly seasoned, and that the different parts of the frames be cut exact in size. It is frequently necessary to shift one or more frames from one hive to another, so that if all hives and frames are not the exact counterpart of each other, there will be no end of vexation and trouble. The width of the timber out of which the narrow frame stuff is to be cut should be $\frac{7}{8}$ths of an inch. The top and end bars (A and B B Fig. 53) should be $\frac{3}{8}$ths of an inch thick, and the bottom bar (C) $\frac{1}{4}$ of an inch. The length of the top bar is $18\frac{7}{8}$in.; end bars, $8\frac{1}{2}$in.; and bottom bar, $17\frac{3}{4}$in. When the end bars are nailed $\frac{5}{8}$ths of an inch from each end of the top bar, as shown in the engraving, and the bottom bar nailed on, the inside dimensions of the frame will

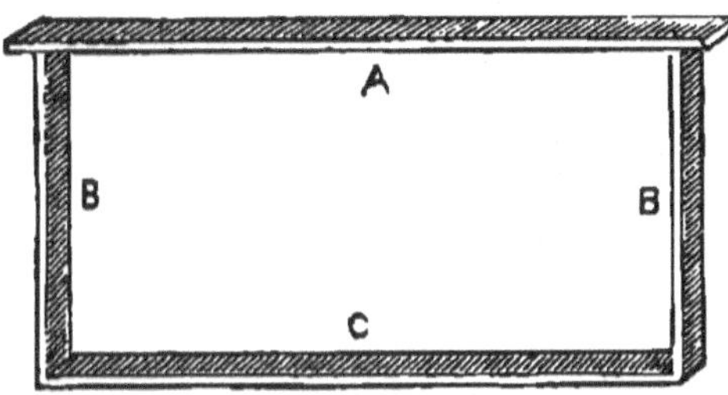

Fig. 53.—NARROW OR BROOD FRAME.

be 17in. long and $8\frac{1}{2}$in. deep, the exact size required for the Langstroth hive. Before, however, nailing them together, a groove must be cut along the centre of the top bar $\frac{1}{8}$th of an inch wide, and of the same depth, for fastening a sheet of comb-foundation in. It would be better to do this before cutting the stuff into short lengths.

BROAD OR SECTION FRAMES.

The top, bottom, and end bars of these frames are exactly the same in length as the narrow frames, and with regard to thickness, should be made out of similar material.

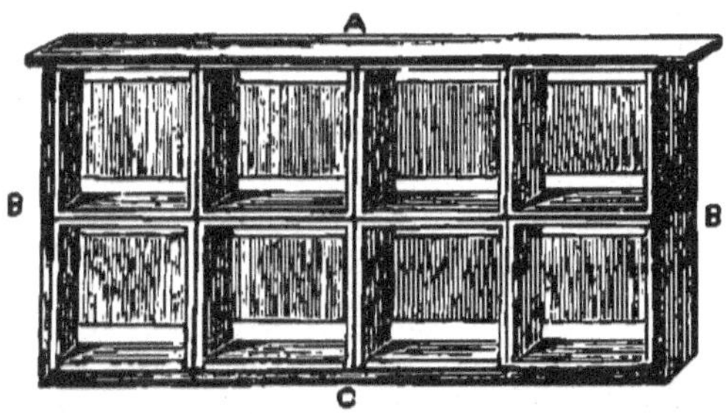

Fig. 54.—SECTION FRAME WITH SECTIONS AND TIN SEPARATORS.

The top and bottom bars should be $\frac{1}{16}$th less than 2in. wide, and the end bars $\frac{1}{16}$th less than $1\frac{3}{4}$in. Sometimes these frames are made too wide, the ends a full 2 inches. The consequence is that with the separators on there is barely room in the hive for the seven frames, and the last one has to be jammed in, a state of things that should not be tolerated. Everything about a hive should work with the greatest of ease. The top and bottom bars being a quarter of an inch narrower than the ends, they should be nailed on in such a way as to allow an equal projection of the end bars on each side of them, so that when the frames are pushed close together in the hive there will be sufficient space between the tops and bottoms of each two frames for the bees to pass to and fro.

HALF-STORY FRAMES.

For raising comb-honey I very much prefer using two half-story supers in place of one full story. My reasons are fully

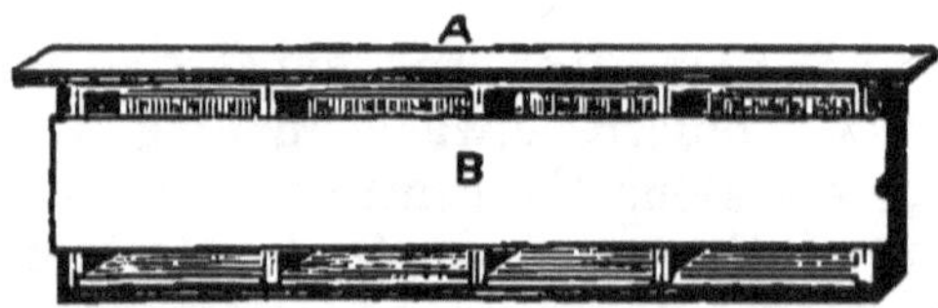

Fig. 55.—HALF STORY FRAME WITH SECTIONS AND TIN SEPARATOR.

explained in another chapter. The section frames for a half-story hive—with the exception of the depth which is only four and a-quarter inches inside—are exactly similar to the larger ones.

FRAME FORM.

To aid us in putting the different parts of the frames together true and square, it is absolutely necessary that we should have a gauge or form for holding the pieces in their proper places while nailing them. The engraving on page 133 represents a useful kind of frame form :—

It is made as follows :—Take an inch board A, 1ft. 8½in. long x 8⅞in. wide, and plane both edges true. Next cut two battens B B, 1⅜in. thick x 2in. wide x 8⅞in. long. Out of one end of each cut a shoulder ⅜in. deep x ⅝in. on, as shown, and nail them on edgewise on the ends of the board A. The length between the battens should be the same as the outside length of frame, viz., 17¾in. The block C may be an inch or more in thickness, 2in. wide, and 1ft. in length, and should be securely nailed to the board A. On the underside of this block screw a piece of steel band 2 inches wide bent to the required form, as shown in the engraving. Three or four inches of the ends of the band should press tightly against the battens B B, to hold the end bars of the frames in their places.

When making the frames, the end bars are to be pressed down between the ends of the band and the battens B B, close up to the board A. The top bar, groove down, is next placed on top, resting on the shoulders of the battens, and nailed to the end bars with two thin wire nails, an inch long, at each end. Now turn the form upside down and nail on the bottom bar; by lifting the frame top bar first, it will come off the form square and true. A narrow strip of thin steel stretched between and riveted to the lower ends of the band D D, will be found a handy contrivance by which the pressure on the end

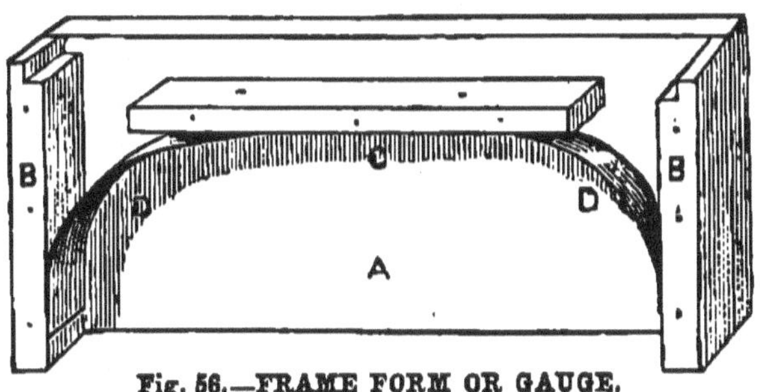

Fig. 56.—FRAME FORM OR GAUGE.

bars may be removed. A slight pull on this will draw the ends of the band towards the centre, when the frame may be taken off without trouble. Both brood and section frames may be made on this form, but it will save time when putting the latter together if a thin strip of wood an eighth of an inch thick is tacked on the upper part of the board A, just where the top bar touches; this will guide the bar into its proper position on the end bars. For half-story frames a shallow form suitable to their depth will be required.

NUMBER OF FRAMES TO A HIVE.

The hive I have described and given instructions for making will take ten narrow frames or seven broad ones. There are a few bee-keepers in America who have adopted an eight-frame Langstroth, but I feel safe in saying that the brood chamber of such a hive would be too small for the climate of any of the Australasian colonies.

The inside of the hive is 14¼ inches wide, which allows each of the ten frames to be three-eighths of an inch apart—the right distance. They may be spaced very quickly by standing on one side of the hive and moving the farthest frame along the rabbets with a forefinger and thumb at each end, till your fingers touch the side of the hive. Now move the next till your fingers touch the first frame, and so on with the whole of them; after a little practice a person can become quite expert at it.

When working for extracted honey the same kind of frames are used in the super as in the lower hive, but for comb-honey the broad frames and sections are used in the super, seven of which, when placed close together as already explained, nearly fill the body, just leaving sufficient space for moving the first one on the rabbets a little to facilitate its removal. Cases, or racks as they are sometimes termed, for holding the sections on the hive instead of frames, are often used. I shall have more to say concerning these in another place.

MATS FOR COVERING FRAMES.

Mats answer two purposes—for keeping the bees below the tops of the frames, and preventing a chilling draught through the hive. No perfectly satisfactory material has yet been introduced that will answer the purpose of a good mat that the bees will not gnaw holes in after a time or propolize fast to the frames. No material should be used that is not porous or that will not absorb the moisture given off by the bees. I have used different kinds of mats, but the cheapest, and I believe the best I have made, was from coarse packing stuff to be obtained at most general merchants. I have had some of these mats in use for the last three years without being gnawed through. I cut them a trifle larger than the space above the frames and leave them with the raw edges.

SECTION BOXES.

There is no way in which comb-honey can be raised equal to the section box system. Not many years ago comb-honey was raised and exposed for sale in large slabs weighing several pounds each, but owing to the messy job of retailing it and the flies attracted by it, very few respectable shopkeepers would have anything to do with comb-honey. In contrast to this, section boxes full of nice white comb-honey may now be seen in nearly every grocer's window in districts where the advanced system of bee-culture has been introduced. They were first brought into use in America some thirteen years ago and have since been adopted in England and many parts of the European continent by progressive bee-keepers; they were introduced into these colonies in 1878.

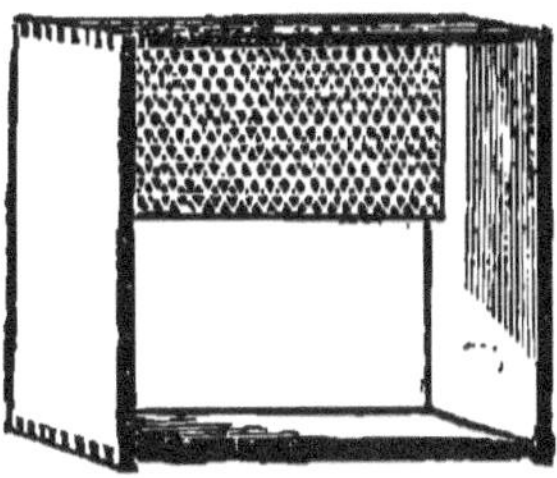

Fig. 57.—ONE-POUND SECTION BOX WITH STARTER OF COMB-FOUNDATION.

There appears to have been almost as many different sizes and forms of section boxes adopted by different bee-keepers since they were first brought into use as there have been different hives, but the most popular at the present day is the one-pound section box, Fig. 57. The two-pound section makes a very nice package but is not nearly so handy nor saleable. Half-pound sections have lately been adopted by some bee-keepers in America, but I am inclined to think they will never become very popular either with bee-keepers or consumers.

REQUISITES OF GOOD SECTIONS.

They should be made of the cleanest and lightest coloured wood obtainable. The material should not exceed one-eighth of an inch in thickness and should be cut smooth. They should be capable of being put together quickly and without nails or glue. And last though not least, they should not cost much.

TO MAKE SECTIONS.

I feel quite satisfied that it will never pay any person not provided with suitable machinery to attempt to make section

boxes. Nothing destroys the nice appearance of comb-honey
more than a clumsily-made section box. Before the neat dovetailed sections now made by Messrs. Bagnall Brothers were
obtainable in these colonies I used to get my material cut by
a cabinet-maker who had a very fine-toothed circular saw, but
as the different parts of the boxes had to be nailed together
stouter material was required for the sides to hold the nails.

The size of the one-pound section—outside measurement—
is 4¼in. by 4¼in. The width of the top, bottom, and sides
should be the same as the section frame, viz., top and bottom
a sixteenth less than one and three-quarter inches, and the
sides a sixteenth less than two inches. When cut for nailing,
the sides should be four inches long by three-sixteenths of an
inch thick, and the top and bottom four and a quarter inches
long by an eighth of an inch thick. A small form on the
principle of the frame form should be used when nailing them.
Half-inch brads I found best for driving, not being so liable to
split the wood as a larger nail. For two-pound sections, suitable for the Langstroth frame, the tops and bottoms require to
be 8½ inches long, but the sides are the same as for the
one-pound box.

<h2 style="text-align:center">ONE PIECE SECTIONS.</h2>

Section boxes made out of one piece of timber are now very
largely used in England and America, and I think might be

Fig. 58.—ONE PIECE SECTION.

introduced by manufacturers into these colonies with advantage.
The pieces, after being ripped from a block of wood that has
been previously shaped to the required form and dovetailed at
the ends, are run across three saws placed the proper distance
apart. The saw cuts run nearly through the pieces, leaving
sufficient wood, however, to hold the joints when they are
folded up.

<h2 style="text-align:center">PUTTING DOVETAILED SECTIONS TOGETHER.</h2>

When first using dovetailed sections I found something more
than hand pressure was required to put them together firm and
strong; I therefore made a small lever cramp, shown in Fig. 59

(No. 1), for this purpose, which answered very well indeed. The base board of the cramp is 5in. wide, 18in. long, and 1in. thick. On one end of this a block, 3in. wide, 4¼in. high, and 2in. thick, is securely nailed. A lever, 2ft. long and 3in. wide, tapered to form a handle, is hinged to the block with a strong strap hinge. Another block, similar to the one on the base board, tapered on the back, is nailed to the lever a bare 4¼in. from the stationary block. If the cramp is made correctly a one-pound section box will barely go between the blocks when the lever is down.

Fig. 59.—SECTION CRAMP AND FORM.

The sections should be put together by hand and then placed between the blocks ; by pressing the lever down the pieces are forced into their places.

CLINCHING THE DOVETAILS.

To make a good finish of them the dovetails should be clinched. For this purpose I use the form, Fig. 59 (No. 2), made by nailing two blocks, similar to those on the lever-cramp, on a short base board, just far enough apart to allow a section box to slide between them. The form is to steady the section while tapping down the ends of the dovetails at each corner. Complaints are sometimes made that the joints of dovetailed sections are not firm enough to keep the boxes from twisting out of the square, even when carefully handled, but I find that when put together as I have described they are equally as firm for all purposes required as when nailed.

SEPARATORS.

In raising comb-honey it is most desirable to have the section boxes uniformly filled, the combs of an even thickness throughout, built with perfectly flat faces, and not projecting beyond the edges of the sections. To this end temporary partitions or separators are generally placed between each two rows of section boxes while in the hive. Were the bees not confined to each particular box by these divisions or walls we should be likely to find the combs built very irregular and bits of wax stuck about the edges of them.

Separators are usually made of tin, but sometimes of very thin wood; tin appears to be the best, as the bees are not so likely to attach comb to it as to wood. They should be made of very light tin, cut three-quarters of an inch longer than the outside dimensions of the frames and $3\frac{1}{2}$in. wide; the ends should be bent at right angles, to hook, as it were, round the end bars, and be lightly tacked to keep them in place. Care should be taken to put them on perfectly flat, and to leave an equal space of a quarter of an inch at the upper and lower parts of the sections to allow the bees to pass in and out and from one box to the others (see Fig. 54).

DISPENSING WITH SEPARATORS.

Much thought has been given lately by some leading bee-keepers to the question of dispensing with separators altogether. No doubt it would be very desirable to do so could we have our comb-honey raised in as good form without them. Not to speak of a saving of expense, there are other objections to their use. For instance, the sections being divided off from each other, and thus to a large extent cutting off continuous communication through the super, tends in a very great measure to prevent the bees entering the boxes as readily as they otherwise would, hence the greater tendency for a colony to swarm when supered for comb-honey. Then, again, there is the space occupied by these separators which might be more profitably filled with honey.

The only method which appears practicable at present to attain this end is to use narrower sections. Nearly all who have experimented conclude that with the two-inch sections separators are indispensable, but with sections running seven to the foot, or a width of slightly under 1¾in. to each box, they have obtained satisfactory results without separators. With only one bee space between the combs, instead of two, as there must be with separators, the narrower sections when well filled are said to weigh about the same as the two-inch ones. A Langstroth hive would take eight rows of the smaller size in the place of seven of the larger boxes. Opinions are very much divided upon the matter at present; I would therefore advise those who might wish to give the system a trial not to go to much expense or trouble at first in making the necessary alterations in their present appliances.

SECTION RACKS AND CASES.

Hitherto I have only mentioned the broad frame system in connection with sections, but there is another method of

Fig. 60.—PRIZE SECTION RACK.

A. *A wedge for jamming the sections together.*   B B. *Tin separators.*
C C. *Glass in outside boxes.*

placing section boxes on a hive which dispenses with the frames. What is termed a rack is formed, consisting of a light framework of wood, across which thin laths are nailed three-eighths of an inch from the bottom at equal distance apart; on the edges of these the sections rest. The rack is set on top of the frames, and takes the place of an ordinary super, a deep cover fits over all, and rests on the hive. Fig. 60 shows a rack filled with prize-sections, that was at one time thought a great deal of in America. The rack and case system for sections has been adopted by a great many bee-keepers in the United States, but there are a few leading men who still adhere to the frames. Some five years ago I gave the above kind of rack a fair trial through one season. At that time I was raising large quantities of comb-honey by the frame method. I soon discarded the racks, and have never used them since. The objections I had to them were :—1st. I could not tier them up, a fatal objection, in my opinion, to any system for raising comb-honey. 2nd. The trouble and expense of blocking up the outside sections with glass or wood to prevent bees getting out. 3rd. When removing some of the boxes the bees would crowd outside, and so get in the way when the cover was to be put on ; and 4th. I could not examine one of the central boxes without disturbing nearly all the others.

THE HEDDON SECTION CASE.

Improvements are continually being made in racks as well as other appliances connected with bee-culture, consequently

Fig. 61.—THE HEDDON SECTION CASE.

there are more convenient ones in use now, but no rack, in my opinion, is equal to the Heddon case for sections (Fig. 61). It is much easier to make than a rack, and to my mind very much handier to use. It will be seen by the engraving that the case is very similar to the body of a half-story hive, divided by partitions crosswise into four compartments, each wide enough to take a one-pound section box. Mr. Heddon, who has raised comb-honey on an extensive scale, was, as he says, "one of the pioneer opposers of wide frames and separators," but having been persuaded to give them another trial, he says :—

"I did so by making 350 wide frame supers (one story or tier of sections high), and used these side by side and over and under 300 of our cases, all worked on the tiering up plan. I used separators of both tin and wood, about 1,800 of wood and 300 of tin."

In his report of the trial to *Gleanings* of May 15th, 1885, he sums up the results as follows :—

"1. My case is the best style of surplus receptacle I know of *to use without separators*, and admits of no improvement by *me*. 2. I would rather abandon separators altogether than use them in two-story supers. 3. I do not think there is any system of using separators equal to the wide frame *when used one story high*. 4. I would use no system of surplusage (either for comb or extracted honey) that did not give me the advantage of tiering up. 5. I prefer tin to wooden separators. 6. There are many advantages in the use of separators, and many in the non-use of them, and each person must be governed by his own special circumstances. 7. Their use or non-use need not affect the *quantity* of surplus secured. I mean to continue the use of separators, and I am now perfecting a different style of super, which I think will aid us greatly in their quick and easy manipulation as well as their usefulness."

Mr. Heddon's opinions, as above quoted, are worth noting, and with the exception of the seventh clause I quite agree with them. I feel convinced that separators are a hindrance to the bees entering the supers in the early part of the season, and also to the rapid filling of the sections at all times. I see no reason to doubt that the Heddon section case is better than frames for use with narrow sections without separators. Professor Cook and others speak very highly of it. I will first give some explanation regarding an appliance used with the section case, and then show how this and a case may be made adaptable to the Langstroth hive.

HONEY-BOARDS.

These are appliances to be placed on top of the frames of the lower hive when putting on a super, and between every two supers when more than one are required on a hive. They are used for the purpose of preventing the building of comb between the upper and lower frames. Every bee-keeper who has had frame hives in use even for a short time will have experienced the nuisance of having the lower part of the frames of a super fastened to the upper part of those in the lower hive ; I have tried several means to prevent it, such as greasing the frames, and allowing but the smallest bee-space between them, but to no purpose. It appears, however, that with the use of a properly constructed honey-board the desired result is obtained. It is curious to note how various appliances that have been discarded come into favour again. Honey-boards were in common use at one time with frame hives, but until quite recently a person known to be using them after they dropped out of general favour would have been looked upon as non-progressive ; now it has been decided that we cannot well do without them.

The principle on which the most improved honey-boards are made to act, is to divide the bee-space between the upper and lower frames ; in other words, there is a greater space left between them, and this is divided into two equal parts by the board, so that there are two bee-spaces between the frames in place of one. It does not prevent the bees fastening the board

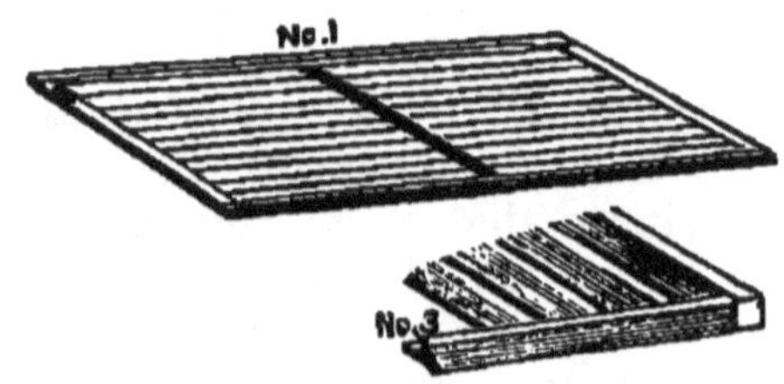

Fig. 62.—HEDDON'S HONEY-BOARD.

No. 1. *Slatted board complete.* No. 3. *Section of same.*

to the lower frames with bits of comb, but it is claimed by those who have used them that the upper bee-space is left entirely clear, so that the frames can be lifted out or the whole super taken off at any time without trouble. Although honey-boards would no doubt be useful on all hives, still, I have felt the most need of some such appliance on hives supered for comb-honey. I find broad frames more frequently " bridged " than narrow ones, perhaps for the reason that the latter are shifted oftener during the extracting season.

HOW TO CONSTRUCT HONEY-BOARDS AND SECTION CASES.

The above engraving of Heddon's honey-board is taken from *Gleanings* for January 15th, 1885. This board (No. 1, Fig. 62) consists of an outside frame, with thin slats running from end to end, tacked on ⅜ths of an inch apart (No. 3, Fig. 62).

The slats are ⅜ths of an inch below the upper edge to give the top bee-space, but there is no allowance made in the board itself for a bee-space below. This is secured by making the frame large enough to rest on the outside edges of the hive, the tops of the frames being about bee-space below the edge. With the hive in general use here (already described) we could not very well, nor wisely, I think, break the junction between the hive and its super with the frame of a honey-board, unless the frame could be made to fit in a similar manner. I see nothing to prevent the honey-board resting on the frames themselves ; we could then make the section cases to fit the lower hive, or each other, the same as our hive bodies do at present.

For strength I would make the frame ¾in. deep by ⅝in. wide, of the exact dimensions of inside of hive, viz., 18½in. by 14¼iu. Before nailing it together, run a groove ⅛in. wide by ¼in. deep, along the centre of the depth of the end pieces. Next cut nine slats, ¼in. by ⅞in. wide, ½in. longer than the inside dimensions of the frame. Nail the two ends (grooves inwards) and one side together, slip the ends of the slats into the grooves, and nail on the other side of frame. Space the nine slats ⅜in. apart and fasten them with small nails at each end ; cut six slats, ⅜in. square, to nail on (three on each side) across the others at equal distances apart, to keep them firm and prevent them twisting. The sides of the frame might be bevelled on their inner edges, to give room for the bees to get up into the sections at the sides of the case. The honey-board is now complete, and may be placed on the frames either side up, as there is ⅜ths of an inch bee-space on both sides below the long slats.

A half-story body, used with the Langstroth hive, can be converted into a Heddon section case for use with the above honey-board by merely putting in the divisions, its measurements for this purpose being correct. The length of the inside is 18½in., and as this space is to take four 4¼in. sections and three division boards (see Fig. 61), we must make the latter of half-inch material, or rather less, so that the sections may slip into their places readily. The divisions should be cut 4¼in. wide and 14¼in. long (the width of the case), and nailed in the body, so that they will divide the length into four compartments of a full 4¼in. each. The upper edge of each division board should be a ¼in. below the top edge of the case. Three strips of stout tin, or, what would be better, thin galvanised iron, 14¼in. long by 1in. wide, are now required; down the centre of each punch a few holes, and nail them along the bottom edges of the divisions, allowing the strips to project a quarter of an inch on each side; these projections are to rest the sections on. Two more strips are required for the ends of the case, which must be bent along their length, so that they may be tacked on in their proper places, allowing only ¼in. to stand out as a support for the sections. The case is now complete.

Each compartment will take eight 1¾in. sections or seven 2in. ones, and just leave sufficient space for blocking them off with a thin piece of wood. If separators are to be used one will be required for each section. When the honey-board is placed on the lower frames, and the case or half-story put on in the usual way, there will be the requisite space, ⅜ths of an inch, from the frames to the board, and the same from the board to the bottom of sections. The ⅜in. pieces across the honey-board might be put on so as to come directly under the divisions of the case; or, instead of the wood, strips of tin could be tacked across the slats; this would leave free communication all over the honey-board. To use honey-boards with our extracting hives we would have to make the boards larger, so that the outside frame could rest on the edges of the hive and support the super to give the necessary bee-spaces.

AXIOM.

"BEES MAY ALWAYS BE MADE PEACEABLE BY INDUCING THEM TO ACCEPT OF LIQUID SWEETS."   *Langstroth.*

CHAPTER VII.

THE HONEY-EXTRACTOR AND MANIPULATION OF EXTRACTED HONEY.

NEXT in importance to the movable comb-hive itself, as an apiarian appliance, ranks undoubtedly the honey-extractor. By its means we are enabled to obtain the liquid honey in perfect purity from the comb, in the form best suited for storing and for transport, and without injury to the combs themselves. These can in this way he made to do duty over and over again, a matter which has an important effect upon the quantity and the cost of the honey produced each season by one colony of bees. Without the extractor the improved form of hive could not have developed half its real advantages. It would, of course, have enabled us, as it does now, to raise comb-honey in the best condition, but the importance of honey as an article of general consumption and of commerce could never have been anything like what it is at present if we had been obliged to follow the old system of obtaining it in a liquid state from the combs.

STRAINED OR PRESSED AND MELTED HONEY.

Formerly, when it was required to separate the honey from the comb, the bee-keeper had his choice of two methods—the one consisting in squeezing or pressing the honey out of the comb in its cold state, the other in melting or boiling down the honey and comb together and separating the wax, etc., which would settle on the surface, as soon as the mass cooled. By both these processes the comb must of course he sacrificed or reduced to the state of melted wax, and it will easily be understood that by either process the original delicate flavour of the honey would be partially or wholly destroyed. Those who have had any experience in separating honey from the combs

in any other way than by using the "extractor" will know
what a disagreeable business it is, and will not wonder that no
one should have been found formerly to undertake that portion
of a bee-keeper's duties as a matter of choice.   Those who still
adhere to the box hive and sulphur pit, as well as those who
obtain honey from bush hives, are obliged to strain or melt
the honey in that way.   Such people, it may be supposed, will
not be over nice as to whether or not a little more or less of
bee-bread, brood, dead bees, and other rubbish may be squeezed
or boiled with the comb ; and, indeed, if they should be ever
so particular they cannot prevent some such mixture taking
place.   I venture to say that no one who has had the opportunity of tasting honey taken with the " extractor " would care
to eat the old sort of strained honey again.

INVENTION OF THE CENTRIFUGAL EXTRACTOR.

To the Austrian Major von Hruschka, of Dolo, near Venice,
we are indebted for the invention of the honey-extractor in
the year 1868.   Like many other important inventions, the
idea seems to have been suggested in a very simple manner.
It is said that Herr von Hruschka's son was amusing himself
twirling a small tin pail, tied to the end of a string after the
manner of a sling, while his father was engaged taking some
honey, and happened to give him a piece of unsealed honey-
comb.   This the boy put in his pail, and afterwards continued
to twirl it round.   The father subsequently chanced to notice
that one side of the piece of comb was quite clear of its honey.
He turned it in the pail, swung the latter round as the boy had
done, and found the other side of the comb emptied of its store.
He grasped the idea at once, set to work, and gave the world
the first honey-extractor.   Although many improvements have
been made in the details of construction, the principle of the
extractor remains the same as in the first one made by Von
Hruschka : the honey is thrown from the combs by centrifugal
force.   There are several kinds made and sold at present,
differing only in trifling details.   The engraving on the next
page shows the form, taken from an American pattern, now
generally used in these colonies.
This is called a " double " or " two-comb " extractor, because
it receives two comb frames at a time.   It consists of a strong

metal (usually tin) cylinder, about 17in. in diameter by 24in. in height. It contains a square basket of tin work, with wire cloth covering on two opposite sides, sufficiently large to take two frames of the Langstroth hive, one on each side, placed on end, and resting against the wire cloth. A spindle, to which the basket is attached, passes down the centre, the lower end working in a bearing raised some height above the bottom of the cylinder, so that there is a space for honey to accumulate below the bottom of the basket. The upper end of the spindle works through a bearing exactly in the centre of the

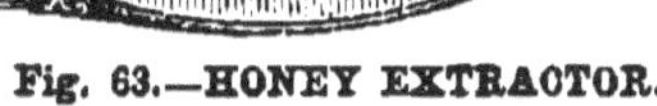

Fig. 63.—HONEY EXTRACTOR.   INSIDE BASKET.

cylinder and attached to the inside of its rim by a light casting; to the top of the spindle is attached a handle with gearing to multiply the motion, so that by turning rather slowly with the hand a considerable velocity is got up on the outsides of the basket. When combs are to be extracted, the capping of the cells having been removed with the uncapping knife hereafter to be described, one frame is placed against each wire cloth side of the basket and made to revolve until the honey from the outer side of each comb is thrown out, through the wire cloth, and, striking against the inside of the cylinder, trickles down to the bottom, to be withdrawn by the tap or honey-gate shown in the figure. The frames are then lifted out and turned in the basket until the second side is emptied like the first. In placing the frames in the baskets care should be taken that the bottom bar shall be *going foremost* when the basket is set revolving, as the honey cells are generally built by the bees with a greater or less slant upwards, and they are more easily emptied by the centrifugal force when the mouth of the cell is turned away from the direction of motion.

As there is a considerable strain on the machine when two heavy combs are made to revolve at a high speed, both the

Fig. 64.—FRAMEWORK FOR TWO-COMB EXTRACTOR.

cylinder and the interior basket require to be strongly constructed, and combs of nearly equal weight should be operated upon at the same time in order to equalise the strain on the central gear. If it be desired to extract from only one comb upon any occasion, a piece of board or some such counterpoise should be placed in the opposite side of the basket.

To make the extractor firm and steady, and to raise it a sufficient height off the floor (if it should be so placed) so that a vessel can be put under the tap to draw off the honey, it should be encased in a framework of wood placed upon feet, as shown in the above figure engraved from a photograph.

No regular apiary should be without one of these machines, unless a still larger and more expeditious one (described further on) be considered necessary. But people who keep only very few hives, and do not care for a little more trouble with the comparatively few combs they may require to extract in the season, may find it convenient to use a

SINGLE-COMB EXTRACTOR.

A simple and cheap implement of this sort was introduced in 1875 by Mr. Abbott in England, where it is extensively used by cottagers and others cultivating bees on a small scale. It is known by the name of the "Little Wonder;" can be had for about one-third of the price of a two-comb extractor, and is of the construction shown in Fig. 65.

The can, or body, is made of tin, two broad straps of the same material are soldered to it and screwed to the handle. A frame of wire netting (*a, b*) is made to fit inside, against which the comb is placed after being uncapped. The iron pin at the lower end of the upright rod being firmly fixed in the floor, a circular motion is given to the machine by a sway of the hand which grasps the loose portion of the handle on the top. As the revolutions increase in rapidity the honey is thrown out of the comb into the can, from which it may be drawn off by the opening on top.

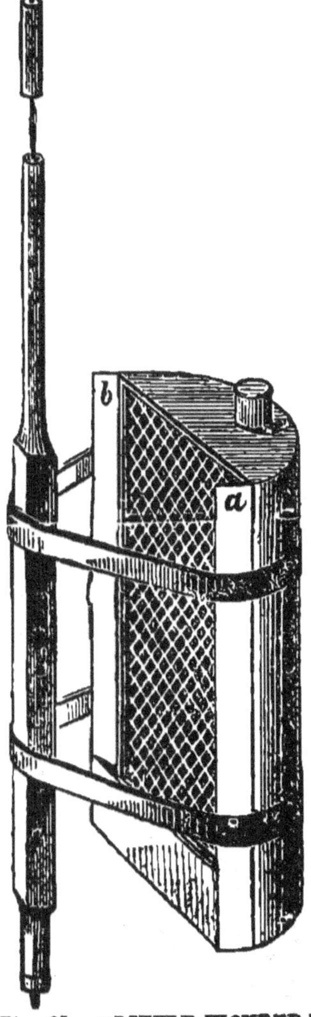

Fig. 65.—"LITTLE WONDER."

SIX-COMB EXTRACTOR.

In order to facilitate the work in large apiaries extractors have been made to take three, four, six, or even eight combs at a time. The engraving on next page represents one that I have had made for use at the Matamata Apiary.

The following description of this machine was given

in the *New Zealand and Australian Bee Journal* in November, 1883:—The case or cylinder inside of which the framework supporting the comb baskets revolves, is made of 1½in. timber, lined with stout tin. It is hexagonal in shape, 44in. in diameter at its widest parts; each of its six sides measuring 22in., making its circumference 132in.; height, 33in.. The above are outside measurements. Wood is the most suitable material for the case, as it keeps the machine firm while at work.

The outer part of the framework upon which the baskets hang is cast metal; cast in pieces to form each of the six sides.

Fig. 66.—REVERSIBLE SIX-COMB HONEY-EXTRACTOR WITH ONE BASKET DETACHED.

These pieces are shaped thus II, the top and bottom bars being 16½in. long, and the depth of the frame 19¼in. The bars are 1in. in width, by $\frac{3}{16}$in. thick. At the corners, where the sides join each other, socket-pieces are riveted in the angles at top and bottom. These pieces, as well as securing all the sides together, answer as parts of the hinges for the comb baskets.

The comb baskets, which are made of tinned wire cloth, are 2in. wide by 19in. deep, secured at top and bottom by two bands of metal, like the bars of the framework. At one end of each of the bands arms project two inches; each being furnished with a pintle, which is made to fit in the socket-pieces, and so form hinges like those of a field gate. These hinges allow of the baskets being turned so as to take each other's place, and thus bring the opposite side of the comb to the front, after one side has been extracted. The baskets turn towards the centre.

The spindle, or journal, is a $\frac{1}{2}$in. round iron bar, 37in. long. A short distance from the top and bottom ends are two six-sided nuts. From each of the six sides of each nut the supporting arms of the framework project; these are made of $\frac{3}{8}$in. round iron, their ends being screwed on to the framework. The lower part of the spindle works in a socket fastened on to the bottom of the case, while the top passes through a curved bar of iron which is screwed on to opposite sides of the case.

Fig. 67.—COWAN'S AUTOMATIC BASKET.

A handle 10in. long fits on to the spindle above the bar of iron, and is made secure by a screw-nut. At one side of the bottom a honey-tap is fastened in, and the extractor is complete.

The only alteration I found necessary to make, after some trial of the machine, was to attach a gearing with multiplying power, having the driving handle at the side instead of working direct from the spindle, as in the figure. This enables the operator better to regulate the speed and keep the basket revolving at an uniform rate. With this machine one man can do nearly three times as much work as with a two-comb extractor, and with but little more labour.

Mr. T. W. Cowan, who has done much to improve apiculture in England, has invented an arrangement, shown above, by which the baskets of the extractor are made to turn automatically by simply reversing the motion of the gearing. A sort of rachet movement is fixed in the upper arms, by which the reversing is effected.

PREPARING COMBS FOR EXTRACTING.

As a rule the combs intended for extracting are left in the hives until the cells are sealed or capped. This is a sure indication that the honey is in good condition, as the bees will not seal up any which has not been properly "ripened." In cool

Fig. 68.—ROOT'S UNCAPPING KNIFE.

and moist weather, when the nectar brought in by the bees is very thin, it is hardly safe to extract the honey before it is sealed, as it may, if not artificially ripened afterwards by evaporation, show a tendency to ferment. In hot weather, however, when the honey is pretty thick, it may safely be extracted as soon as the cells are filled and before the bees have had time to seal them over. When this can be done it saves much valuable time, some labour, and some injury to the combs, inseparable from the process called "uncapping." In most cases, however, some portion, if not all the surface of the

Fig. 69.—BINGHAM AND HETHERINGTON KNIFE, WITH CAP-CATCHER.

comb, will require to be uncapped before being put into the extractor, and for this purpose an uncapping knife is necessary. These are made of various forms, two of which are shown above.

The blade of the knife is of steel, thin, and sharpened on both sides. The Bingham and Hetherington knife takes more of the form of a trowel, except that it is much thicker in the blade and has bevelled edges. After using both kinds I much prefer the latter. The cap-catcher can be fixed on the knife to prevent the cappings, after they are shaved off, falling on the comb, but I have not found the need of using it, for by giving the comb a slight inclination forward the cappings drop clear. With the aid of one of these knives the operator, after a little practice, can shave off the cappings very easily and cleanly, with the least possible injury to the tops of the cells. While this is being done the comb must be held on end over an un-capping box or can, so formed as to retain the wax cappings, to be afterwards melted down, and to allow the honey, which is sometimes unavoidably cut off with the cappings or which may trickle from the comb during the operation, to pass through a strainer into a receptacle provided for it. Various contri-vances are adopted for this purpose. The following (Fig. 70) is one I have generally used, and can recommend :—

Fig. 70.—DADANT'S UNCAPPING CAN.

It is somewhat like the cylinder of a honey-extractor, but made in two parts—the upper one, to the bottom of which the strainer is attached, slipping a short distance into the lower one. The cone rising from the bottom of the can gives sup-port to the strainer. Those I have in use are made of galvanised iron, are 20 inches diameter, and 30 inches high.

A flange 2 inches from bottom of upper can forms a support for it by resting on upper edge of lower one. The strainers are made of stout wire-cloth ten meshes to the lineal inch, and permanently fixed to upper can. A tap at bottom and two strong handles to each can make them complete ; cost, 30s. A light wood framework is made to hook on upper rim, as in the figure, on which to rest the frames of combs while uncapping them.

Sometimes it is advisable to extract the honey out of a piece of broken comb, or a whole comb not built in a frame. For this purpose a so-called broken-comb basket of the form shown below (Fig. 71) is convenient.

The two pieces of wirecloth are joined by hinges at C C. A wire, with two bent ends, B B passes through the tin frame of one piece, and can be easily turned to hook into the other

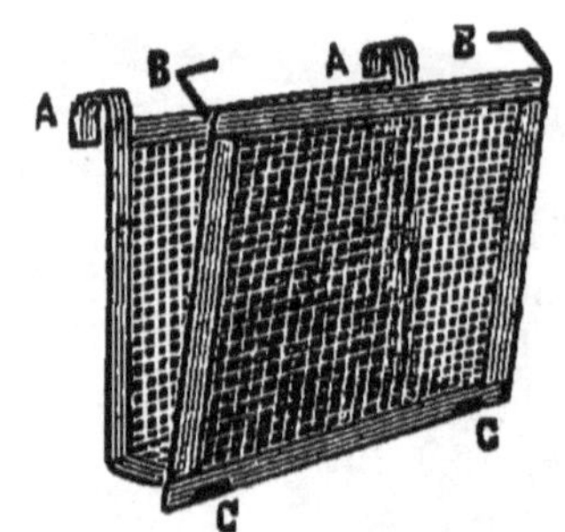

Fig. 71.—BROKEN-COMB BASKET.

frame at A when the comb to be extracted has been placed between the two ; the whole is then suspended on the top bar of the extractor-basket by means of the hooks A A.

MANIPULATION OF EXTRACTED HONEY.

The honey as it flows out of the extractor is by no means in a fit condition to be filled into the vessels in which it is to be stored or sent to market ; no matter what care is taken with the uncapping and extracting, there will be some pieces of wax mixed with the honey, and perhaps some larvæ or dead bees. If it be passed through a suitable strainer, all these foreign substances may be removed, or if it be collected in a tank and allowed to settle long enough, until they, being lighter than honey, accumulate on the surface, the clean honey may then be drawn off from the bottom of the tank, through a tap or honey-gate. To obtain the honey in the best possible condition, it is desirable that both these processes—the straining and the settling in the tank—should be gone through.

The most convenient form of fitting up the extracting house has been a subject of many inquiries and much thought amongst bee-keepers; but of all the plans I have seen suggested, that of Mr. T. J. Mulvany, in the *New Zealand and Australian Bee Journal* of October, 1884, is the most simple, least expensive, and best. The idea of the double tank is an excellent one. I have adopted it myself, and find it most convenient. The following description is in nearly his own words.

ARRANGEMENT OF EXTRACTING-HOUSE.

Much may be done to save time and labour by a proper arrangement of the extractor, strainer, and tanks in the extract-

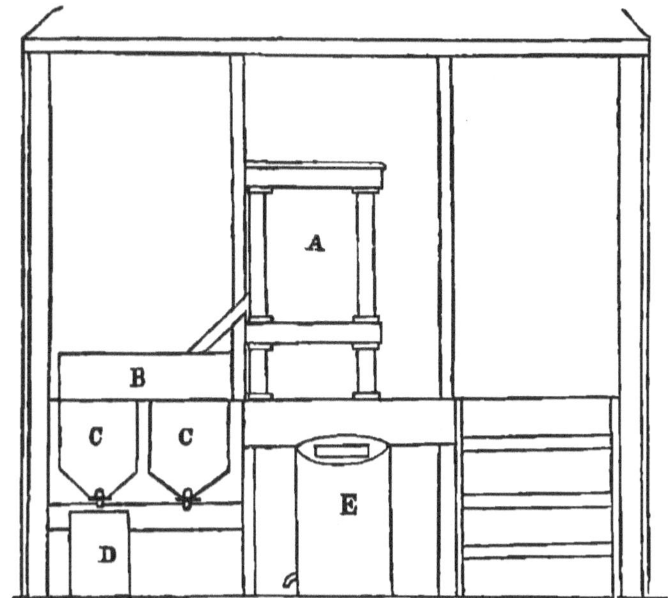

Fig. 72.—CROSS-SECTION OF EXTRACTING HOUSE WITH VIEW OF EXTRACTOR. STRAINER, AND TANKS.

ing house. The two sketches, Figures 72 and 73, show, in elevation and in plan, the arrangement above referred to, by means of which the honey is allowed to flow direct from the extractor into a strainer, and thence into a tank, the whole process being automatic, so that the pure honey can ultimately be drawn off from the tank without further trouble, direct into the packing tins or other vessels.

In both figures, A represents an ordinary two-comb extractor, fixed on a platform about 2ft. 7in. above the level of the floor at one end of the extracting-house; B, the strainer; C C, a double tank; D, the position of a 60 lb. tin or other vessel ready to be filled from the tank; and E, the uncapping can.

The strainer is a vessel of strong tin, stretching across the double tank, on the outer edges of which it rests by means of

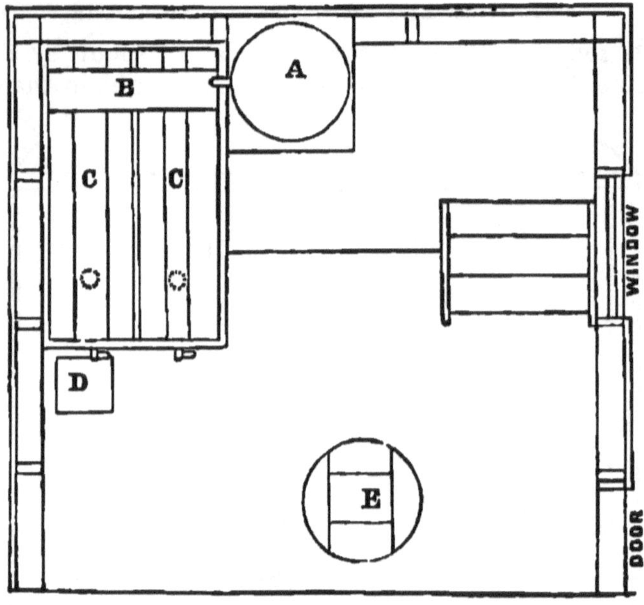

Fig. 73.—GROUND PLAN.

flanges attached to its ends. The bottom, for one half its length, is close, the other half having a fine wire gauze (sixteen wires to the lineal inch) let in and soldered like a milk strainer. By simply turning the strainer end for end, it can be made to work into either of the two divisions of the tank. The honey, as it flows from the extractor, passes first through a rough strainer, consisting of a frame of wood, two inches deep and half an inch thick, the bottom covered with perforated zinc, fitting loosely in the top of the deep tin strainer, on the sides of which it hangs by means of bent lugs. This coarse strainer catches all dead bees, larvæ, and large pieces of wax, and is easily lifted out and cleaned without disturbing the fine strainer, through which the honey passes more slowly into the tank.

The tanks can, of course, be made of any size that may be considered desirable. That shown in the sketches was considered large enough for an apiary of a hundred hives. It was only 4ft. 2in. long, by 2ft. 6in. wide, and 1ft. 4in. deep (outside measure), made of inch boards, with a division board in the centre; each division lined with strong sheet tin, soldered so as to be quite water-tight. Each division holds upwards of 500lb. of honey. The tank is made in two divisions in order to admit of the honey of any one day's extracting remaining to settle all the next day, even if extracting be going on every day. If more than 500lb. of honey be likely to be extracted in any one day, the tank should be made larger. The process of double straining and settling in the tank tends materially to ensure the complete "ripening" of the honey ; and being, as already observed, automatic, it saves all manual labour, all waste of honey, and ensures perfect cleanliness.

I have two double tanks at Matamata, of the following dimensions (inside measure) : 6ft. x 4ft. x 18in. deep; each division will hold over 1,800lb. when level full of honey of the specific gravity of 1·488. For a short time last season I found I required both in use at one time, and had over 5,800lb. in the two tanks. If the extracting house is large enough, two such tanks can easily be placed *across* the room, with the extractor stage *between* them ; so that when one tank is full, the extractor will only require turning round to the other. They are made of 1¼in. timber, lined with stout tin.

Golden Rule.

"Keep your colonies strong."

Oettl.

CHAPTER VIII.

COMB FOUNDATION.

THE third great improvement introduced of late years, taking rank only after the inventions of the movable frame-hive and the honey-extractor, is that of furnishing the bees with the foundation or septum of the combs which we wish them to build. By this means we are now enabled to dictate to the busy little workers exactly where a comb is to be built, and whether it shall contain worker or drone cells ; to secure its

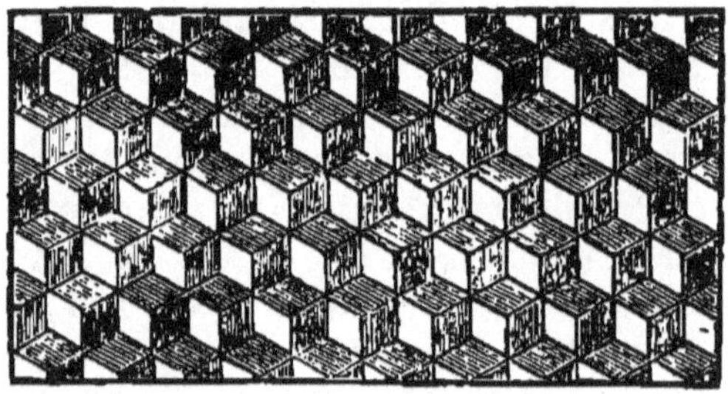

Fig. 74.—COMB FOUNDATION.

being built quite straight, and with an even surface ; and to save the bees a great deal of time in the secretion of wax just at the period when their labour can be best employed in the storing of honey.

HISTORY OF THE INVENTION.

There would be no use in fitting up a hive with movable bars or frames, unless we could secure the building of the combs along the line of the bars or within the frames. The bees, if left to themselves, would be just as likely to build their combs across such lines, connecting all the bars or frames together, and thus rendering them quite useless. It was necessary to give them a fair start, at all events, and the early practice was to attach a strip of old comb to the bottom of the bar or the under side of the top bar of the frame. These pieces of comb were called "starters," and it was no doubt with the idea of furnishing such starters in a convenient manner that apiarists first commenced experiments in order to produce strips of manufactured comb-foundation. As soon as any success had been achieved in that direction, it could not long escape the notice of intelligent apiarists, that great advantages would ensue if the bees could be furnished with an unerring guide for the whole extent of the comb, and with a considerable portion of the wax required for its construction. It was not, however, until numerous experiments had been made, and a considerable time had elapsed after the first successful attempt to make narrow strips of comb-foundation, that the beautiful sheets, with which most bee-keepers are now familiar, were produced.

It appears that so early as the year 1840 or 1842 a German, named Kretchmer, used strips of linen coated with wax and passed between engraved rollers to give them the impressions of the bottom of the cells. His son, E. Kretchmer, of Iowa, writing to the *American Bee Journal* of December, 1878, says: "Comb-foundations were made by my father in Germany in 1842; they were made by a pair of engraved rollers, and starch was used to prevent the wax from adhering to the rollers." This sort of foundation does not seem to have been a success, and it was not until the year 1857 that another German, Herr Mehring, of Frankinthal, introduced impressed sheets of wax, instead of waxed linen. These wax sheets were four or five times as thick as the partition in the natural comb, which is, as Prof. Cook informs us, only $\frac{1}{180}$ of an inch in thickness; they were pressed between metal plates, which gave them impressions corresponding to the rhomboidal bases of the natural cells; and this foundation soon found a very general use in Germany. In 1874, a poor German, named Friedrich Weiss, residing in America, invented the machine which brought the foundation to something nearly like what is now made. Professor Cook gives to him the whole credit of the invention, and to Mr. A. I. Root the no less laudable merit of having, by his energy and enthusiasm, brought both the roller machine and the foundation into general use.

ADVANTAGES DERIVED FROM ITS USE.

It has already been stated (in Chapter III.) that bees require to consume a large quantity of honey—taking the mean of experiments, about fourteen pounds— in order to secrete one pound of wax. Assuming the honey to be worth only fivepence per pound, each pound of wax thus secreted represents a value of nearly six shillings. By supplying wax foundation, which costs less than half that price, we save more than half the cost of the material. A still greater advantage is the saving of time to the bees, and the opportunities thus given them to store a much greater quantity of honey. Not only can they store, instead of making into wax, upwards of twenty pounds of honey, represented by some pound and a half of foundation supplied to the ten frames of a hive, but they can have the ten combs fully built out in one-fourth of the time that should be devoted to the building of entirely new comb; and all the bees that would be so employed are set free to store honey instead. We may reckon that in an ordinary season a fair swarm will work out the ten sheets of foundation in a Langstroth hive in one week; without the aid of the foundation it would take them four or five weeks. I have had swarms that worked out and filled with honey and brood all the sheets in a *two-story hive*, and threw off a good new swarm, within three weeks from the date of hiving. The time thus gained may make all the difference between profit and loss in a short honey season. Besides this saving of time and gain in honey, we secure straight and even combs, such as are rarely, if ever, built without the aid of foundation; we can control the building of drone-comb, and consequently the breeding of drones within such limits as may be deemed advisable; and it will be found that, even without the precaution of wiring, the combs so built will be much stronger and will withstand the strain of extracting much better than those built without foundation. It is also of very great value in the case of swarms hived late in the season, which are thereby enabled to build their comb in a short time, and put themselves in a better condition for the winter.

PRINCIPAL POINTS OF GOOD FOUNDATION.

Good comb-foundation must, in the first place, be made of nothing but pure beeswax; various substitutes have been tried, but they have all failed. The nearer it approaches nature in the size and shape of the cells the better. The cells should have a thin base, with high and soft rudiments of the side walls. I have hitherto advised the use of foundation running about six square feet to the pound ; but I now think, after further experience, that for brood and extracting combs, where wire is not used, a little heavier, say five square feet to the pound, is better. The lighter kind will do where wire is used.

COMB-FOUNDATION MACHINES.

There are two methods for impressing the wax sheets—one by passing them between engraved rollers, as shown in Fig. 75 ; the second by heavy pressure between flat plates in a

Fig. 75.—A. I. ROOT'S 10in. ROLLER MACHINE.

machine like a small hand printing press. There are many who prefer the pressed to the rolled foundation ; but after making both kinds, I cannot see any superiority in either. Of the roller machines there are five different sorts made, known as the Root, the Dunham, the Vandervort, the Pelham, and the Van Deusen machines. The first four make the natural based foundation, while the last impresses the sheet with a flat-bottomed cell. This latter also has its admirers.

M

I have used both the Root and the Vandervort, and have seen samples of foundation made in the other machines ; but I prefer that made in the first-mentioned to any other. The flat-bottomed foundation is used mostly, I believe, for section-boxes.

With the given press the sheet of wax is placed between two copper plates, which are made to close like a book. The

Fig. 76.—THE GIVEN PRESS.

plates are then put under the press, and with the aid of a compound lever are subjected to a tremendous pressure, which gives the sheet the desired impressions. This press has one advantage over the roller machines : it can imbed wires in the sheets of foundation while they are being impressed. A frame already wired is placed over the lower plate and a sheet of wax on the wires; the plates are put through the usual process, and the sheet comes out wired and fastened in the frame. The Given press numbers among its friends many of the leading bee-keepers of America.

DRONE CELL FOUNDATION.

All the machines now manufactured, whether of the roller or press description, are made to impress worker-sized cells, unless specially ordered for drone cells. It was at first thought that the larger cells would be advantageous in the surplus boxes, and a number of drone comb machines were sold; but it was found that the use of such foundation encouraged the breeding of too many drones and the consequent occupation of the surplus boxes by the queen—two things which it is important to guard against. I had given this matter some thought when ordering my first machine in 1878, and decided not to order a second one for drone cells, as I anticipated that the use of drone comb would result, as it has proved to do, in a disadvantage. I find no difficulty in obtaining as much good drone comb as I desire by the following method. During the height of the honey season place a few frames, with only starters of foundation, in the top boxes; the bees will, at that time, work them out at once with drone comb and fill them with honey before the queen has an opportunity to take charge of them for the purpose of egg laying.

PROCESS OF MANUFACTURE.

The wax used for this purpose must be not only pure beeswax, as already stated, but must also be as clean as possible; any dirt mixed with the wax tends to make the foundation brittle. When melting the wax a double boiler should always be used, the inner one for the wax and the outer one for water, in order to prevent the wax from burning; burnt wax is of no use for foundation. To make rapid work two such double boilers should be used, one in which the wax is melted from its cold state, the other to be kept supplied with melted wax at the proper temperature and used as a dipping boiler. The next things needed are a tub of cold water, two or more dipping boards the length of the sheets required, and a thin knife. The boards should be made of very thin wood—or stout galvanised iron will answer—and made so that they can be reversed when dipping. Wax melts at a temperature of about 145° Fahr., and the contents of the dipping boiler should be kept at very little over that temperature. The boards, after being soaked in the cold water and drained, are at first just slightly touched over with soapy water to give them a start ; care should be taken to use no more soap than is absolutely necessary, as it is said the bees dislike it. Take a board and dip it overhead in the wax ; lift it out and let it drip ; as soon as it has ceased to drip, quickly reverse the board and dip it overhead again ; and repeat the process until the sheet is of the desired thickness. Two or three times is sufficient for stout foundation, and once or twice for thin. After the last dip plunge the board into the cold water, and if everything is right the sheets will peel off without trouble. If the wax is too hot the sheets will break, if too cold they will stick to the board. According as the wax in the dipping boiler is used up it should be replenished from the melting boiler, which must be kept on a good fire. For the Given press the dipping boards are made about the size of the inside dimensions of the frames, and are used in the same manner as the others.

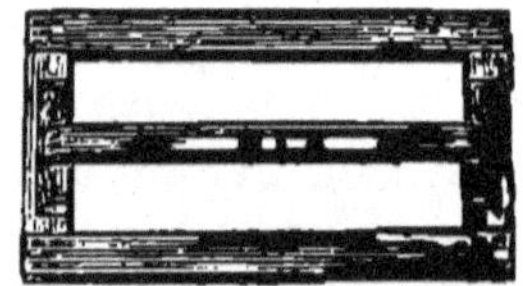

Fig. 77.—GAUGE FOR TRIMMING FOUNDATION.

After all the sheets are dipped it is better to leave them until next day before putting them through the machine. They should be nice and pliable, or else they will break when pressed or rolled. If the sheets are placed in warm water it will soften them. A little thin starch is a very good thing to put on the rollers in order to prevent the wax from sticking to them ; tins for holding the starch are supplied with the machines. For the Given press it is better to brush the plates over occasionally with concentrated lye. As the sheets require trimming to fit the frames after coming from the machine, a gauge should be made for that purpose as shown above.

By laying five or six sheets squarely together, and the gauge on the top, so many can be trimmed at once with the aid of a sharp butcher's knife. The sheets of foundation, when not to be used in wired frames, should be cut a quarter inch shorter and half an inch shallower than the inside dimensions of the frames, but for wired frames they should be made to fill them as nearly as possible.

TO FASTEN FOUNDATION IN FRAMES.

For this purpose a board is required, three-eighths of an inch in thickness, to fit easily inside the frames. Two thin battens should be nailed on the back, projecting a little at the ends, as shown below.

Fig. 78.—COMB-FOUNDATION BOARD.

This board is to steady the sheet of foundation while it is being fastened in the groove of the frame. The following engraving and explanation will make the operation clear to the reader :—

Fig. 79.—MODE OF FASTENING IN FRAME.

In cold weather slightly warm the sheets before handling. Cut one edge perfectly straight with a sharp knife and straight edge. Hold the board, A, in the left hand; lay the frame, C (inverted), against the projections B; the board itself will then be within the frame. Lay the sheet of comb, D, against the board and press into groove in frame. Now, by elevating one end of the frame (as shown in the figure) and pouring a little melted wax in groove at upper corner, it will run down to E, and fasten the sheet securely. As soon as firmly set, support the sheet while reversing the frame, and do the same on the other side. When done, hang in an empty hive for safety. Use stoutest sheets in lower hive. For melting the wax required for fastening the sheets the most convenient appliance we can possibly have is a wax-smelter, made upon the principle shown in the following sketch :—

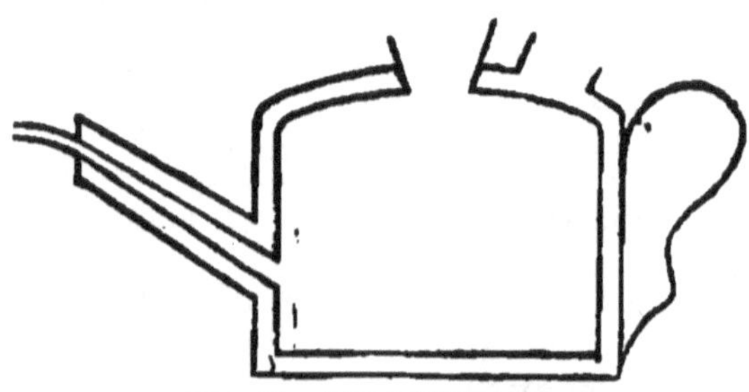

Fig. 60.—WAX-SMELTER.

It is double jacketted, the space between the two containing hot water, and the wax thus kept from cooling in the inner vessel. These smelters are generally made of tin, but are better (though of course more expensive) if made of copper. A lamp or slow fire can be used to rest the smelter upon and keep up the temperature during the intervals of using it.

TO FASTEN FOUNDATION IN SECTIONS.

The Parker comb lever is an indispensable appliance for fastening either starters or small sheets of foundation in section boxes.

The lever is fastened at the sides to the block in such a way that it can be pushed forward or drawn back. A stop is nailed to the block, against which the edge of the top piece of the section box is placed; the front edge of the lever is now pushed forward to the centre of the section top, the edge of the starter or sheet of foundation placed under it; the end of the handle

being then raised, the front of the lever presses upon the edge
of the foundation, the rest of which is at the same time to be
bent upwards to its proper position in the box; the end of
the lever is then drawn backwards, with pressure, and the
foundation remains fastened in its place. The machine should

Fig. 81.—PARKER'S COMB-LEVER.

be screwed to a table or bench, and the under edge of the lever
moistened with a little honey, to keep it from sticking to the
wax.

WIRED FOUNDATION.

In order to prevent sagging or bulging of the foundation,
owing to undue heat and the weight to which it is sometimes
exposed during the operation of comb building, and especially
to prevent breakage of the comb when thick honey is being
extracted, it has been found desirable to strengthen the septum
in some manner. Different substances, such as wood, vegetable
parchment, strong paper, linen, vulcanite, and wire, have been
tried as a base for foundation, but none of these, except the
last, have been found to answer the purpose. Capt. Hetherington, of Cherry Valley, New York, was, I believe, the first to
use wired foundation some few years ago, and already it is
coming into very general use. Very thin wire is first secured
in the frames, and afterwards imbedded in the foundation,
either by hand or by the Given press. The following answer
to a correspondent, in the *New Zealand and Australian Bee
Journal*, expresses nearly all I have to say as to a choice between
hand and machine wiring :—" As soon as a demand sprang up
for wired foundation in America a Mr. Given invented a

machine by which sheets of wax are pressed into frames already
wired for the purpose, and at the same time giving the sheets
the ordinary impressions of the base of the cells. Several
manufacturers at once commenced making and selling frames
of wire foundation, but it was soon discovered by purchasers
that this method of obtaining it was too expensive, as, in
packing, one frame of comb would occupy nearly as much
space as $1\frac{1}{2}$lbs. of ordinary foundation, consequently the
freight on a few sheets of the wired article came very
heavy. It was also found that many of the sheets broke
away from the wires in transit through rough handling. The
method now generally adopted by bee-keepers is to purchase
the ordinary foundation and wire it themselves. This is by
far the cheapest and best plan, and the one I would advise
those who desire to use wired foundation to adopt." Professor
Cook says also, "Some, even with the Given press, prefer to
put the foundation into the wires by hand."

The following instructions will enable any one to fix the
wiring in his frames : - Pierce the top and bottom bars of the

WIRING THE FRAMES.

Fig. 82.—WIRED FRAME.

frames, before putting them together, with holes two inches
apart, commencing a half inch inside the end bar. The holes
should be exactly along the centre of the bars, the number
required being nine to each. A small lever press, with a set of
short steel awls firmly fixed in it, could be made to pierce the
nine holes at one stroke. The wiring is done after the frames
are put together, No. 30 *tinned* wire being best for the purpose.
When imbedding the wires by hand, fasten one end of the wire
to a tack driven in near one end of the top bar, and pass the
other end through the holes (as in figure), fastening it in the
same manner at the other end of the frame. Care should be
taken, when tightening the wires, not to draw the frame out
of the square.

IMBEDDING THE FOUNDATION.

Cut an inch board (A in the figure) a little larger than the size of the frame ; on this screw another piece, B, three-eighths of an inch thick, cut slightly smaller than the inside of the frame, letting the grain of each board cross that of the other,

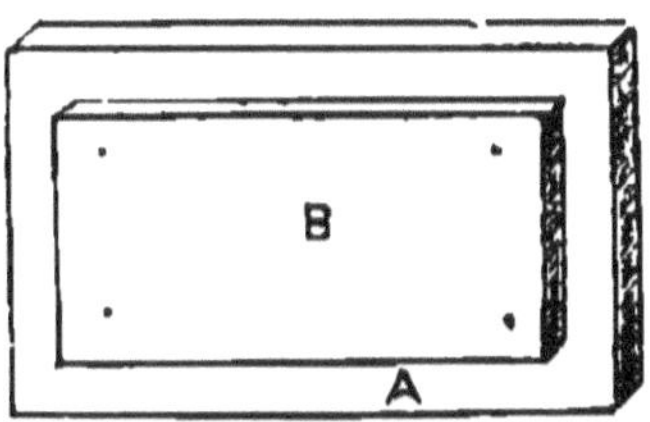

Fig. 83.—WIRING BOARD.

which will prevent twisting. Lay a sheet of foundation on the board B, and a wired frame over it, resting upon the lower board A. One edge of the sheet should be *close* against the top bar of the frame. The wires can now be imbedded in the centre of the sheet by passing over each a button-hook in which a shallow groove has been filed. The sheets should be slightly warmed. A still better implement for imbedding the wires is shown here (Fig. 84). The lower curved part, upon

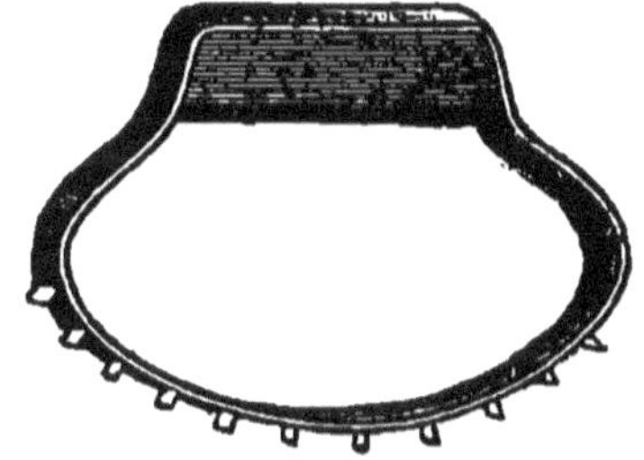

Fig. 84.—EASTERDAY'S WIRE IMBEDDER.

which the points are fastened, is just long enough to reach from the top to the bottom bar of the frame. It is grasped by the upper part in the hand, is used on the wires with a rocking motion, and imbeds them very rapidly. Mr. Root says he considers it does the work " quicker and easier than any other plan hitherto tried."

TO SECURE STRAIGHT COMBS.

It is of very great importance to have the combs built perfectly straight, and within the frames. To secure this the hives should stand level, as already explained, so that the frames may hang plumb. When wires are not used and the frames are not hanging vertically the lower parts of the sheets will project beyond them, and if much out, they are very likely to be fastened by the bees to the next frame. There is a very simple device, however—shown in the following engraving— which will prevent the sheet getting out of place and help to support it while being worked out.

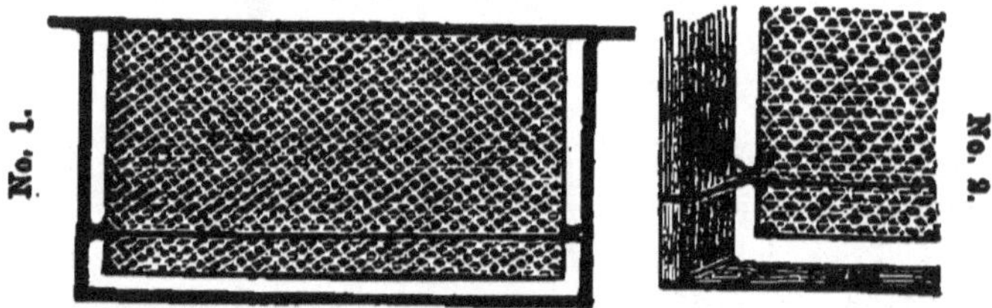

Fig. 84a.—DEVICE FOR SECURING STRAIGHT COMBS.

No. 1 represents a frame of comb-foundation. No. 2 is a section of No. 1. About 1½in. from the bottom of the sheet a thin piece of twine is passed round one end bar of the frame, then crossed (as shown) and one part taken along each side of the sheet, crossed again, and tied at the back of the other end bar. The twine need only be tied sufficiently tight to prevent it slipping down the end bars. There is another method of fixing the twine, i.e., by boring a small hole in each end bar and passing the double part of the twine through one, putting a small toggle in the bight, and fastening the two ends after passing them through the other hole. This simple arrangement will be found of great value in securing straight combs. The twine may be removed after the combs are fastened to the end bars.

CHAPTER IX.

MANIPULATION OF BEES AND FEEDING.

THE common, but erroneous, idea prevailing amongst those who have paid little or no attention to the nature and habits of the honey-bee, that to go near one is to run a risk of being attacked by it, may, I think, be attributed to the fact that they have had instilled into their minds while young an idea that the bee is an enemy they have to fear. It is not an uncommon occurrence for a mother, on seeing her infant near flowers on which there are bees flitting about, to say, " Oh! come away, my child; there's a bee, it will sting you," and she immediately takes the child away from the supposed danger. This lesson, to dread the bee, thus early inculcated, is never forgotten.

Again there are many people who believe that bees have a special aversion to them, that they cannot go within fifty yards of one, as they will sometimes tell you, without its making for them and " declaring war;" and no amount of persuasion will convince them that they may be mistaken. A person who has this idea firmly fixed in his mind is likely to act in such a way when a bee is near him as to invite its attack, and so condemn the bee for what he has himself, unconsciously perhaps, been the cause of. It is my opinion that if there are any such people that bees attack without apparent provocation they are very few indeed, and that no one, as far as my experience teaches me, is more liable to be stung than another provided they both act in a like manner. People of a nervous temperament as a rule make very poor manipulators.

HANDLING BEES.

There are certain fixed rules to be observed when handling bees if freedom from stings is to be secured. The most important are: 1st, to avoid jarring the hive; 2nd, to avoid breathing into the hive or upon the bees ; 3rd, to avoid making any quick movements about the hive ; 4th, to be careful not to stand in the line of flight to the hive ; 5th, let all manipulation, as far as possible, be conducted during fine weather and while the bees are flying ; 6th, the operator should act in a fearless but gentle manner ; and 7th, never strike at a bee, but when one gives warning of stinging bow the head slightly, if unprotected with a veil, and walk slowly away.

Bees are more irritable during cloudy or showery weather, owing, perhaps, as it is said, to the peculiar electrical conditions of the atmosphere. Queenless colonies are more easily provoked to anger than when in their normal condition. So marked is this as a rule that I can often detect them on first opening the hives. The scent of the poison from their stings will excite bees, and the crushing of one while manipulating will usually make the rest very angry. For the protection of the face the manipulator can wear a

BEE-VEIL.

This should be made of some light material, such as leno or tarlatan, long euough to hang from the brim of a hat to the

Fig. 85.—WIRE-CLOTH BEE-VEIL.

Fig. 86.—TARLATAN BEE-VEIL.

lower part of the chest, when there will be ample to tuck under the collar of the vest or coat (Fig. 86). A piece of strong elastic run through a hem round the top will keep it tight and close round the hat, which is the better for having a broad brim. Some bee-keepers prefer having a piece of wire cloth sewed into the veil (Fig. 85), on account of the wire being less obstruction to the sight and not confining the breath so much as the other material. A bee-veil should always be worn by a beginner ; it gives him a sense of security, and therefore more confidence. My first veil was made about three years ago. I never felt the need of one till I commenced to handle hybrids pretty extensively.

BEE-GLOVES.

These I have never worn, therefore I cannot say whether they are useful or not, but as it seems to me that I require the most perfect freedom with my fingers for handling the frames, I should think gloves of any kind on the hands are an encumbrance. Indiárubber gloves are usually sold for this purpose, but a pair of thin woollen gloves covered with cotton ones, and the whole dipped in a strong solution of soda and water, have been highly recommended. I should think that if gloves be worn at all the extreme joints at least of the fingers and thumbs ought to be left free.

QUIETING BEES.

During the season, when honey is being gathered rapidly, bees as a rule can be handled without showing any signs of displeasure, but at other times it may be necessary to proceed with caution, and have at hand some means of quieting them should they resent our interference. Smoke is one of the best bee quieters we have ; a few puffs will generally cause them to gorge themselves with honey, in which condition of superabundant fulness they may be handled with impunity.

SMOKERS.

One old-fashioned method of applying smoke to bees was to make a tight roll of cotton rags, which, being lighted at one end, was held in the hand while the smoke was blown by the mouth into the hive, but now we have a more pleasant and convenient method. A tube or fire-box of tin is attached to a small bellows worked with a spring (Figs. 87 and 88); a few live embers are first placed in the tube, then the material for burning, which may consist of cotton rags, dry rotten wood, dried cow-dung, or any other material that will burn and produce a dense smoke, when a few puffs of the bellows will set the smoker in working order.

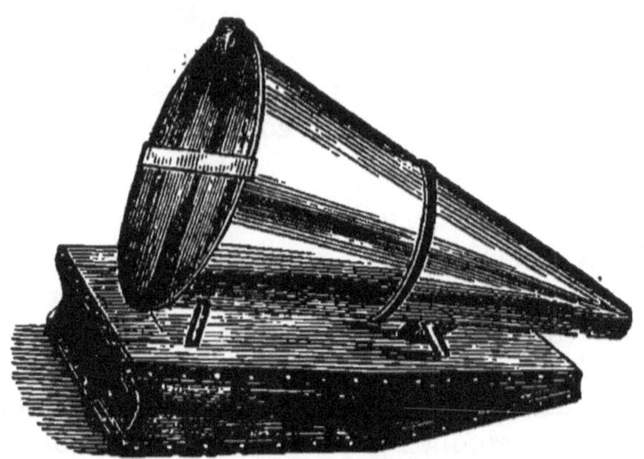

Fig. 87.—CLARK'S COLD-BLAST SMOKER.

Two of the best smokers now in use are made on different principles. In one the blast of air from the bellows is blown through the fire; in the other the draught enters the tube between the fire and the mouth of the tube. By this arrangement cold air is mixed with the smoke as it leaves the tube. The latter is termed a " cold-blast " smoker (Fig. 87) and the former a " direct draught " smoker (Fig. 88). Each has special

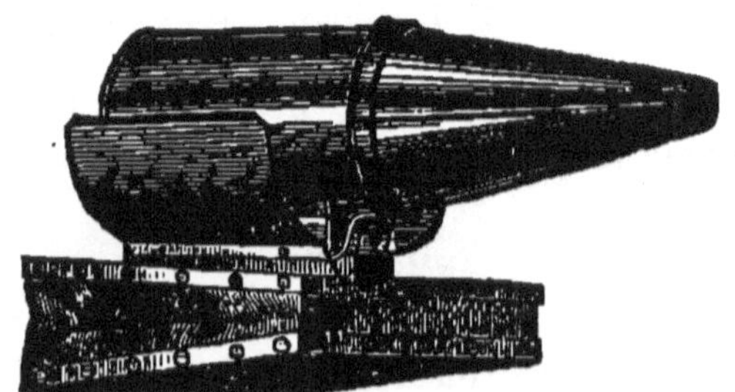

Fig. 88.—BINGHAM'S DIRECT-DRAUGHT SMOKER.

advantages not possessed by the other. The smoke blown from the " cold-blast " smoker being mixed with cold air instead of hot is an advantage I have no doubt appreciated by the bees;

but with the " direct draught," the air going through the fire must necessarily be hot when it leaves the smoker, and after using it steadily for some little time the smoke is apt to be very hot ; herein I prefer the " cold blast." On the other hand, the advantage of the Bingham direct-draught smoker over all others in use, with perhaps the exception of King's (which I have not seen), is that it can be set down without the fire going out immediately ; in fact, it will burn away for hours, or until the fuel is exhausted. Those who have been troubled with their smoker going out every now and then as soon as they laid it down for a minute will be able to appreciate this advantage. I don't know anything more annoying than while manipulating bees, especially if they happen to be hybrids, to find your smoker out that you only a minute before laid out of your hand. The peculiar feature about the Bingham (patented in America) which insures the constant burning of the fire is the non-attachment of the small tube directing the blast from the bellows to the fire-box itself. There is about a quarter of an inch space left between the end of the air tube and the fire-box, so that a current of air is always rushing through this opening into the fire.

I have both the above smokers in use at Matamata, and although I like the principle of the " cold blast " I prefer the Bingham for the reasons given. Could the principles of the two be combined in one I think it would make a perfect smoker.

FUEL FOR SMOKERS.

For the cold-blast smoker I find very dry half-rotten wood or dry cow-dung as good as anything I have used ; but for the Bingham it is better to use good dry sound wood, cut into lengths of about four inches by one-half inch square. To set the smoker going, drop in first a few live embers and cover them with small fuel, give a few puffs to start the fire, and fill up with the larger pieces. A correspondent to the *New Zealand and Australian Bee Journal* recommended soaking the fuel in a solution of saltpetre (an ounce to the pint of water) and drying it carefully before using it ; a match will then be sufficient to light it.

As it would be difficult to explain clearly how to make a smoker without a number of illustrations, and more difficult still for an ordinary bee-keeper to make one even were instructions given, it will be sufficient to mention that they can be purchased of Messrs. Bagnall Bros. and Co., or their agents, at the following prices : Clark's cold blast, 3s., by post, 4s. 6d. ; Bingham's direct draught, 4s. 6d., by post, 6s.   At these prices it would not pay any bee-keeper to make one or two for his own use.

HOW TO OPEN A HIVE.

During the honey season it is usually only necessary to be careful when removing the cover and mat, and if the bees "boil up" over the top of the frames to give them a puff or two of smoke.   But at other times, or with bees inclined to be vicious, it is better to blow a little smoke in at the entrance half a minute or so before removing the cover ; this will give the bees time to begin filling themselves with honey.   After the cover is removed a few more puffs on top will make them quiet.

When removing any of the frames it is better first to move two or three of the side frames a little to make room for taking out the first one, when any of the others can be taken out without trouble.

COMB-HOLDERS.

A small comb-holder, similar to the one shown in Fig. 89, to hook on the side of the hive, is very handy for hanging a

Fig. 89.—COMB-HOLDER.

frame on when taken from the hive.   It is made of stout folded tin.   Another kind, easily made, is shown in Fig. 90.   This is very much like the body of a hive cut in half lengthwise, with a bottom nailed on.   A double one can be made by extending the ends and putting the division in the centre of them ; a hand-hole cut in the top of the division is necessary for carrying it.

CURES FOR STINGS.

I am inclined to believe that there are as many "infallible cures" for stings as there are for rheumatism—every person seems to have one. The first thing to be done is to remove the sting—not, however, as most people would do, with the thumb and finger, but by a scraping process with the thumb or finger-nail. If the former plan were adopted the contents of the poison bag would be squeezed into the wound, but by the latter method the sting is more easily and quickly removed without expressing any poison.

I must here plead ignorance of the best or even a good remedy to stop the irritation caused by a sting. I have tried a few on myself, but I cannot say whether they had any effect or not; certain it is that they did not stop the irritation at once, as they were supposed to do. The poison being an acid,

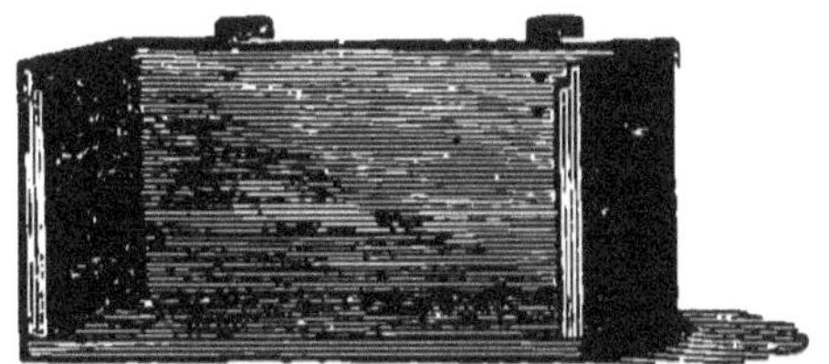

Fig. 90.—SIMPLICITY COMB-HOLDER.

we naturally expect that an alkali will neutralise its effects, therefore ammonia should be about the best remedy if it is applied immediately. A writer in a recent number of the *British Bee Journal* recommends the use of strong carbolic acid, such as is used for disinfecting drains. His method of applying it is as follows :—

"Immediately after we have been stung we extract the sting, touch the place with a stick or glass rod which has been dipped into carbolic acid, and then apply a drop of water; and the union of the two causes a paralysis of the nerves which control the lymphatics, and so prevents the poison being taken up into the system. Only the smallest possible quantity must be used, as the acid, being a caustic, when mixed with water, leaves a scar, which, however, will disappear in a few days, unless an excessive quantity is used."

Some recommend an onion, cut and applied to the wound, but whatever is done should be done quickly, before the poison has time to get into the system, and be careful not to *rub* anything in.

It is very rarely that any serious effects ensue from a sting beyond a little pain, aud perhaps a swelling of the part around the wound, but there have been cases where death has resulted from the sting of a bee, the immediate cause being either suffocation, a shock to the nervous system, or blood poisoning, where the person has been in a bad state of health. The following from a correspondent to the *British Bee Journal* gives the treatment for extreme cases :—

"In the *Journal* for December 'E. H. B.' wishes to know the treatment to be adopted when a man or woman gets into a state of coma from a bee sting, which, of course, will depend on the cause of the coma.

"The bee poison, when taken into the system, gives rise to a form of blood poisoning, but as the amount of poison even from twenty stings is so small, the symptoms, except in those cases where the person stung is in a bad state of health, never become serious.

"Death has resulted from perrons having been stung inside the mouth or the throat, by eating fruit, honey, etc., in which a bee has been overlooked, and the swelling caused by the bee poison, closing up the windpipe, has killed the patient by suffocation.

"In these cases the only remedy would be to make an artificial opening in the windpipe, an operation difficult even for an experienced surgeon.

"In those cases where death is imminent from the 'shock,' the best remedies would be stimulants, either in the form of brandy, whisky, etc., sal-volatile internally, or by injection under the skin, strong coffee or tea, together with galvanism, and hot cloths applied to the head and chest.

"The after treatment would consist in building up the patient's strength by means of stimulants, strong beef-tea and soups, milk and eggs, with quinine, bark, and mineral acids.—GEORGE WALKER, L.R.C.P., Wimbledon, 22nd December, 1882."

FEEDING.

There are times and seasons when a little attention given to the matter of feeding stocks will tend greatly to increase the profits from the apiary. A prudent apiarist will no more neglect feeding his bees when they require it than he would his horse, cow, dog, or any other of his domestic animals. Feeding is resorted to for one of two reasons, viz., either to supply a colony if it has fallen short of the necessary food, or for the purpose of stimulating breeding. The seasons when food is more likely to be required than at any other time are autumn and early spring.

FEEDING FOR WINTER.

In the autumn examination of the hives, when preparing them for winter (see Chapter. XIV.), the first point is to see to the supply of food. In most districts 25lbs. of honey will be ample for winter stores; in others again, like Matamata, where there is comparatively no forage between clover and clover, 35lbs. is little enough for an ordinary colony. I believe it to be poor policy to run the bees too close at the end of the season for the sake of getting a little more honey; for although we may get a greater price for the honey than the syrup will cost for feeding, still when we come to add the expense in time and labour of feeding, to say nothing of the many inconveniences attached to it, and the risk of starting robbing, we shall find very little profit in the extra honey that has been taken. A colony well supplied with food in the autumn will keep up breeding later than one with a scant supply, and thus come out stronger in spring. When food is required for winter stores it should be supplied before cold weather sets in, and is best given as rapidly as the bees will take it. Where there has been a short supply for winter it will most likely be necessary to feed again just as breeding commences at the end of winter; in short, food should be given at any time it is required.

STIMULATIVE FEEDING.

It has already been stated that the activity of the queen as regards egg-laying in spring depends upon the amount of honey being gathered. As the season advances, and the worker-bees are enabled to gather more than is sufficient for present requirements, so egg-laying increases until the queen is depositing to the full extent of her capacity. Where early spring forage is abundant breeding will go on rapidly and the colonies will gain strength much earlier in the season than where it is scarce, unless we make up for the want by supplying the bees with food artificially. It is a fact that rapid breeding does not so much depend upon the amount of food already in the hive as upon the amount being *stored*; so that when we wish to stimulate breeding at a time when honey in the fields is scarce we must feed.   For stimulating purposes the amount of food given does not so much matter as supplying it regularly ; half a pint of syrup or less given every twenty-four hours is ample.

WHAT AND WHEN TO FEED.

Next to sealed honey, a good syrup made from white sugar is the best food we can give. Candy is very good, and when run into a frame is handy for hanging in the centre of the cluster in winter. I have wintered bees on it, but I think on the whole a fairly thick syrup is best even in winter if placed convenient for the bees. Our winters in any part of Australasia are not so severe, however, as to prevent bees reaching food in any part of a hive.

Recipe for Syrup.—To every pound of sugar add half-pint of water, put it into a saucepan and boil for a few minutes ; keep stirring. This when cool is ready for use.

Candy.—Take, say, 10lbs. of sugar, put in a little water (about three half-pints), mix well, and boil, keeping it well stirred to prevent burning. Boil until it is ready to sugar off. You can determine when this point is reached by putting some in a saucer ; or test it, as confectioners do, by dipping your finger in a cup of cold water, then in the candy, and back into the water again. When it breaks like egg-shells from the end of your finger it is just right. Take it off the fire at once, and as soon as it begins to harden round the sides keep stirring till it gets quite thick. Very great care must be taken to prevent the food from burning ; it is said burnt sugar is poison to bees. The candy can be made into cakes by pouring it into plates previously greased, or it may be poured into a frame by fastening the frame down on a flat board on which a sheet of paper has been spread to prevent the candy sticking to the board. The frame should rest on the board closely all round, so that the candy will not run underneath it. As soon as the candy is ready, pour it into a frame—a Langstroth frame will hold about 8lbs. If made according to the directions it will be firm, dry, and opaque when cold, and will stick to the frame, so that it may be suspended in the hive like a frame of honey. It should be warmed a little in cold weather before being put in a hive. Cakes of candy can be placed on top of the frames under the mats, care being taken not to uncover the bees.

The best time to feed syrup is just before dusk. Feeding is sure to cause some excitement amongst the bees, and if it is given in the daytime numbers of them will fly out in an excited manner and so attract the attention of other bees, which may cause robbing. Candy may be placed in the hive at any time. The food for stimulative purposes should be given every evening.

HOW TO FEED SYRUP.

Food should always be given to a colony in such a manner as to be out of the reach of strange bees. A number of different kinds of feeders are made for holding syrup ; some for placing on top of the frames, others for hanging on the hive, but only one that I know for placing outside. Undoubtedly the safest plan is to place all food within the hive, where it can best be protected by the inmates. Perhaps the most simple and handiest feeder is the " Simplicity " (Fig. 91), for standing

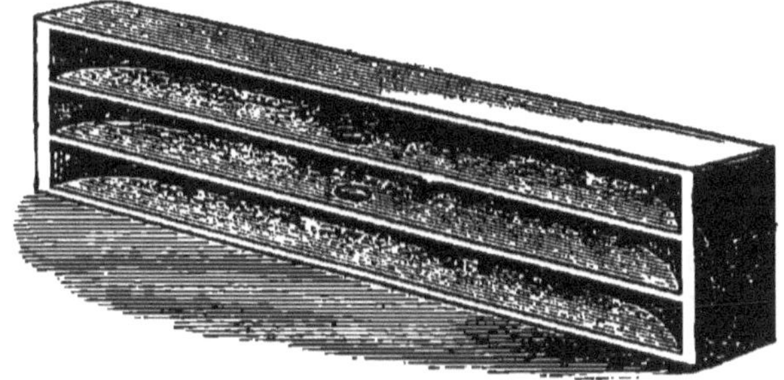

Fig. 91.—" SIMPLICITY " FEEDER.

on top of the frame under the mat. As usually made, it is cut out of a solid piece of wood 1ft. long, 3in. wide, by 2in. deep. The grooves are cut with a circular saw. I have made them by cutting the grooves with a plough and nailing ends on afterwards. The partitions afford a foothold for the bees and prevent them getting drowned in the syrup. In wide feeders a thin float should be used. In top feeding the mats must be so arranged as to leave no part of the frames uncovered.

Empty combs can be utilised as feeders, and I am very much in favour of these, as they can be hung in the centre of the cluster of bees. They can be filled by laying them on an inclined board and allowing the syrup to drop through a perforated vessel held about a foot above the combs. They should be hung up for a time after being filled, till there is no drip from them, before being placed in the hives. Another kind, for hanging in the hive, answers the purpose of a division board at the same time. A simple but at the same time a very useful feeder can be made out of an empty broad frame by nailing three bottom bars lengthwise in the centre of it, so as to form

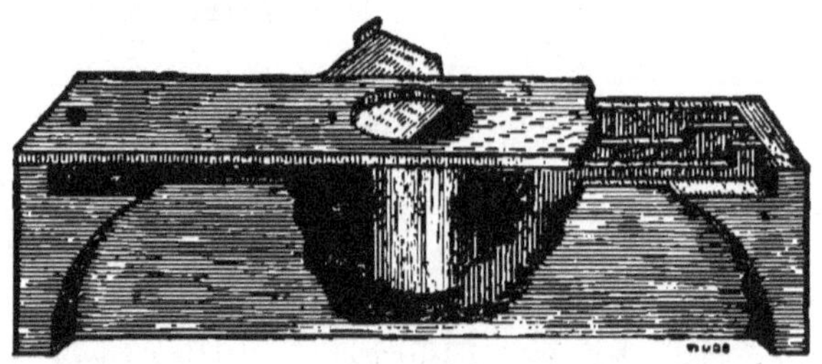

Fig. 92.—GREY'S ENTRANCE FEEDER.

a trough. Grey's feeder, for placing across the entrance to the hive, is shown in Fig. 92 ; it can also be used inside like the "Simplicity." Any or all of these feeders can be purchased for a small sum from hive manufacturers, and it is the better way for those who wish to make their own to buy one to serve as a pattern. An ingenious person will find many things that he can utilise as feeders. All wooden feeders should be given a thin coating of hot wax before being used, to keep them clean and prevent them absorbing the syrup.

Rule.

"SEE THAT YOUR BEES ARE NEVER SHORT OF FOOD."

CHAPTER X.

TRANSFERRING.

ALTHOUGH the operation of transferring bees and combs from a box to a frame hive may appear a formidable undertaking to the beginner, he will soon find that it only requires a little confidence to get over all the difficulties easily. After he has succeeded with the first case, he feels astonished that he should have felt half afraid to commence.

IMPLEMENTS REQUIRED.

Transferring may be done at any time during the summer, but the best time for it is in the spring, before the combs are heavy with honey. A few tools are necessary, and these should be ready at hand before commencing. A hammer and chisel will be required to knock the box asunder, a long knife to cut the combs, a lighted smoker, an empty box, a hive with frames, and a box or barrel turned bottom upwards to serve as a bench. Some transferring wires (Fig. 93, No. II.) will also be required for fastening the combs in the frames.

Fig. 93.—TRANSFERRING WIRES AND CLASPS.

These I usually make out of No. 16 wire in the following manner :—Lay a frame on its side, and cut the wires an inch longer than the outside depth of the frames, and then make a bend in the wires half an inch from each end, so that they can be made to grip the top and bottom bars of the frame. About thirty of these may be required for each hive. A few clasps (Fig. 93, No. I.) made of tin strips are also handy. A transferring board, of the construction shown in the following figure, is a most useful appliance.

It is easily made, and very convenient when working. The spaces between the bars admit of the transferring wires being fixed on the lower side of the frame as well as on the upper side as it lies with its enclosed comb upon the board. Any honey that drips from the combs during the operation can be caught on a dish placed underneath the board.

A fine warm day should be chosen, and I find the morning the best time for this work. Everything required being at hand, blow a few puffs of smoke into the entrance of the box,

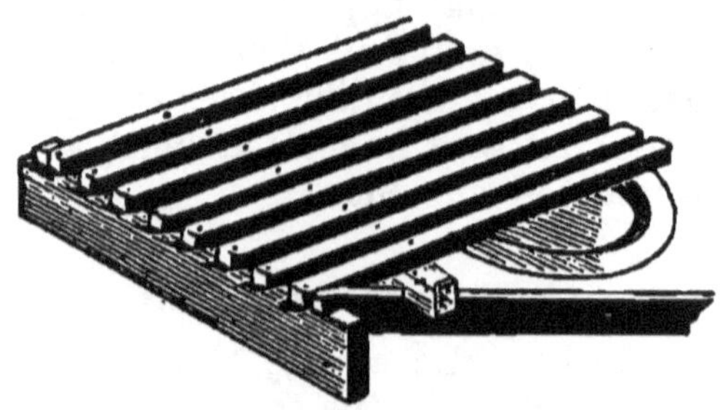

Fig. 94.—TRANSFERRING BOARD.

and after a minute or two turn it bottom upwards, just at the back of the place where it stood, and place the empty box over it. A cloth may be tied round the junction of the two boxes, to steady them and to keep the bees confined, if the operator is at all timid. An empty box may be placed where the old one stood, to catch any bees that return from the field during the operation.

DRIVING.

The bees are now to be forced to leave the old box and their combs and to cluster with their queen in the empty box which has been placed on the top, just as a natural swarm does when newly hived. This process of forced swarming is called "driving." It is done by rapping on the sides of the box hive in such a way as to frighten the bees until they fill themselves with honey and retire from the apprehended danger, as it is their instinct to do under such circumstances. This rapping on the sides of the box may be done with the hands, but better with a pair of sticks, beginning gently and gradually increasing the force of the blows, but not so violently as to endanger the breaking down of the combs inside. When the drumming has been continued without intermission for a period ranging from ten to twenty minutes the bees will be nearly all clustered in the upper box. When this has taken place the upper box may be lifted off and placed on the old stand, free ingress and egress being allowed for the bees in front.

FIXING COMBS IN FRAMES.

The combs are now ready to be transferred. If there are robber bees about—which may soon be known by seeing them settling on the combs—take the box of combs into a room or shed—if no bee tent (see Chapter XV.) be at hand—where the robbers cannot get in. With your long knife cut the combs free from the sides of the box, and take an end and a side off

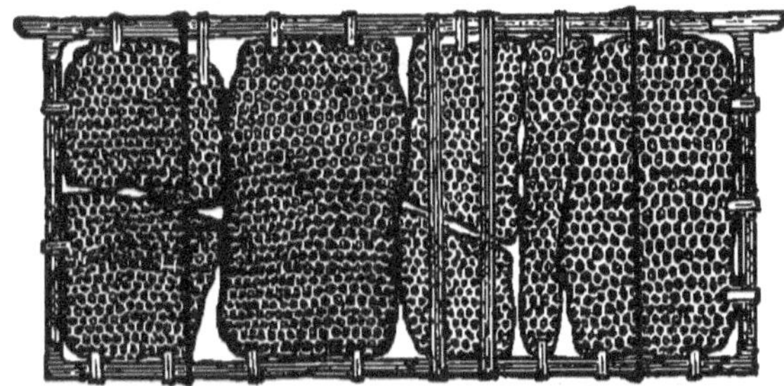

Fig. 95.—PIECES OF COMB TRANSFERRED TO FRAME.

carefully. Now cut the combs out separately without breaking them, and select the straightest containing brood in the first instance. Lay a comb on the transferring board, with a frame over it, and if it be larger than the frame, cut it just a trifle larger than the inside, so that by springing the frame a little it will go over and grip the comb so cut. If this be neatly done the comb will require no other fastening. According as each frame is finished, hang it in the hive.

When more than one piece is required to fill a frame, select only such as are straight, containing brood or honey, and secure them in the frames with the wires and clasps, as shown in the above illustration.

With the aid of the transferring board (Fig. 94) the wires can be put on both sides of the frame without moving it

Proceed thus till all the straight worker combs *containing brood or honey* are transferred; take no others, unless there happen to be any perfectly straight worker comb empty large enough to fill the frames in one piece, when these may be used also. Under no circumstances should crooked or drone comb be transferred. The hives may be filled up with frames of comb taken from other hives or with frames of foundation, the brood combs being kept in the centre. Place the hive, as soon as it is ready, where the box-hive stood; raise the front a little, and shake the driven bees out of the box, so that they can enter the hive, which they will do at once, and if all be well, and no accident have happened to the queen, they will proceed without delay to fasten the combs in the frames. In the course of a day or two the wires can be removed.

Other methods are sometimes recommended for securing the combs in the frames, but having tried most, and having transferred some hundreds of colonies with their combs, there is no plan I have found to answer so well as the wires I have described. They are easily put on, and may be taken off without lifting the frames from the hive.

I have often transferred without driving the bees into an empty box, merely turning the box-hive up, driving the bees with smoke to one end, then tearing off the other end, cutting out and transferring the combs and shaking the bees into the new hive, which had in the first instance been placed where the old box-hive had stood. The first described process takes a longer time, but it makes cleaner work, and is the proper plan for a beginner to adopt.

MR. HEDDON'S NEW PRACTICE.

Mr. Heddon recommends doing away altogether with the transferring of the combs into frames. He drives the bees into an empty box, as already explained, and then shakes them at once in front of the new hive, already filled with frames of comb or of comb-foundation, and placed where the box-hive had stood. The old box-hive with its contents is then set aside for twenty days, until all the young bees have emerged from the cells. If the weather turns cold the box should be placed in a warm room until a sufficient number of bees have emerged to keep up the temperature of the box. The young bees are now drummed out as before. If a queen should be present she is killed, and the bees are either united with the others or taken to form a separate colony. If there be any honey left in any of the combs, it is extracted, and all the old combs are melted down.

OPEN DRIVING.

There is a method known as open driving practised by experienced bee-keepers and experts when competing at bee-shows, at which prizes are given to those who shall in the neatest and quickest manner drive the bees and capture and exhibit the queen. In open driving the cloth round the junction of the boxes is dispensed with, and one end of the upper or receiving box is elevated in such a way as to permit the passage of the bees to be seen by the operator and spectators. A sharp look out is kept, and the queen captured while ascending.

RULE.

"ONE STRONG COLONY WILL BE MORE PROFITABLE THAN SEVERAL WEAK ONES."

CHAPTER XI.

INCREASE OF STOCKS—NATURAL SWARMING—DIVIDING.

THE question of the increase of stocks in an apiary, how to promote, control, or even to prevent it, is one of importance in all cases, and one which must be treated by each bee-keeper according to the special objects he has in view—the first point to be determined being

WHAT RATE OF INCREASE IS DESIRABLE.

It may be desirable to stock a new apiary as quickly as possible, or, if opportunity offers, to dispose profitably of a number of colonies ; in which case the largest increase compatible with the formation of none but strong colonies must be worked for, regardless of a honey harvest. If the object be to form an apiary gradually, and with the least outlay of capital, the best plan will be to try for such a moderate annual increase of stocks as may be consistent with securing a fair return in honey at the same time ; or, lastly, if the apiary has attained the full extent intended, the bee-keeper will naturally want to obtain the greatest possible quantity of honey from a fixed number of hives ; and it then becomes a question whether that end may be best secured by suppressing *all* increase, as far as it is possible to do so.

CIRCUMSTANCES WHICH AFFECT A DECISION.

Having settled the preliminary question of large, moderate, or no increase, the bee-keeper will next have to consider the circumstances with which he has to deal and which may exert an influence upon the results he wishes to attain. These may be said to consist chiefly in : 1, the climate, as it affects the habits of the bee and the commencement and duration of the swarming season ; 2, the sources from which the chief honey harvest is to be derived, when they come into blossom, and consequently, when the chief take of honey should commence ; and 3, the race of bees he keeps, and their natural tendency to increase by swarming, which tendency is greater in some races than in others.

The bearing of these circumstances upon the matter will be apparent when it is borne in mind that in some climates, where the bees are inactive during nearly half the year, and where the honey season is probably very short, the ordinary swarming period may occur just at the commencement of that honey season. If the bees should swarm just at that time, unless recourse be had to the most improved system of management, the swarm or new stock will be taken up in building comb, and the parent stock in rearing a young queen and recruiting its strength, at the very time when all the workers should be employed gathering in the honey harvest, and thus the best part of the season may be lost. On the other hand, in countries where the climate is mild and the queens begin to breed early in the spring, the swarming season may occur a month or more before the beginning of the regular honey harvest ; and in such a case, if a colony casts off an early and strong swarm, then even, if left to nature, the new stock may have ample time to build the necessary comb, and the parent stock to be recruited, so that both may be in a position to take full advantage of the honey season, and probably to collect twice as much honey as the original colony could have done if it had not swarmed. It is obvious that the same rules would not apply to both these cases, and there are intermediate phases which would require to be treated according to the peculiarities of each.

MODES OF ATTAINING THE OBJECT.

Whatever rate of increase may be desired, and whatever the circumstances of the case may be, there are two modes by either of which the bee-keeper may seek to attain his ends— the one by " natural swarming " (that is, by availing himself of the natural tendency of the bees to cast off swarms, and using the art of modern apiculture to control that tendency as far as possible in the direction he aims at), the other by "dividing," or, as it is commonly but not very appropriately termed, "artificial swarming." Each of these methods has certain advantages; the modern system of movable frame hives, the use of comb-foundation, and the aid of queen-rearing give wonderful facilities for dividing colonies in a way which was quite impossible with the old straw skeps or box hives. With them it was only practicable to force a colony to swarm by "driving" the bees, and that process was properly enough called artificial swarming; but the modern process of dividing is quite a different thing, as we shall see presently. The same improvements, however, which render this artificial mode of increase so practicable, also give great facilities for controlling to a great extent the operations of natural swarm-ing. There has been considerable difference of opinion even among experienced bee-keepers as to which mode of increase should be practised. Many leading men still think that the natural swarming system possesses advantages which more than counterbalance some very obvious inconveniences which attend it; while on the other hand there are many among the most advanced apiarists who are in favour of the dividing method, which undoubtedly is capable of being carried out with extraordinary success by experts. Some writers go so far as to say that no intelligent bee-keeper with a full know-ledge of his business should allow a natural swarm to take place in his apiary. I do not like to lay down such dog-matical rules. I am firmly of opinion, after practising both methods of increase pretty extensively, that except under special circumstances or in the hands of expert apiarists, natural swarming, controlled according to modern practice, is productive of as good results as any known method of artificial increase; and I am sure it will be wiser for the beginner to follow its practice at first, and until he shall feel himself capable of carrying out the non-swarming or dividing method with perfect ease and certainty. I shall therefore say what has to be said with respect to both systems, leaving it to the judgment of the reader to determine whether the one or the other should be adopted, either as a general rule or under special circumstances.

NATURAL SWARMING.

The natural instinct for swarming, with which bees are endowed, is an admirable provision for the propagation of their

race and its spread over any district favourable for their existence. When we recollect that it is only by means of the queens
and drones that the race can be propagated ; that these queens
and drones cannot exist by themselves, or without the workers
of a colony ; that the queens require to be renewed every two
or three years, in order to keep up strong stocks, while the
workers are only for a season, and the drones for a few months;
and finally, that only one queen can be tolerated at a time in
any colony, we cannot fail to be struck with admiration at the
beautiful manner in which the swarming instinct is adapted to
this state of things.

CAUSES OF SWARMING.

The queen and a comparatively small stock of workers pass
through the winter months under various conditions, according
to the climate, as will be noticed under the head of "Wintering."
In spring, as soon as the temperature and the forage are such
as to set the workers in a state of full activity, the queen commences her great work of egg-laying, and as we have seen, she
is capable of laying from ten to twenty thousand per week ; at
all events, after five or six weeks she will, if all be right, have
laid so many eggs, that if all were hatched, and developed into
workers, there would be far more than the number requisite
for, or that could be reasonably employed in, an ordinary hive.
One half of the number will have come to maturity at the end
of the six weeks, and the other half will be out in three weeks
more. The hive will be already rather overcrowded, and some
twenty or thirty thousand bees can be well spared to form a
good swarm, there being fully as many in the egg and larva state
ready to supply their place within three weeks. In this state
of affairs instinct leads the workers to build several queen
cells, and the queen, as soon as the queen cells are closed, to
lead off the swarm, all filled with honey and ready to commence
comb-building at once in their new home, wherever it may be.
The bees left behind in the old hive have *several* young queens
maturing, lest one or more of them should fail ; they have also
the drones, usually flying at this time, or maturing in the
drone-comb, and an ample stock of workers also maturing by
degrees—therefore all the elements of their future strength.
If the old queen has left with her swarm just when the first

queen cells were closed, then the first young queen will emerge in eight or nine days, and in the meantime the stock will have been recruited by a large number of young bees. If they still feel themselves over-strong, or are still actuated by a desire for swarming, the first young queens may go off with one or more after-swarms or "casts;" if not, the first out will remain in possession of the hive, and all the others will be destroyed in their cells. In five or six days more the young queen will probably be fertilised, and shortly after will begin to lay eggs. This is the natural course of the swarming, which provides for a multiplication of the self-sustaining stocks or colonies, and at the same time for a succession of young queens.

THE SWARMING SEASON.

The time of year when first swarms may be expected depends upon the climate, season, and strength of the colonies. In New Zealand, in the northern part of the province of Auckland, they sometimes issue as early as September; at the Thames it is usually the middle or latter part of October before the earliest come off; while further South it is later still. Swarms should be expected all through the months of November and December, and even up to the middle or end of January.

As to the Australian colonies, I have taken care to obtain the best information I could upon this as well as other apicultural matters, from bee-keeping correspondents, to whom I have already expressed my acknowledgments, and the general result is as follows. In Queensland, with its nearly tropical climate, the swarming season continues from mid-August to mid-April; the great difficulty there is to check the swarming tendency and keep the stocks strong enough to collect a fair return of surplus honey. Mr. Fullwood mentions cases of increase in one year elevenfold. In New South Wales the ordinary swarming season is given as October to December; but in some parts it begins in September and lasts till February. In South Australia, as Mr. Bonney informs me, the regular swarming season is September and October, but in good seasons the bees swarm again in December and January. He has known stocks to increase by natural swarming in eighteen months, from one to thirteen. In Victoria the season varies, but in some of its northern parts swarming has been known to commence as early as August. In Tasmania the season is similar to that in the southern portion of the Province of Auckland—from mid-October to mid-January.

SYMPTOMS OF SWARMING.

It is generally known amongst experienced bee-keepers, that first swarms give but little outward indications by which the apiarist may know when they are likely to issue. True, the bees may hang clustered outside the hive, which is thought by some people to be a sure sign; but although this may be caused in some instances by an over-crowded hive, it more often occurs through bad ventilation and want of shade, and can by no means be taken as an indication of the immediate issue of a swarm. I have known cases where the bees have been hanging outside the hive in large numbers for weeks prior to swarming, keeping the household constantly on the alert lest they should lose them.

This waste of time on the part of the bees during perhaps the best part of the season should not be allowed, and could not occur with movable comb-hives and proper management. Mr. Langstroth says, "There are no signs from which the apiarian can predict the certain issue of a *first* swarm. For years I spent much time in the vain attempt to discover some *infallible* indications of first swarming, until facts convinced me that there can be no such indications." One of the surest signs in the interior of the hive is the presence of newly built queen cells, and this can be easily ascertained at all times by the use of movable comb-hives.

ISSUE OF THE SWARM.

The actual process of swarming is so admirably described by Langstroth, whose closeness of observation and clearness of description are equally inimitable, that I cannot withhold from the reader a passage from the chapter upon natural swarming in his treatise upon "The Hive and Honey Bee." He says, at p. 112 :—

"I have repeatedly witnessed, in my observing hives, the whole process of swarming. On the day fixed for their departure the queen is very restless, and instead of depositing her eggs in the cells, roams over the combs and communicates her agitation to the whole colony.

The emigrating bees usually fill themselves with honey just before their departure; but in one instance I saw them lay in their supplies more than two hours before they left. A short time before the swarm rises, a few bees may generally be seen sporting in the air, with their heads turned always to the hive; and they occasionally fly in and out, as though impatient for the important event to take place. At length a violent agitation commences in the hive; the bees appear almost frantic, whirling around in circles, continually enlarging like those made by a stone thrown into still water, until at last the whole hive is in a state of the greatest ferment, and the bees, rushing impetuously to the entrance, pour forth in one steady stream. Not a bee lurks behind, but each pushes straight ahead, as though flying for dear life, or urged on by some invisible power in its headlong career.

"Often the queen does not come out until many have left; and she is frequently so heavy, from the number of eggs in her ovaries, that she falls to the ground incapable of rising with her colony in the air. The bees soon miss her, and a very interesting scene may now be witnessed. Diligent search is at once made for their lost mother; the swarm scattering in all directions, so that the leaves of the adjoining trees and bushes are often covered almost as thickly with anxious explorers as with drops of rain after a copious shower. If she cannot be found, they commonly return to the old hive in from ten to fifteen minutes, though they occasionally attempt to enter a strange one, or to unite with another swarm."

In a case of the sort last mentioned a careful search should be made in the neighbourhood of the hive and between it and the place where the swarm settled, when the queen may be found on the ground, and probably surrounded by a number of her bees, and may thus be saved and returned to the hive. The attempt to swarm will then most probably be repeated the next day, if the weather should prove favourable.

OBJECTIONS RAISED AGAINST NATURAL SWARMING.

The three great objections usually raised against natural swarming are, first, the uncertainty of the time when a stock may swarm, and the consequent necessity of watching for it; second, the risk of being unable to capture the swarm if it should settle in some place not easily accessible, and the certainty of losing it if it should abscond; and third, the probability of two or more swarms issuing about the same time in a large apiary, and uniting together in one cluster. I think none of these objections, practically speaking, very formidable. As to the first, a careful and observant apiarist, from his periodical outward inspection of his hives, will be able to form a pretty correct idea of the strength of the colonies, and the probabilities of their swarming, without constantly examining the interior. It is well known that first swarms issue, as a rule, only on fine sunny days, and not earlier than nine or ten o'clock in the morning, nor later than two or three o'clock in the afternoon, unless, indeed, they have been kept back by unfavourable weather for several days. The necessity for watching is therefore, as a rule, confined to that portion of the fine days in the swarming season. If there is reason to expect any of the stocks to swarm, and if the bee-keeper himself is not engaged, during the time referred to, at or tolerably near to the apiary, it certainly will be well to have somebody (an intelligent child will do as well as auy) to keep an eye on the hives during those hours of the forenoon and afternoon. An occasional visit to the apiary every hour or so, to see if a swarm has settled on any of the bushes generally used for the purpose, may be sufficient as far as securing the swarm is concerned; but it is much more satisfactory to *see* from what hive it proceeds, and to watch, and if necessary force its settling in the way hereafter mentioned, in case of its showing any tendency to abscond. Any danger from the second cause may be reduced to a minimum, if care be taken to grow a few shrubs in convenient positions just in front of the apiary; the bees will almost invariably settle and cluster, for awhile at least, upon a tree or shrub in the neighbourhood of the apiary, if they find a few to choose from which are inviting and suitable for the purpose. With regard to the third objection—the voluntary uniting of swarms—it may sometimes cause a little trouble in a large apiary; but it is not *always* objectionable—on the contrary, it is sometimes very convenient in the case of small after-swarms. These latter, however, are, as a rule, to be avoided altogether under proper management. And with regard to the larger swarms which it is wished to keep separate, this can generally be managed by the prompt intervention of the bee-keeper, if on the spot, and by having several swarm-boxes always at hand; or if two large swarms should unite, they may be divided and hived in two lots, as shown further on, but with the almost certain loss of one queen at least.

PREPARING FOR SWARMS.

At the approach of the swarming period, everything requisite to facilitate the hiving of swarms should be in readiness, so that the bee-keeper can lay his hand on the necessary appliances at a moment's notice. All the new hives likely to be required for the season's increase should be placed in position, according to the directions given in Chapter V., great attention being paid to the proper bedding and levelling of the bottom boards. In the next place it is necessary to have some kind of

SWARM-BOX

or other contrivance in which to take a swarm previous to hiving it. Some form of box is generally used, though a round-bottomed bag with a light hoop of stout wire or cane sewn round the mouth of it to keep it open is a very handy device, especially for fastening to the end of a light pole to take a swarm settled above arm's reach. The bag, if large enough, may be put right round the cluster, and with the aid of the hoop the bees can be scraped, as it were, off the branch, and so be taken with ease. Mr. T. J. Mulvany, jun., uses for this purpose a light box about 20 inches long by 10 or 11 wide and deep, to which is attached an open sack of stout calico about two feet long. One end of this sack-like appendage is tacked round the open part of the box, the other end being open and bell-mouthed, just large enough to fit over a Langstroth hive. In each end of the box there is a ventilating hole about 3 inches in diameter, over which is tacked perforated zinc. The mode of using it in taking a swarm will be described further on. As a receiving box when driving bees, or for carrying a swarm in, I do not know of a more handy contrivance. I have had one in use—that Mr. Mulvany kindly presented me with—through two seasons, and I have found it of great service. The main points to look to in whatever form of implement may be chosen are lightness, strength, durability, and handiness. With regard to the latter feature, each individual bee-keeper will have his own ideas; and I need only here remark that I find a plain box made out of $\frac{5}{8}$-inch material 18 inches square by 10 inches deep to suit me very well.

TAKING AND HIVING SWARMS.

To many people the taking of a swarm and hiving it appears rather a dangerous operation, whereas there is nothing connected with the work that need excite the least alarm or prevent any person undertaking it. It has already been stated that the bees fill themselves with honey just previous to swarming, and as " bees gorged with honey never volunteer an attack," they may be handled at this time with little risk of their stinging; and so long as they are not unnecessarily hurt they may be tumbled and shaken about to an astonishing degree without attempting to defend themselves. Usually swarms cluster in such a manner that they may be shaken into a box or other receptacle with the greatest ease, though it may occasionally happen that a swarm will settle in a very awkward place for taking it; however, there are few difficulties in this way that may not, by the exercise of a little ingenuity on the part of the operator, be successfully overcome.

First swarms, headed as they are by laying queens, are not long before they commence to settle. If it is a still, hot day the bees will choose the shady side of a tree or shrub, and, if a windy day, the lee side. Should they, however, cluster in a spot where the hot rays of the sun are beating down upon them, or the wind is blowing the cluster about, it may be taken for granted that the swarm will not remain long in that position, and if compelled to rise again there is no knowing where the bees may make for ; at any rate there would be but little chance left of the owner securing them. A swarm *should always be taken as soon as possible after it has settled*, and the sooner it is located in the hive to commence work the better. My own plan of taking and hiving swarms is as follows : When I find the swarm settling, I take a bottom board and a hive already fitted with frames of comb or foundation—the latter is temporarily removed from its permanent position—and place them close under or near the cluster, the hive being set as level as circumstances will admit and the frames spaced correctly. The front of the hive is propped up a little on the bottom board, and some kind of cloth—a sack ripped open will do—is spread in front, one edge resting on the bottom board. By this time the bulk of the bees will probably have settled. I now take a light, convenient-sized box similar to the one described—several of which are always at hand—and shake or brush the bees into it, getting as much of the cluster in as I can, and turning the box partly on its side in my hands I empty a *few* bees out of it on to the

bottom board at the entrance to the hive. I continue shaking
out a few at a time till the bees commence running into the
hive, when the box may be emptied in front by degrees. It
is as well not to shake all the bees in front at once or they
may rise again and give extra trouble. A few bees will be
sure to rise and settle again in the same place, but unless there
is a large cluster no notice need be taken of them, for when
they find the queen not there they will soon go in search of
her, and hearing the joyous hum of their companions entering
the hive will soon join them. If by any chance the queen
should not have entered the hive, the fact may soon be known by
the bees running about in all directions in search of her, and
led by them the bee-keeper will soon discover her whereabouts,
when he must take steps to get her in, and the bees will soon
follow. After all the bees are in the hive, it may be shifted
at once to its permanent stand from whence it was removed,
or it may be left till evening. I prefer shifting it at once.

I have found the above plan of hiving swarms the best of
any I have yet tried. Some recommend leaving the hive in
its permanent position, and carrying the swarm to be shaken
out in front of it; but with me this plan has not been at all
successful. A large number of bees are sure to rise again and
fly back to the place where they were brought from, when
they are lost to the new colony, whereas, if they are hived at
the spot, those that rise will soon find their way back to their
queen.

Mr. Mulvany describes his mode of hiving swarms as
follows :—

"The mouth of the sack, or the box itself with the sacking drawn
back, according to circumstances, is held around or just under the
cluster, and the bees shaken or brushed into the box, which is then
laid on the ground, mouth downwards, but partly raised, until all
stragglers have found out and joined the queen in the box. When all
the bees are secured, the sacking is drawn forwards, gathered in one
hand, and tied with a string as near the mouth of the box as possible.
The swarm can then be carried anywhere, accurately weighed (which
I consider one of the conveniences of the plan), and either hived at
once or hung up in a shady place till required. When brought to the
hive, which is already in its place and provided with frames of foundation, comb, or honey and brood as desired, the cover is lifted off, the
end of the sacking drawn over the opened hive, the string untied, and
the bees shaken down *on top of the frames*. The entrance of the hive
may be stopped loosely with cotton wadding, and the bees left to

themselves till they settle down on the frames, which they usually do in a very short time. The swarm-box is then taken off, the mat and cover placed on the hive, the entrance made free, and the new colony goes to work at once. Swarms taken at the apiary are hived immediately when taken—any brought from a distance, and which may have been a day or more in the box, are hived in the evening or after dark. In this way we never have any trouble from the bees deserting the new hive."

Should a swarm settle in the midst of a bush or on the ground, a cloth should be spread as near as possible, and the box placed on it in the most convenient position for brushing some of the bees into it. As soon as the main body of the bees begin to make for the box they may be left to take their own time, care being taken *always* to see that hives and boxes containing new swarms are shaded from the heat of the sun. When a box has to be left, as in the above case, till the swarm clusters in it, it is better to let it remain till a little before dusk before hiving the bees. I have found in the majority of cases that when swarms have been disturbed by trying to hive them before evening, after they have been left long enough in the boxes they were taken in to settle themselves, they have risen again, and sometimes have gone straight away. In the case of a swarm having settled and no one at hand with sufficient courage to hive it, a sheet thrown around it will usually prevent the bees absconding till assistance can be obtained.

When two swarms issue at the same time, and cluster together (a circumstance which, as already mentioned, is liable to take place in a large apiary), and we are desirous of keeping down increase, they may be hived together, if sufficient room is provided by putting on the supers. But in the event of there being more than two, and the bees are too numerous to be put in one hive, the cluster should be divided into parts, the size of a good swarm each, and put into two or more separate hives, into each of which a frame of brood and eggs has been placed. Should none of the queens be seen, it may be ascertained in a short time which of the hives, if any, are queenless, by queen cells being started over the eggs or larvæ provided them, when, if any are at hand, we may, to save time, give them either a queen or a nearly mature queen-cell.

ABSCONDING SWARMS.

There are few things connected with bee-keeping more vexing than to have swarms leave for parts unknown. Absconding swarms may be divided into three classes, viz., those that go off without clustering, those that leave after being hived, and established colonies that leave their hives in despair. With regard to the first, in my experience the cases have been very rare indeed in which swarms have left without first settling somewhere in or near the apiary. I cannot call to mind half a dozen altogether. But I have known many to leave after being hived, more especially with those hived on the plan of carrying the swarm some distance to the hive immediately after it is taken; this is the principal reason for my adopting and advocating the plan of carrying the hive to the swarm. I have had very few leave when hived in this way. Sometimes a swarm will *not* stay, no matter how carefully it has been hived, or what inducement there may be for it to remain. There is no accounting for this. They do not always go straight away, but sometimes give you an opportunity to hive them again and again, only to repeat the same process of leaving their hives. If the queen of such a swarm could be captured and have a wing clipped, it would most likely prevent any further trouble. As to the last class, where established colonies leave their hives *en masse*, this is caused in almost every case through starvation, and is only carried into effect as a last resource, after the colony has dwindled down and apparently lost all hope of surviving. Of course, no one working on the lines of advanced bee culture will permit such a thing as this to happen.

Nucleus colonies (see Chapter XII.) are sometimes known to leave their hives in a body and follow their queen on her " wedding flight," though I do not remember ever having had one that did so. I think this is only likely to occur where very small nucleus hives and colonies are used for queen-rearing, and I would therefore recommend having none but fair-sized ones, such as described in Chapter XII., and keeping the bees well supplied with brood until the young queens commence to lay.

It is a good practice to give every newly hived swarm a frame of eggs and larvæ either from the parent stock or from another hive; it will not be so likely to leave when this is done, and the bees are provided with the means of raising a queen, should any accident in hiving them have deprived them of their own.

CLIPPING THE QUEEN'S WING.

For the purpose of obviating some of the inconveniences connected with natural swarming, many leading bee-keepers have adopted the plan of clipping a wing of the queen *after she has been impregnated*, to prevent her flying with the swarm. Many reasons can be given in favour of this method, while there is only one that I know of which some consider against it.

When a swarm issues, the clipped queen, although she cannot fly, will make an attempt, and consequently fall to the ground close to the hive. The bee-keeper, being at hand, must pick her up and cage her. As soon as the bees are all out, remove the hive to another stand some distance away, and place an empty one, fitted with frames of comb or foundation, and a frame of eggs and brood, in its stead. Open the entrance as wide as possible, and lay the caged queen down close to it. In the meantime the swarm may have settled; but when the bees discover that the queen is not with them they will not be long before they return and naturally enter the new hive. While they are going in, release the queen, and see that she goes in with them. When two or more swarms issue at the same time and form one cluster, they will separate in a short time, and return to their several stands; so that if the bee-keeper has secured the queens and changed the hives before their return, he will have done his hiving with little trouble. Care must be taken to clip a wing of *every queen*; for if there should happen to be one queen in the cluster, the bees will not return.

The objection to this method sometimes put forward is the necessity of having some one constantly on the alert to secure the queen and change the hives. In a large apiary it would be necessary to see the swarm coming out of the hive, or else it would be difficult to find the queen. It is quite certain that if the bees return with or without their queen, they will make another attempt next day, and if they have to return to the old hive again, the probability is that if they find her they will injure or kill the queen, and swarm out with the first young one that emerges; so that some person should always be at hand during the swarming season. On the other hand, it can be said that where the bees are allowed to swarm at all, some one must be about pretty often, if not all the time, as already noticed in this chapter. The facilities afforded for hiving swarms, and the certainty of not losing any, will no doubt be considered by most bee-keepers sufficient to more than repay for the little extra care required in watching the apiary.

PROCESS OF CLIPPING.

Take hold of the queen by the wings with the thumb and forefinger of the left hand—be careful not to hold her by the body—place her on a board (still keeping the wings between the thumb and finger), so that she can stand upon her feet—this will keep her legs out of the way—now pass the point of a small pair of scissors under one of the front wings and clip.

AFTER-SWARMS.

It has already been explained in another part of this chapter that in the ordinary course of things a second swarm may be expected about eight or nine days after the first, and in some cases a third and even a fourth may issue within a few more days, unless the necessary steps have been taken to prevent it. These are what are termed after-swarms, and are invariably headed by virgin queens which have been reared from one lot of queen cells. After-swarms are much smaller than the first, and unless a large increase is required they should always be suppressed. It often occurs when a number of young queens emerge about the same time that all save one will accompany the last swarm; hence it is that we often find four or five dead queens about the front of a hive shortly after placing one of these swarms in it. To an experienced bee-keeper it is not difficult to judge correctly, when a swarm is in the air, whether or not it is headed by a virgin queen. An impregnated queen leading off a swarm usually acts in a business-like manner, and quickly makes choice of a place for settling, and as quickly alights; but a virgin queen often leads her bees and their master dancing attendance upon her for a considerable time while she is undecided where to alight, and then probably she will suddenly lead the swarm off a long distance before she makes up her mind to stop. When a swarm remains in the air longer than usual, and has no appearance of settling, some fine dirt, sand, or a spray of water should be thrown among the bees to disorganise them ; this will generally have the desired effect. It is said that a blank charge fired at or near the head of an absconding swarm will cause it to settle.

PREVENTION OF SWARMING.

By carefully following out certain rules based upon our knowledge of the nature and habits of bees we can control or regulate swarming to a very great extent, but there is no system at present known which will give us such complete control as would enable us to *prevent swarming altogether*, when we desire to do so. We may indeed do as some have advocated, that is, take away the queens for a time, and so stop breeding, but I would much prefer to put up with the inconvenience of a few extra swarms rather than adopt such a doubtful method.

The best aids we can have towards the prevention of swarming are roomy and well-ventilated hives, and if need be shade. At the same time very much will depend on the knowledge and tact of the bee-keeper, that is, his knowing what to do and how to do it, and his doing the proper thing at the right time.

There are times when everything depends upon the ability of the bee-keeper to adapt himself to circumstances, when his judgment for the time being is all he can trust to, and if this fail him, it may result in the loss of the greater part of the season's crop. Such a time is when the honey season has been retarded by bad weather, and suddenly gives promise of a better state of things towards the last few weeks of it, when just sufficient honey has been gathered to keep up breeding. The colonies are strong, and it only requires two or three days of favourable weather to start them swarming. The bee-keeper is aware that the season, however favourable it may turn out, can only last for a short time ; he knows his colonies are liable to have a bad attack of the " swarming fever," and also that his chance of securing a crop of honey depends entirely upon his being able to keep the majority of his bees steady at work instead of swarming.

If done at the right time, much may be accomplished in the desired direction by the use of the extractor and giving more room by putting on an extra super, or even two, if they should be needed ; and as a great deal will depend upon keeping the brood-nest at its normal temperature, the extra supers should be put on immediately above it. Abundant ventilation can generally be secured in the hives herein described by pushing them forward till the front overlaps the alighting boards an inch or two ; should more be required, however, the covers can be raised an inch or so, for while plenty of honey is coming in there need be no fear of robbery. Swarming may be kept down to a great extent in the early part of the season by enlarging the hives in good time before the main honey harvest sets in ; but, as I have before remarked, all operations connected with the prevention of swarming require good judgment and foresight on the part of the bee-keeper, that he may hit on the right time to perform them.

It may here be incidentally remarked that some bee-keepers make use of drone and queen excluders at swarming time— among them Mr. Alley, who uses the excluder shown in Chapter XII. When a swarm issues, the queen is trapped behind the excluder, where she can be caught, and the process of hiving is carried out the same as with a clipped queen.

PREVENTION OF AFTER-SWARMING.

This, as compared with the previous question, may be considered an easy matter. We have seen that in the ordinary course of events a second or after-swarm may be expected in eight or nine days after the first issues, and that there are several embryo queens maturing in the hive. But should unfavourable weather set in about the time for the first leaving, it would be kept back, and may be prevented from issuing till near the time for the young queens coming to maturity. I have even known cases where, owing to the prevalence of bad weather, after all preparations had been made for swarming, the young queens have been destroyed, and swarming given up for the time. At all events, we can reckon, as a rule, that the first young queen will not emerge from her cell in less than eight days from the time the first swarm issues. Now if we see that all but one of these embryo queens are removed, that

is, all the queen cells but one, and only allow this one to come
to maturity in the hive, there cannot be any after-swarm, as
this queen will be required in the hive. It would not, however,
be correct to remove the cells *immediately* after the first swarm
leaves, as will be presently shown. The old queen would be
laying up to within a very short time of her leaving the hive ;
consequently there would be eggs in the cells at that time.
Supposing the queen cells to be cut out during the first day or
two after, the bees would be almost sure to build others, and
thus frustrate our plans ; but if we let them remain for about
five days before we remove them, the larvæ would have grown
large by that time, and there would be little likelihood of other
cells being started. Choice of a good cell should be made for
the one that is to remain in the hive, and the others can be
utilized in forming nuclei. (See Queen-rearing.)

SUPPLYING THE OLD STOCK WITH A FERTILE QUEEN.

There is another method by which we may both prevent
after-swarms and the delay caused by waiting for the young
queen to emerge and commence laying, which takes up a considerable time. If we allow eight days for the queen to emerge,
we may add another five before she shall be impregnated, and
six more before she commences to lay, making nineteen days
in all that the hive will have been practically queenless. By
rearing and keeping on hand a supply of young queens for the
purpose, we may prevent the loss of the greater part of this
time by cutting out *all* the queen cells and supplying the stock
with a fertile queen at once, making use of the cells as previously
advised. By this plan we can gain at least sixteen or seventeen
days, which with a young and vigorous queen would mean probably over 30,000 eggs, that is provided she has facilities for
depositing them. Another advantage, there is no risk of the
hive becoming queenless by the loss of the young queen during
her " wedding flight," as often occurs. On no account should
the bees be overcrowded for want of room.

PREVENTING INCREASE OF COLONIES.

This is a question of the best method for dealing with swarms
so as to prevent as far as possible an increase in the number
of colonies. It is often the case that the bee-keeper has as

many stocks as he can well look after, and wishes to keep down increase; he cannot prevent swarming altogether, and would like to make use of the swarms, but without increasing the number of his colonies. In this, as in most operations connected with bee-keeping, various methods are advocated by different bee-keepers to attain the same end. I have been very successful with the following plan : Hive the swarm in the manner already described, and after cutting out all queen cells from the parent stock, place the hive containing the swarm as a super on the hive it came from. In the course of a day or two examine the hive, and if there are any eggs and larvæ in the super, shift the combs containing them below, taking care to provide plenty of room. Should there be a very large quantity of brood in the hive, some of it might be given to other and weaker colonies. When this method is carefully carried out, I find it very rarely fail to have the desired effect.

DIVIDING.

Colonies may be increased by dividing at any time during the swarming season, but it should only be attempted with strong ones, and then only during a fair honey flow, such as would be likely to cause them to swarm naturally. If it be intended to increase solely by dividing, then it should be carried out just at the commencement of the main honey harvest, when the colonies are very populous, and just before they make preparations for swarming. If it be desired to double the number of stocks, and there are already young fertile queens in nucleus hives on hand, then the nucleus plan may be adopted as follows : On a fine warm day cage one of the young queens from a nucleus hive, and transfer the frames with the adhering bees to an ordinary hive; hunt up the queen of the hive about to be divided, and place her, with the frame she is on, in an empty hive for the time being, that it may be known where she is. Next move the hive into which the nucleus colony has been placed alongside the one to be divided and lift four of the central frames well supplied with brood and the adhering bees, and hang them alternately with those of the nucleus colony; also shake the bees from two or three frames into the new hive, and after filling the remaining space with empty combs or comb-foundation, place it in the position formerly occupied by the nucleus hive, and hang the caged queen between two of the central frames. The old queen with her frame can now be placed in her hive again, and the hive filled with frames of comb or foundation in the place of the frames removed. To prevent too many bees returning to the divided colony, I have often blocked up the entrance with wire cloth—taking care to allow plenty of ventilation—till the evening of the following day, and then liberated both bees and queen, but usually the latter is at liberty before that time with the cage I use. In the course of a few days the surplus boxes can be put on. If less increase is desired, the nucleus colony can be made up by taking frames of brood and bees from two or more old colonies in the same way, instead of from one. It should of course be seen to that each hive is supplied with sufficient food.

When there are no spare queens or queen cells at hand, a different plan must be adopted. In this case one frame of brood, with the adhering bees *and queen*, should be taken from the old hive and placed in the new one ; then move the old hive to a new stand, and put the new one in its place. A frame of honey, if it can be spared, without bees, may also be removed from the old hive to the new one, and another taken from some other hive can be put in, and both be filled up with empty combs or foundation. Most of the old bees from the removed hive will return to the old stand, but the young bees left and those gradually maturing will be sufficient to do the necessary work of the hive and raise a queen for themselves.

Dividing should only be done during fine weather, and great care must be taken to see that the hives are kept snug, and that the brood does not get chilled. Spare queen cells may be utilized in dividing.

AXIOM,

" A MODERATE INCREASE OF COLONIES IN ANY ONE SEASON WILL, IN THE LONG RUN, PROVE THE EASIEST, SAFEST, AND CHEAPEST MODE OF MANAGING BEES."

Langstroth.

CHAPTER XII.

QUEEN REARING.

NECESSITY OF THE PRACTICE.

IN order to obtain good results from the apiary we must, as a matter of course, have good bees and plenty of them; and in order to have good bees we must first of all have good queens to breed from. That there is often a vast dissimilarity in the characteristic qualities of different colonies in the same apiary, no one who has had even a short experience will deny. How often do we find the bees of some colonies constantly irritable and disposed to sting, while those of others may be handled with impunity; or some giving a good return of honey, while others are doing little or nothing. Such cases may be seen in every apiary where a careful breeding of queens has not been systematically carried out. That it is possible, on the other hand, by means of such a system, to develop to a greater extent the good qualities, and to breed out the bad ones in the honey-bee, is no longer a matter of doubt. Such advanced apiarists as Alley, Heddon, Doolittle, and others in America, who have gone about the work in a conscientious as well as a scientific manner, have undoubtedly succeeded in developing a superior strain of bees. There is nothing to cause surprise in all this, when we consider the analogous case of the results obtained by select breeding of horses, cattle, and all our domestic animals. The breeder of bees has one advantage as compared with breeders of horses and cattle—he has not to wait so long for the results of his experiments; the bee-keeper can do as much with his bees in the way of crossing and improving races in from four to five years as the cattle breeder could probably accomplish with his stock in twenty or thirty years. Considering the many advantages to be gained by cultivating the best qualities in our bees, I am induced to look upon the rearing of select queens as one of the most important branches of modern apiculture, and I would therefore advise every bee-keeper to make it one of his special studies.

But independent of all considerations about improving the breed, the modern system of apiculture cannot be carried out in its entirety if this important branch of it be neglected, therefore it becomes imperatively necessary for bee-keepers of the present day to rear and keep a stock of queens on hand sufficient for their needs. It has been shown in the preceding chapter that in the successful practice of either the natural or artificial methods of increase a supply of queens is required, and some spare ones should also be kept ready to make up for losses that may, and in a large apiary certainly *will*, occur during the season when surplus honey is being taken. The sudden loss of a queen at this time would cause a delay of about 24 days before the hive would be furnished with a laying queen again, that is, if the bees have to rear one for themselves from a newly-hatched larva, and it is easy to understand what effect this would have upon the colony, and how necessary it is that we be prepared for such contingencies.

A WORD CONCERNING DRONES.

When endeavouring to improve our bees by cross-breeding we must of course be as particular about raising select drones for mating purposes as about the queens themselves. As the mating takes place in the air (see Chapter III.) and is not, at least as yet, under our control, our only security is to have our select young queens mated when *only* select drones are flying. The periods of the year when we are most likely to succeed in this way are the early spring and the late autumn; in the former by managing to breed our select queens and drones in advance of all others, in the latter by making the colony which produces the best drones queenless before the drones are killed off, and thus secure that these shall be flying when there are none alive in the other colonies. At other times throughout the season there will of course be drones from all the hives upon the wing.

ENTRANCE GUARDS.

It is, however, claimed by some breeders, that with the aid of entrance guards, or " drone excluders," the drones which are not

P

intended for mating purposes may be restrained from flying.
These guards (Fig. 96) are made of perforated zinc, and fitted

Fig. 96.—JONES'S ENTRANCE GUARD.

so as to cover the entrance to the hive. The perforations,
being $\frac{5}{32}$ of an inch wide, are large enough for the worker bees
to pass through, but too small for the drones. Another kind
of entrance guard is shown below. It is an invention of Mr.
Alley's, and answers the double purpose of a trap as well as an

Fig. 97.—ALLEY'S DRONE EXCLUDER, DRONE AND QUEEN TRAP.

excluder. This guard has an upper compartment, into which
lead two wire-cloth cones, seen through the side openings in
the figure. The drones, after failing to make their way out
through the perforated zinc, finally force their way up through
the cones ; but not being able to return the same way they
become prisoners in the upper compartment. Should workers
go up through the cones, they can still make their way out
through the perforated zinc on top. From various conflicting
reports which have come under my notice, I am inclined to
doubt the efficacy of all these guards, as they are now made,
for the objects intended.

HOW TO SECURE CHOICE QUEEN CELLS.

During the past seven years I have paid great attention to the rearing of queens, both for home use and for sale. I have tried several methods for raising queen cells, but none have given me so much satisfaction as the one I first saw described in *Gleanings in Bee Culture* for August, 1880, by Jos. M. Brooks, and which I have since practised. It is very similar to Mr. Alley's method, explained in his " Handy Book," a copy of which should be in every bee-keeper's library.

To secure good queen cells early in the season, we should select, as soon as breeding has commenced in early spring, two or more, as may be required, of our best colonies, and work them on in advance of the rest by slow feeding, or, if need be, by giving them frames of emerging brood from other colonies, taking care to keep them covered up well. As soon as the one chosen for raising drones is sufficiently strong, insert a clean empty drone-comb—to be obtained in the manner explained in Chapter VIII.—in the centre of the brood-chamber. Note the time when the drone-brood is capped, and in eight or nine days after, place a frame of clean new worker-comb in the

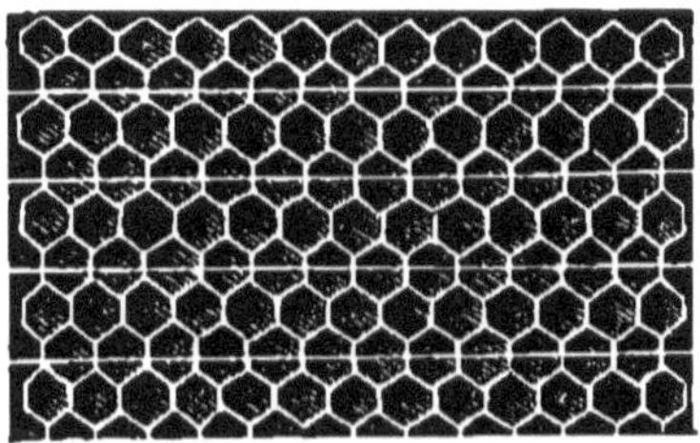

Fig. 98.—COMB CONTAINING EGGS.

centre of the brood-chamber of the hive containing your choicest queen. I would here point out that the cleaner the comb is the better; I find combs built the previous season, that have only contained honey, give the best results. The colony now being pretty strong, with plenty of brood in the combs, the new one inserted will soon be in charge of the queen, and in three or four days will be full of eggs. As soon as the eggs commence to hatch, which will be in three days after they were laid, remove the comb to a warm room,

and if more eggs are required, insert another in its place. Lay
the frame of comb flat on a table or other convenient place,
and with a sharp, thin-bladed knife, dipped in thin starch or
diluted honey to prevent its sticking, cut the comb into strips,
by running the knife along every second row of cells, as shown
by the white lines in Fig. 98, taking care to leave one row of
cells containing eggs intact in each strip. Some empty frames
will next be required, having two thin laths of wood nailed
inside longitudinally, so as to divide the depth into three compartments, as shown below. Next take the strips of comb,
and after destroying the egg in every alternate cell on one side
of the strip—which may easily be done by pressing it with the
head of a wax match—fasten the strips under the top and two
centre bars of the frame with a little melted wax, allowing the
cells in which the alternate eggs have been destroyed to point

Fig. 99.—FRAME FOR RAISING QUEEN CELLS ON.

directly downwards. The object of destroying each alternate
egg is to prevent the cells being built too close together. A
space intervening gives facilities for cutting them out subsequently without injury. Care must be taken, when fastening
the strips, that the wax is not too hot, or else it may melt the
comb and kill the eggs. Having filled as many frames as may
be required (I generally find one comb sufficient to afford strips
for three frames), the next step to be taken is to remove the
queen, every egg, and all uncapped brood from some one or
more strong colonies, and place a frame of strips in the centre
of the brood-chamber in each case. Mr. Alley recommends
preparing the colony by removing the queen, etc., some twelve
hours or so before giving them the selected eggs. Mark the
date and age of the eggs on the frame, and also upon the
cover of the hive. A memorandum book is very useful in
connection with this work for keeping records in. The queen
and brood removed can be utilized in forming a nucleus colony
by caging the queen, removing a strong colony from its stand,

placing the hive containing the brood and caged queen in its place, and shaking the bees from a couple of frames down near the entrance, to secure some young bees with the old ones that will return from the removed hive ; the queen can be released in twenty-four hours.

We have now, by removing the queen, forced the colony to turn its attention to raising others, and by depriving it of its own eggs and larvæ, have compelled it to raise queens from those supplied to it. We have also, by taking away all its uncapped brood, lessened its labours, and thereby obliged it in a manner to give more heed to the matter in hand.

It is often stated that better queens, as a rule, are developed under the swarming impulse than can be raised by the forcing process. The reason given is that the larvæ from which queens are to be reared, when the bees are preparing to swarm, receive the attention of the nurse bees, with this object in view, from the time of hatching, and are abundantly supplied with the " royal jelly "—so much so, indeed, as to apparently have more than they can consume, some usually being found in the bottoms of the cells after the queens have emerged. This surplus jelly being found in a cell is considered a good sign that a strong, healthy queen has developed from it. I have no doubt that this is all correct ; but if these conditions can be brought about by the forcing process, there appears to be no good reason for supposing that the queens raised in that way will not be just as good ; and by the method I am describing this can be effected, as I have proved time after time. The main considerations are to develop the queens in strong colonies, and to let the nurse bees have as little to do as possible, that their whole attention may be devoted to rearing the queens from the selected eggs or larvæ we have supplied them with. The larva of a worker bee several days old can be transformed into a queen, but all breeders agree that such queens are of little use.

In less than twenty-four hours after the eggs have been given to the colony, several queen cells will be started over them. Some colonies will build more than others, but I think we may reckon the average at about fifteen with Italian bees, though I have had as many as thirty-five in a frame. There will be more built when honey is plentiful ; and if little or none is being gathered, the bees should be fed while cell-building is going on. Twelve cells are considered enough for one colony to care for, and this is near the number that is usually found in a hive from which a strong colony has just cast a swarm.   As soon as the cells are forward enough to be plainly seen, destroy all except about ten or twelve of the largest and best-looking ones.

Having now as far as possible fulfilled on our part every condition necessary to ensure the rearing of good queens, we must be content to leave the rest to the bees for a few days. The cells, when fully formed and capped, will have something of the appearance of Fig. 100, though the engraving is rather a flattering one.

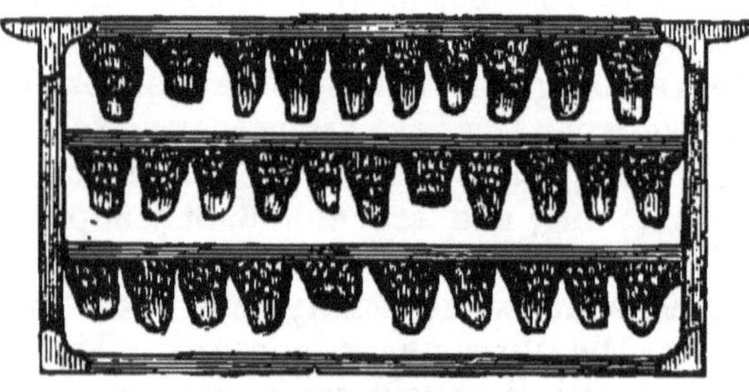

Fig. 100.—FRAME OF QUEEN CELLS.

It will be remembered that the date and age of the eggs— three days—was marked on the frame, so that we can calculate the day when the queens will be at maturity ; that will be on the thirteenth after inserting the eggs.   Being able to know within a few hours when the queens will emerge is one of the great advantages of this system of queen-rearing.   By the old methods, and even when cells are built under the swarming impulse, it is impossible to say correctly how old the embryo queens may be.

As soon as the cells are capped, a frame or two of emerging brood may be given to the colony to strengthen it.   It will have been noticed by those who have had any experience in queen-breeding, that there is often a marked difference between queen cells ; some are long, pointed, and dense-looking, while others are stunted and thin-walled.   The latter are always reckoned to contain poor queens, and it will be well to shun them, and make use of none but well-formed, rough-looking, long, pointed ones.   On the morning of the twelfth day after the eggs were given the nuclei can be formed.

FORMING NUCLEI.

A nucleus colony in connection with queen-rearing is a small colony formed for the special purpose of caring for a young queen during her maidenhood, or until she may be required to do duty in another colony. A nucleus hive, described on page 127, is a small hive suitable for the colony, and is rarely used except for queen-rearing purposes. Some queen-breeders use a very small hive, with much smaller frames than their common ones for keeping their queens in till mated, but for several reasons I consider it best to have but the one frame in both the queen-rearing and the ordinary hives. In the first place, a nucleus colony can be formed in a few minutes from any hive by simply transferring two or three frames and the adhering bees from it to a nucleus hive. Then again, a nucleus colony can be built up at any time, or united with another, where the frames are all alike, with very little trouble. And lastly, we have only the one sized frame to make. I have always used a nucleus hive such as I have described, and would not care to use any other.

The required number of nucleus hives being ready—their entrances covered with wire cloth to confine the bees—take the frame of cells and cut out carefully all but one; then return the frame to the hive until the queen shall have emerged, when it may be removed and a frame of comb or of foundation inserted in its place. Care must be taken that the queen cells are not injured or chilled; a small box, with some soft material to lay the cells upon, is handy to keep them in until they are inserted in the combs. Now go to a strong colony and hunt up the queen. This is sometimes a difficult task with a strong colony of black bees. If you have an empty hive alongside to place the frames in after you have examined them, much trouble may be saved. Having found the queen, place her with the frame she is on in a hive by herself for the time being, and insert a queen cell in each of the other combs as you take them from the hive, remembering that you require some brood, a fair number of bees, and a fair share of honey in each nucleus. I usually put either one pretty full frame of brood, or two that are not so well filled, with the adhering bees, and a frame with honey, which may be taken from another hive, or else a frame of foundation, in each nucleus.   The frames of brood and bees should be taken as equally as possible to form the different nucleus colonies. A strong stock will generally furnish enough for five nuclei.

Something is required to support the combs while the cells are being inserted.   To enable one to work easily and quickly the comb should be about on a level with the shoulders while stooping or kneeling beside the hive.   A stand like that shown below is quickly made and is very serviceable.

Fig. 101.—COMB STAND.

The drawer near the ground will be found handy for keeping queen cells and small tools in.

HOW TO INSERT QUEEN CELLS.

When cutting the cells from the frame, as much as possible of the base should be taken, clear up to the wood.   With the frame which is to receive a cell placed on the support or stand in a convenient position, cut a small hole in the comb just large enough to put in the cell without pinching it in any way. In cutting the part for the base let it fit as nicely as practicable, as shown by the white line in the next illustration.

As soon as the cell is inserted, place the comb with the adhering bees in the centre of a nucleus hive ; put in the other combs as already explained, and put on the mat and cover. Be sure that you have blocked up the entrance with wire cloth so that no bees can escape. Then proceed with the other nuclei in the same manner. When all are finished, take the nucleus hives to a cool shady place, or if they can be put in a dark, well ventilated room or shed, it will be better still. Keep them closed till the evening of the second day after that on which the cells were inserted, when they may be placed where they are to remain, and the bees liberated a little before dusk. By confining the bees in this way for a day or two they

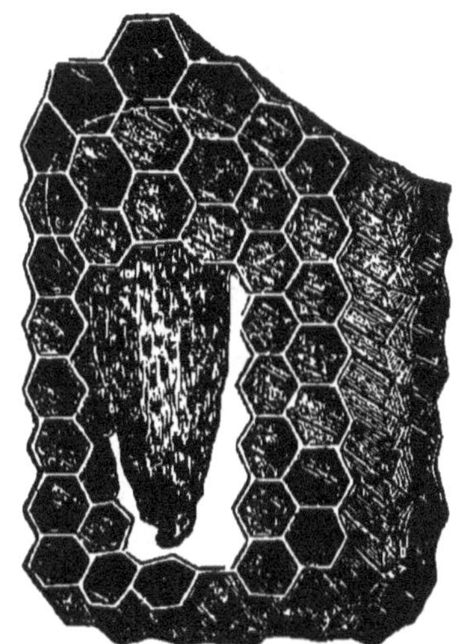

Fig. 102.—INSERTED QUEEN CELL (from which the Queen has emerged).

become reconciled to their new quarters, and very few fly back to their old hive. Before I adopted this plan I sometimes had a deal of trouble on account of so many bees deserting the nuclei.

The above method of forming nuclei and inserting queen cells is no doubt the best to adopt when queen rearing is only carried on on a limited scale, and where the loss of a queen cell would be felt, but where, as in my own case, a saving of time is of greater consequence than the loss of a cell now and again, a knowledge of my plan may be of service.

MY METHOD OF FORMING NUCLEI AND INSERTING QUEEN CELLS.

With the nucleus hives, a few spare combs, and provided with some long pins, I go to a hive, and without troubling to look for the queen—except merely to glance over the combs as I take them out—I insert the cells as quickly as possible. Instead of taking time to fit them nicely, I give a hasty look at the cell, cut a hole in the comb I think will suit, put in the cell and fasten it there by running two pins through the base of it into the comb, one each way—sometimes one is sufficient. Advantage may be taken of a depression iu the comb and so save cutting a hole. In this way I can insert the cells and form the nuclei in a very short time. If the queen should be seen during the operation, she is placed with the frame she is on to one side till all is finished, when she is put back into the hive after contracting it with division boards, if necessary. Should she not be seen it only means the loss of one queen cell, which is more than made up for by the time saved in not waiting to find her. I have often spent a considerable time looking for the queen in a strong colony and then perhaps had to give it up. Professor Cook recommends inserting the queen cells twenty-four hours after the nuclei are formed, but says : "We may do it sooner but always at the risk of having the cell destroyed." I very rarely find one destroyed, and I think the risk is likely to be greater when time is allowed for the bees to commence building cells before giving them one. Occasionally it happens that a nucleus colony will not accept a queen cell even when it *has* been queenless for some little time. When this occurs a cell should be protected in a cage when placed in the hive until the queen emerges, when there is likely to be no further trouble.

MATING YOUNG QUEENS.

To return to our nuclei. We left them just after liberating the bees. At that time the queens would be one day old ; in four or five more they will take their wedding flight just when our select drones are about fourteen days old and flying. If our plans have been carefully matured there would be no other drones flying from our apiary at this time, so that there would be every likelihood of our queens mating as we desired.

Sometimes quite a number of young queens will be lost during their wedding trip, at other times very few. I have never been able to satisfactorily account for this difference. Whether it be that there are more bee-enemies about at one time than another I cannot say, but of this I am certain, that there are a less number lost when the nucleus hives are far apart and located some little distance away from the main part of the apiary. Mr. Alley says that the daughters of some queens are more liable to be lost than others but cannot account for it. In another place he says : " I bred from a queen last season, not one in fifty of whose daughters were lost in mating." Possibly some have a sense of locality better developed than others, and are therefore less likely to miss their proper home on their return from their first flight. At any rate it is a matter worth giving attention to.

When the young queens commence to lay, which they will do in a few days after mating, they are ready to be made use of unless we desire to test them, and when raising them for sale they should always be tested for purity and laying qualities for at least a month. By following up with cell building others may be ready to place in the nuclei when the young laying queens are removed, though there may not be the same chance to have the second lot of queens mated by selected drones unless it can be accomplished by the use of drone excluders, the utility of which, as I have before remarked, I am rather doubtful about. Even then no other bees should be near the apiary.

NECESSARY DISTANCE APART OF DIFFERENT RACES TO ENSURE PURE MATING.

This is a question upon which a considerable difference of opinion exists. Mr. Alley thinks that half a mile is far enough, while many other experienced apiarists consider that a mile, or even two, is rather close. It is one of those still debatable questions connected with apiculture which may be argued on both sides for any length of time without being settled satisfactorily one way or the other by actual proof. I am inclined to think, however, that if we want to make sure of our queens

being mated with drones of our own apiary, there should be
no other bees located nearer than one or two miles off.

QUEEN NURSERIES.

When queen rearing is carried on extensively, that is, as a
special business, appliances termed queen nurseries are often
brought into requisition to aid the queen breeder. By the use
of these, as will be presently seen, much time is saved, and a
proportionately greater number of queens can be reared by
each colony so employed, as the queen cells may be taken
away from them after they have advanced a certain stage and
be brought to maturity in the nurseries, when the bees may
again be furnished with material for cell building, or have
fertile queens given to them. There are two descriptions of
nurseries in use. One is a kind of double cased hive without
a top—usually made of tin. The space between the inner and
outer cases, which extends all round the sides and bottom, is
about an inch wide. This space is filled with water, and a
small lamp is kept burning underneath ; regulated so as to
maintain a temperature within the nursery of about 90°—the
ordinary temperature of a hive when the colony is rearing
queens. They are usually made large enough to hold about
six frames conveniently, which can be suspended within them
in the same manner as in a hive. When queen cells are to be
matured in this nursery, they are removed from the colony in
which they were built, soon after being capped, and placed in
it, and the nursery covered. Several lots of cells can be
maturing in it at the same time, but great care must be taken
to keep the nursery at an even temperature or the embryo
queens will be destroyed.

The other kind of nursery—so called—is a set of small
cages fitted into a frame ; the cells are placed in the cages and
the frame is suspended in a hive containing a strong colony,
which supplies the necessary heat for maturing them. Two
or even three frames of cages would not be too many for a
very strong colony.

The nursery shown in Fig 103 is the invention of Mr.
Alley, and as I have had some experience with it I can answer
for its usefulness. One great advantage in making use of a
colony instead of the lamp nursery for maturing cells is that

there is no risk, nor trouble in attending to the regulation of
the temperature. In Alley's nursery the young queens are
protected when they emerge, and the cages are so made that a
supply of food can be placed in them to serve the queens

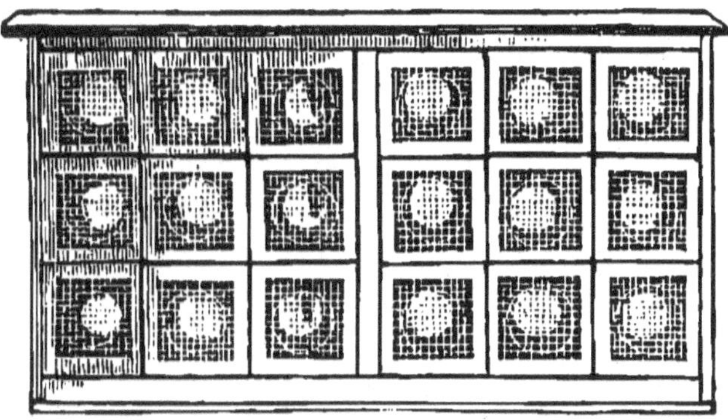

Fig. 103.—ALLEY'S QUEEN NURSERY.

should it be necessary to keep them in the cages for a time
until they can conveniently be disposed of. Of course the cells
can be removed before the queens emerge and be given to
nuclei in the manner before explained.

To make the cages (Fig. 104), eighteen of which fit in a
narrow Langstroth frame, I use a smooth batten $\frac{7}{8}$in. thick by

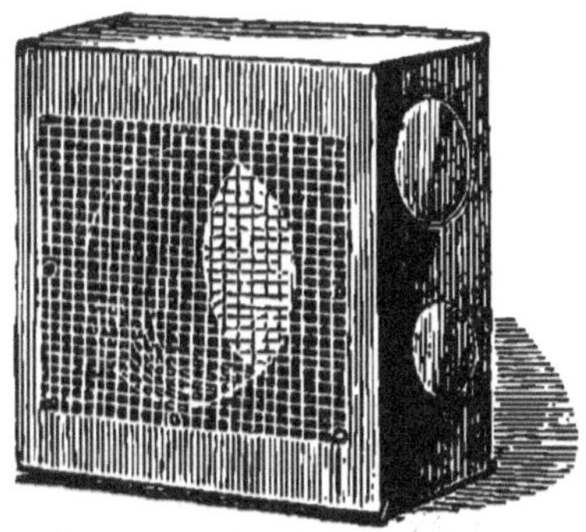

Fig. 104.—ALLEY'S QUEEN NURSERY CAGE.

$2\frac{7}{16}$in. wide and mark it off into lengths of $2\frac{11}{16}$in. for each cage.
Before cutting them off, the holes are bored—I find an expansive
bit best for this purpose as it bores smoothly. The large central

hole in each cage (see Fig. 104) should be $1\frac{1}{2}$in. diameter, and after boring these, turn the batten on its edge and bore one $\frac{11}{16}$in. and one $\frac{5}{8}$in. hole to communicate with each of the larger ones. Cut the battens into the lengths marked for each cage and tack fine wire cloth over both ends of the central holes. Next nail two end bars a' full $\frac{3}{8}$in. thick in a Langstroth frame, dividing it lengthwise into three equal compartments, and the nursery is ready. If the cages have been accurately made six cages will fit nicely in each compartment.

The queen cell is put into the larger hole on the edge, and food in the smaller one. The cell may be fastened by pressing the wax at the base of it against the edge of the hole or by running a pin through the wood and base of cell. The food should consist of honey, which can be given on a piece of sponge. Place the cages in the frame with the holes in the edges toward the top bar, and hang it in the centre of a strong colony—I usually put them in the upper part of a two story hive. The queens as they emerge are protected, have sufficient food to last them for a short time, and are safe till they can be made use of. If they are to be kept in for a day or two more food must be supplied.

INTRODUCING QUEENS.

Next to rearing queens we must know how to introduce them safely into strange colonies. The ordinary conditions to ensure safety are,—that the colony must first be made queenless, that is, the old queen must be removed. In the next place the new one, when first placed in the hive, should be protected in such a way that while the bees can see her and even feel her with their antennæ they are prevented from stinging her, as some would be apt to do before they become used to her. And lastly, the colony should be fed if there is no honey being gathered while the queen is being introduced. There are exceptions to the second clause. In the busy season, when honey is coming in rapidly, if the queens can be changed without much disturbance of the hive, the new one is likely to be accepted just as readily if she is turned loose on the frames, as she would be were she protected for a day or two. I have often introduced them in this manner with success. On the other hand, I have had great difficulty with some colonies trying to get them to accept a queen even when introduced in the usual way. Last season, when introducing some queens that had just arrived from Italy, I succeeded with four out of five without any trouble, each taking up their respective stations in less than twenty-four hours, but the fifth one I found balled, and was just in time to save her. I caged her again but in such a way that the bees could not liberate her, and on looking over the combs I found queen cells commenced, which I destroyed. In twenty-four hours I turned the queen loose, when the bees again balled her ; this occurred twice more, when I gave it up with that colony and tried another, to which she was successfully introduced by the following day.

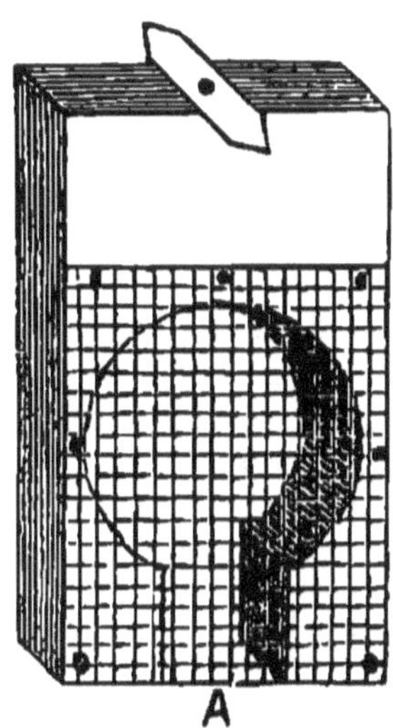

Fig. 105.—ALLEY'S INTRODUCING CAGE.

From what has been stated in a former chapter it will be understood that if we wish to change our bees from one variety to another we have only to change our queens. In this way we may change from blacks to Italians, or vice versa.

I have used several kinds of introducing cages but the most simple and handiest, according to my judgment, is Alley's (Fig. 105). I have used no other for the past two seasons. It is made by boring a 1¼in. hole half-an-inch from one end in a small block of wood 3in. long, 2in. wide, by half-an-inch thick ;

cut out the notch (A) to communicate with the hole, sufficiently large to admit the queen after the wire cloth is tacked on. Cover both sides of the block with fine wire cloth (as in figure), and tack on a strip of tin 2in. long by half-an-inch wide, to the centre of the solid end, in such a manner that it may be turned at right angles to the cage.

Place the queen to be introduced, with a worker or two, in the cage, and plug up the notch pretty firmly with some comb in which there is a little honey to answer as food. Remove the queen you are about to supersede, and queen cells if any, and hang the cage by the tin from the top bars of two of the centre frames. The bees will liberate the queen by gnawing away the plug of wax. Examine the hive in twenty-four hours. If honey is scarce feed liberally while introducing.

DIRECT INTRODUCTION OF QUEENS.

Mr. Simmins, a prominent English apiarist, has made known a system by which he claims to be able to introduce queens successfully without going through the usual process of caging them. A number of leading bee-keepers have reported favourably of the method, while some failures have also been mentioned. Briefly the system is as follows :—The queen to be introduced, with the comb she is on and the adhering bees, are taken straight from the nucleus or other hive that she may be in and placed at once in the centre of the queenless colony. To quote his own words :—

" In manipulating use smoke as under ordinary conditions, not on any account to excess. Never handle the queen, or cause her to become restless by any carelessness on your part. The comb to be inserted with queen and bees should not be taken from one part of the apiary to another openly in the hand ; nevertheless, let it be carried in a nucleus hive, or comb-box having no lid so the bees may be exposed to the light and air. The colony to receive the queen should first have its combs parted to give ample room to insert the queen-comb without crushing, or the bees 'brushing' each other ; let the whole surface of the frames be exposed to the light while obtaining the nucleus, then insert the same and close the hive at once. When no honey has been coming in feed over night the colonies to be operated upon. . . . The original queen is not to be removed until the introduction takes place."

I have not tried the above plan, but I believe it would work very well during the busy season.

INTRODUCING VIRGIN QUEENS.

This is a matter in which I cannot claim to have had special success. To introduce successfully a mated queen is an easy matter compared to introducing an unmated one. Sometimes I have had fair success with a considerable number ; at other times, and under similar conditions as far as I could judge, I have lost fully 50 per cent. However, several experienced apiarists have lately given the matter much thought, and experimented in different ways, and some now claim to be able to introduce unmated queens without difficulty.

The principal feature in Alley's method appears to be making the colony queenless three days before attempting the introduction. He says :—

"In order to introduce such queens (virgins) successfully the colony should remain queenless three days (seventy-two hours) ; then give the bees a pretty good fumigating with tobacco smoke. Remember the bees must remain queenless *three* days at the least, and during the meantime no queen must be near them, otherwise the operation will prove a failure. Virgin queens can also be introduced successfully by daubing them with honey and using no tobacco smoke. . . . This is a much slower process than by fumigating them with tobacco smoke, but just as successful."

In another place he gives the ordinary plan of introducing queens, only laying stress upon the point of keeping the bees queenless for three days.

Mr. Doolittle, who claims to have introduced several hundred without having lost one, says : "To secure the best results the queens should be about three days old when placed in the introducing cages, but a difference of two days either way will make no great difference." After removing the queen which he wishes to supersede, he places a caged virgin queen, well supplied with food, on the top of the frames, and leaves her there for five days, when he opens the cage and closes the hive, leaving the queen to come out of the cage at her leisure. Twenty-four hours afterwards he examines the combs and removes any queen cells that have not been already destroyed by the bees. Both Alley's and Doolittle's plans appear to be specially meant for introducing to nucleus colonies.

Mr. D. A. Jones, of Ontario, stated at the Toronto Convention that he had introduced fifty queens into fifty hives in

fifty minutes with the aid of chloroform. His method is to
fit three pieces of sponge pretty tightly into the nozzle of a
smoker ; on the middle piece he pours a teaspoonful of chloroform ; he then puffs the fumes for a quarter of a minute into
each hive which has been previously prepared by being made
queenless. In about two minutes he returns to the first one
and gives it a few more puffs, and so on through the whole, at
the same time letting the queens run in ; and if it be in the
middle of the day he puffs them a third time after about two
minutes more, in order to catch the bees that are coming in.
He says : " I have during the past season taken the worst
cases of fertile workers and the most difficult queenless colonies
that I ever had to deal with, and I never missed yet."   I
believe this statement refers more particularly to mated queens,
but I suppose the plan would work equally well with virgins.

" SHIPPING " QUEENS.

I make use of this term "shipping" because it is now so
generally adopted that scarcely any other word would be understood to convey the meaning intended, although it is by no
means necessarily connected with the idea of conveyance by
ship or by water. It is the term which Americans apply to all
manner of " sending off " queens, bees, honey or other commodities, whether by road. railway, or sea. As regards the
transport of queens it would be more correctly termed posting
or mailing queens, as they are generally forwarded by post
parcels.
Wonderful progress has been made of late in sending bees
and queens safely over long distances. Not long ago it
was considered risky to send a caged queen on a voyage
occupying more than eight or nine days, but within the
last eighteen months they have arrived safe after travelling
over twenty days. This success is due to the improvements made both in the shipping cages and in the food
supplied to the queens and bees. Formerly hard candy
and water was placed in the cages, but some two years
ago Mr. I. R. Good, an American, invented a much more suitable food which serves the purpose without water. The use of
this, together with the improved cages, has quite revolutionised
the queen trade. I make the " Good candy," as it is called, by

putting some ordinary sugar into a mortar, a little at the time, and grinding it up very fine, then adding a little honey while still grinding it till it is as fine as butter. It must not be made very soft, just so that it can be rolled into pretty firm balls without flattening out much when placed on a board.

Fig. 106.—QUEEN SHIPPING CAGE.

The cage I have been using lately, and found to answer very well, is similar to the one used by Mr. Doolittle (Fig. 106). It is made from a block of wood 3¾in. long, 2in. wide, by 1¼in. thick. A 1½in. hole is bored in the centre of the flat side of the block to within less than a quarter of an inch of going through. Two ¾in. holes are then bored in the edge, passing down within an eighth of an inch of the sides of the larger hole, but not breaking into it; these also go nearly through the block. A little very hot wax is poured into the small holes to give the wood a thin coating; this prevents it absorbing the moisture from the food which is placed in these holes. A small hole is cut with a penknife from the larger into each

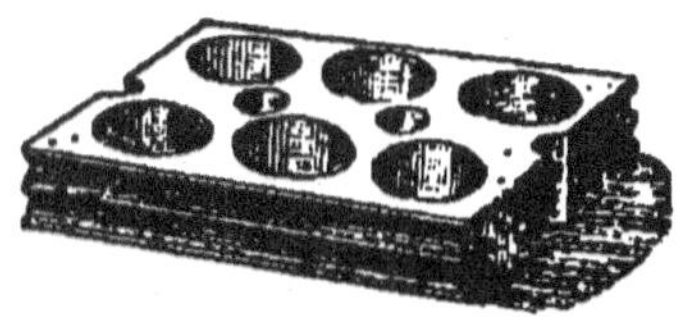

Fig. 107.—BENTON'S SHIPPING CAGE FOR TWO QUEENS.

of the feed holes, and the latter are filled with food pressed in tight and corked. The wire cloth can be partly tacked on, leaving one corner open to put in the queen and bees; about eight bees are sufficient to accompany the queen. To prepare it for mailing run a wide groove a quarter of an inch deep in a three-eighth inch batten the width and length of the cage, and tack this on after the bees are in, groove down, next the wire cloth, to give ventilation at the ends when the cage is wrapped for addressing.

Another shipping cage (Fig. 107), invented and used by Mr. Benton, has been successful in conveying queens safely from Europe to America, the trip occupying twenty-four days. Mr. A. E. Bonney, of Adelaide, writes me that he has used it with great success. Mr. A. I. Root gives the following description of it in *Gleanings*, from which the illustration has been taken :—

"The cage is made of some tough wood resembling pine. The board is planed on both sides, and then cut up in pieces 2¾ by 4 inches. The cage shown contained two queens when it was sent. The holes bored in the piece of wood are 1⅜in., and are bored nearly through. They come so near each other that passages are cut with a penknife connecting them ; that is, each three are connected in this way. One of the holes at each end row of three is coated with melted wax, and then filled with the usual candy made of powdered sugar and honey. No water is used, but abundant ventilation is given in several different ways. Only one of the three holes is ventilated, however, viz., the one in the end opposite the one containing the candy. The middle hole in each row has no ventilating passages. It would seem that this affords the bees an opportunity of choosing one of the holes that is much ventilated, or the central, where there is but little ventilation. A few bees could keep pretty warm in one of these round holes, especially if they choose the one without ventilating-holes.

"The cage is so made that, even when cramped in the mail bags, the holes cannot well all get closed. It is for this reason the grooves are made in the side of the piece of wood. Ten holes about the size of an ordinary darning needle are pricked through these side grooves into this end hole. The spur of the bit makes another hole. . . . You will observe two smaller-sized holes near the centre of the block. These smaller holes are connected with a hole about as large as a gimlet would make, shown partly at one end of the block. This also comes out where the block is grooved or cut in. Five small holes are made through into this gimlet-hole, so these ventilating-vestibules at each end of the block are both ventilated from two sides in such a way that the ventilating-holes cannot well get stopped up. A cover of wood about an eighth of an inch thick is tacked over the holes when completed. I do not know how friend B. gets in his queen and bees unless he lays the wooden cover on the block so as to partly close the holes, and then puts in the bees and queen one by one, after which he slides the cover on and fastens it with wire nails."

It conduces to safety to cage the queen and bees the evening previous to mailing them. There is one matter connected with mailing queens that should be borne in mind, that is, the bees and queens should be put up in such a way that there can be no possibility of them stinging any person handling the cage, nor the contents injuring anything in the mail bag. In the postal regulations, any article likely to injure the contents of the mail bag or do bodily harm to the officers is not allowed to be sent by post. Some years ago I wrote to the Postmaster-General of New Zealand, and sent him a specimen queen-cage, asking his permission to mail queens, and received his consent; but this would assuredly be withdrawn if any complaints were made by the officials. Bee-keepers in America some four years ago were put to great inconvenience and loss owing to queens being prohibited to be sent through the mails, due to the negligence of some one not putting them up properly, and it was only after some considerable time and difficulty that Professor Cook managed to get the prohibition withdrawn. It therefore behoves all to be very careful in this matter.

SHIPPING WHOLE COLONIES.

It is often necessary to deliver valuable queens accompanied by either a nucleus or a full colony, and therefore a few words with reference to the safe packing of such colonies when intended for travelling long distances will not be out of place here.

The main requisites are, to have the frames well secured so that they cannot move ; to see that the box or hive is well ventilated ; and that sufficient food and water are supplied.

Shipping boxes made for the purpose out of light material are much handier and less expensive than the ordinary hives. I make mine out of half-inch timber. The sides are 11in. deep and 19in. long ; the ends 10¼in. deep and wide enough to take the number of frames required. Fillets of wood half-an-inch square and the length of an end of a frame are nailed inside on each end, far enough apart to allow the frames to slide between them easily. A small notch is cut out of one end for an entrance, and covered with a piece of tin. Nine holes of one inch diameter are bored on each side, six in a line two inches from the bottom and three the same distance from the top edge, and these are covered with wire cloth on the inside. The sides are nailed to the ends, and kept flush at bottom ; this leaves the ends three-quarters of an inch short on top. A batten ¾in. thick and 2in. wide is then nailed on the outside at top of the ends to come flush with the top of the sides. The shoulders of the frames rest on the top of the ends, and are secured from moving by the fillets. The bottom is then nailed on and the box is ready for the bees.

The greatest difficulty I have had in shipping bees has been to ensure a plentiful supply of water on the voyage. I have tried two methods, but neither has been as satisfactory as I could have desired. First I arranged a sponge so that the bees could get at it through a piece of wire cloth, and pasted instructions on the box to water it every day—but this was often neglected. Next I made flat zinc bottles with a tapering neck ; these were filled and placed neck downwards on a piece of flannel in a small shallow zinc pan tacked to the bottom of the box, the bottle being secured with tacks to the side. This plan was also found frequently to fail. I think the heat of the boxes caused the water to evaporate too quickly ; at all events the water soon disappeared. I now believe the better way will be to adopt a combination of both plans ; that is, to put in one or two bottles of water and to arrange a sponge on the outside in the manner already described. The risk will at least be diminished by such precautions.

Old tough combs are the best to use for shipping purposes, not being so liable to break as new ones. No uncapped brood should be put in. nor too many bees ; neither should heavy combs of honey be used ; it is better to have the food fairly distributed through all the combs. The top should be screwed on for greater convenience of taking it off afterwards without any hammering or unnecessary jarring of the box. Full instructions for keeping the box in a cool dark place, etc ;, etc., should be pasted on top of each box, and if at all possible let the sender, or some very careful person acting for him, see to the shipping personally.

AXIOM.

"QUEENLESS COLONIES, UNLESS SUPPLIED WITH A QUEEN, WILL INEVITABLY DWINDLE AWAY, OR BE DESTROYED BY THE BEE-MOTH, OR BY ROBBER BEES."   *Langstroth.*

CHAPTER XIII.

SURPLUS HONEY—MODE OF SECURING AND MARKETING.

THE provident instinct which induces the honey-bee to store up honey for future use, and the restless industry with which it utilizes every opportunity of doing so, are the inherent qualities of the insect which enable man to turn its labours profitably to his own account. Bees, like all other animals, are actuated by a natural desire to "increase and multiply," which is perhaps more strangely evidenced in their case than in most others, because out of twenty thousand or more that form a colony there is only one, the queen, which can be supposed to have a truly parental instinct. It is, however, clearly with the object of supplying food for the rearing of future generations, and for the use of all during a time when little or none can be gathered, that these busy workers are led to collect on all favourable occasions so much larger quantities of honey than what can be required for immediate use. The aids afforded to the bees in a state of domestication under the improved system of culture enable them to store much larger quantities than they could do if living in their natural homes, and by preventing as far as possible the multiplication of swarms, the great bulk of the surplus honey, which would otherwise be devoted to the formation of new colonies, can be utilized for the benefit of their owner without detriment to the bees.

SPRING MANAGEMENT.

With the advent of spring come some of the chief duties of the apiarist. The object of his labours at this time will be to see that his stocks are progressing favourably, and that nothing is left undone to get them into good condition for taking every advantage of the honey season from its commencement. Much of the season's success depends upon good management in spring. Whatever may be the desire of the bee-keeper— whether it be to take all the honey he can, to increase his colonies at the expense of honey, or to secure a share of both— the management in early spring will be the same.

On the first favourable opportunity, after breeding has commenced, an examination of the stocks should be made and the condition of each noted for future reference. The first point is to see that they have a plentiful supply of food, any deficiency to be made up in the manner explained under the head of " Feeding"; and the next, to confine the bees on to as few combs as they can cover. The object of this is to conserve the heat of the bees as much as possible for the benefit of brood-rearing. Hives containing less than will cover seven frames should be contracted by division boards (Fig. 108), and where there are not sufficient bees to cover four or five frames, unless the colony has a young and valuable queen, it would be better to unite it with another (see page 249). It should be borne in mind that—other circumstances being equal—the more bees there are to cover the brood the more rapid will brood-rearing proceed.

DIVISION BOARDS.

No bee-keeper should be without a supply of these on hand, as they are often most valuable in winter and early spring.

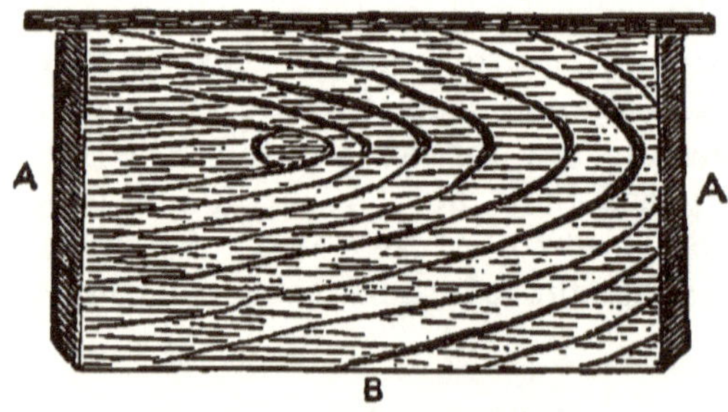

Fig. 108.—DIVISION BOARD.

With the aid of a pair of close-fitting division boards I have been able to keep up brood-rearing in small colonies all through he winter. A hive can be contracted to any size, and be made as snug as possible in a few minutes, by the use of these boards. I make them out of an inch board 9⅜in. wide, cut off in lengths of 14¼in. (the inside width of a hive). The two ends, A A, and bottom, B, are bevelled so as to leave a thin edge all round to come in contact with the hive. This enables a person to fix them in more readily without injury to the bees. A top bar of a frame is nailed on the upper edge of each, so that when they are placed in the hive they should touch the bottom board and sides of the hive, and the top bar should rest lightly on the tin supports, the same as the frames, to keep them steady and in place. When contracting a hive for a small colony, remove or place at the side all the unoccupied combs, leaving just as many in the centre as are sufficient for present requirements; then place a division board on each

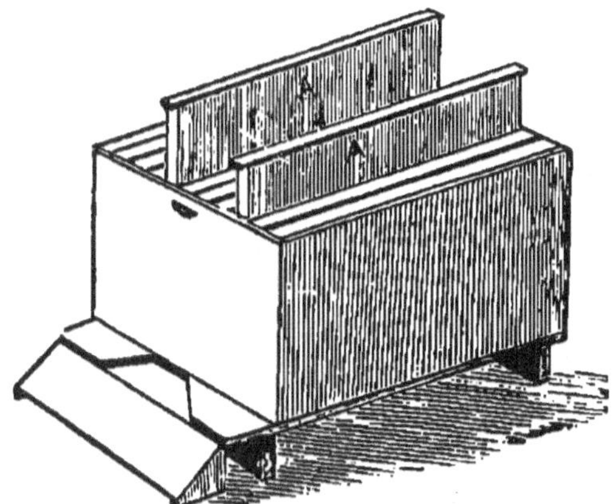

Fig. 109.—HIVE WITH DIVISION BOARDS.

side of them, as shown in Fig. 109.—A A are the division boards raised in order to show them. Some bee-keepers use chaff division boards, made by tacking calico or other light material over both sides of a close-fitting broad frame and filling in between with chaff, but the one I have described is easily made, will last for years, and is all that is required, at least in any part of Australasia. An extra mat or so should be placed in each hive and everything done to make them snug and warm.

An examination of the hives should be made once a week, or as often as may be necessary; if food is being given to any of the colonies they may require looking through every day or

two.   Where spring forage is scarce and stimulative feeding is
requisite to quicken brood rearing, it should be started about
six weeks prior to the commencement of the main harvest, the
object being to get the hives full of young bees at this time, as
everything depends upon having strong colonies to gather the
honey while it lasts.   No fixed time can be given for commencing to work up the colonies ; this must be left to the
judgment of each individual bee-keeper, as it depends upon the
flowering of the particular vegetation from which the main
supply of his honey is secured ; it should not, however, be less
than six weeks before that period.

SPREADING BROOD.

As brood-rearing progresses and plenty of young bees are
emerging from the centre combs, the brood nest may be enlarged by spreading the frames and placing a clean empty comb
in the centre, provided there are plenty of bees to cover all the
brood.   The empty comb, from its position will be taken
charge of by the queen almost at once, and in a short time will
be filled with eggs, whereas an outside comb would be very
sparingly occupied at first.   Division boards may be moved
as required to enlarge the brood nest.   This spreading of the
brood, however, requires great care and judgment ; if carried
too far, a cold night may drive the bees from the brood in the
outer frames, and the result would be that such brood would
get chilled and die.   An empty comb may be carefully inserted
at intervals as the number of bees increase.   If the caps of the
cells of a frame of honey be bruised with a knife just
sufficiently to start the honey running, and the frame be inserted
in the brood nest, it will have the same effect with regard to
stimulating the queen as syrup feeding.   From some of the
strongest colonies a frame of emerging brood may occasionally
be given to the weaker ones to equalize their strength.

PUTTING ON SURPLUS BOXES.

Having followed out the foregoing instructions and worked
the stocks into good condition by the commencement of the
main honey harvest, they may, if increase is the principal
object in view, either be allowed to swarm, or be divided, as
the case may be, before putting on the surplus boxes ; but if
honey only, or honey with a moderate increase is desired, these

boxes should be put on before the bees make the slightest preparations for swarming.

My own method is to place the surplus boxes on when the
hives are fairly full of bees, plenty of young ones emerging, the
weather warm, and honey commencing to come in pretty
steadily, taking care to get them on before any queen cells are
started. To induce the bees to start work at once in them I
lift one or two of the side frames from the lower hive containing honey only, with the adhering bees, and place them in
the centre of the top box, with an empty comb between them,
replacing those from below with empty combs, which may be
put toward the centre of the brood nest. By this plan
swarming is kept back, and in the course of a day or two, if
the weather is favourable, comb-building and honey storing
will be going on above. As the season advances and more
honey is being gathered, and the top box is getting well
stocked with workers, if I wish to keep swarming down, I place
another super with empty combs or frames of foundation, *next
the lower hive under the super already on*, and commence to
extract from the combs wholly or partly sealed, always taking
care to keep the lower super well supplied with empty combs.
After a while the hive contains an enormous force of workers,
and when it does throw off a swarm it is an extra large one—
one that is large enough to occupy a hive and surplus boxes at
once ; while if the queen cells are cut out of the parent stock
and it is supplied with a young laying queen it will scarcely
feel the loss of the swarm. Or even if nothing more be done
than to prevent after swarms issuing, the colony will not be
long before it is in a very strong condition again.

The same method applies when raising comb-honey, but to
induce the bees to enter the sections some half-worked ones
should be placed in each super ; if these are not to be obtained
a clean new comb may be cut into sizes to fit the sections and
fixed in them.* If the broad frame system be adopted, then
half-story supers should be used, and as soon as the bees
are fairly started to work in the first one another should
be placed underneath. The same applies to section cases,

* Mr. J. B. Mason, in a late issue of the *American Bee Journal*, says, that
if a section or two with the "young wax-workers," that is, the bees just commencing to draw out the combs, be transferred from some other hive to the
new super, it will cause the bees to take charge of it at once.

which I have recommended as being preferable to racks. Much closer attention to ventilation and supplying extra room in the surplus boxes is needed to keep down swarming when working for comb than for extracted honey.

REVERSING FRAMES.

A system of reversing frames—which at present bids fair to become generally adopted—has within the past year or two been tried by a number of prominent bee-keepers in America. The objects sought to be accomplished by the system are, first, to ensure the frames being perfectly filled with comb and fastened all round, and second, to prevent the combs in the brood-chamber being clogged with honey, or, as Mr. Heddon puts it, "an aid in supplying a brood-chamber for breeding purposes only, and the surplus arrangement above to possess nearly all the honey." Should these results be obtained—and late reports leave little room for doubt—it will be a great step in advance in the practice of apiculture.

A large proportion of combs are not built down to the bottom bars of the frames, neither are they fastened the whole of their depth down the end bars—consequently there is both a waste of room and a want of strength in the combs. If the frames are reversed, that is, turned upside-down after the combs have been built in the first place, the bees, for their own safety, are compelled to attach them securely both to the end bars and what were formerly the bottom bars, thus filling the frames solid. The next advantage, and which more nearly concerns us here, is the storing of the surplus honey in the surplus boxes; it is claimed that by the proper use of the system the bees may be induced to enter the surplus boxes more readily, and even in a manner be compelled to transfer the honey from the brood-chamber to the surplus receptacles. As a rule, the brood-combs are about one-third filled with honey, which is always stored along the upper part, next to the top bar, with the brood in the centre and lower part. When the top boxes are about to be put on, if the frames are reversed, the honey and brood being in an unnatural position with regard to each other, the bees, to rectify matters, will remove the honey, with a view to storing it above, and as the upper parts of the combs now contain brood, there is no alternative but to store it in the surplus boxes, and thus more breeding space is secured below.

Mr. Heddon, who appears to be a thoroughly careful and observant apiarist, has had several thousand reversible frames in use for the past two seasons, and states in his report to *Gleanings* for July, 1885 :—

"I h•ve to report not only practical success in their manipulation (the second year of their use), not only as far as gluing, etc., is concerned, but the generally conceived advantages of reversing are more than realised. The comfort of frames solid full of comb, and that comb nearly solid full of brood, is pleasing to the eye of the apiarist. I find the three-fourths space between the lower half of the end of the frame and the hive a great advantage. A few hives that were overlooked, and became clogged with honey, crowding out the queen, had to have the brood-combs extracted. Before replacing them with the bees we reversed them, giving plenty of surplus room above, and this reversing prevented any further clogging of the brood-frames. I am pleased beyond expectation, and never expect to use anything else but reversible frames for either comb or extracted honey production."

HEDDON'S REVERSIBLE FRAME.

Scores of devices have been suggested for making frames reversible without altering the hives; but Mr. Heddon's (shown in the following engraving) appears to be as simple and practicable as any, and the fact that he has found it to answer so well is a strong recommendation in its favour.

Fig. 110—HEDDON'S REVERSIBLE FRAME.

On reference to the figure it will be seen that the frame consists of an ordinary top bar, to which are attached two short end bars, with a rectangular frame pivoted between them; —it is very simple and easily made. For a Langstroth frame the rectangular frame should be 17 inches long by 8¾ inches deep, outside measure, and made out of three-eighth-inch material, one of the longest sides having a groove to fasten comb-foundation in when wires are not used. The short end bars to which the frame is pivoted should be about 5¼ inches long and tapered at their lower ends. These are fastened to a top bar in the usual way of making an ordinary frame. All that is necessary now is to fasten the end bars to the centre of the ends of the rectangular frame in such a manner that the latter will turn under the top bar.

QUEEN EXCLUDER HONEY BOARD.

This device is for preventing the queen entering and laying in the upper boxes. It is usually made of a sheet of perforated

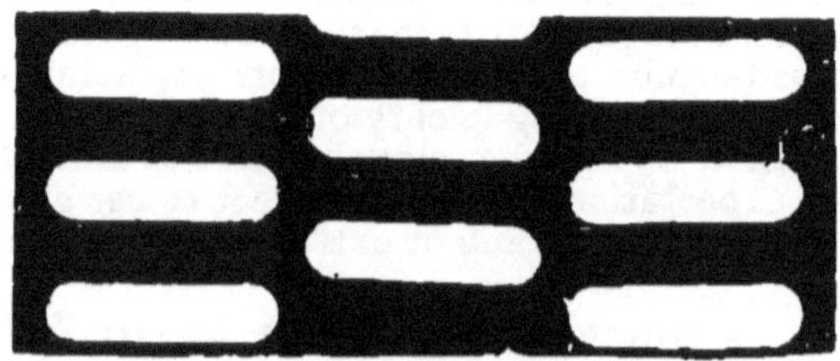

Fig. 111.—QUEEN EXCLUDER.

zinc (Fig. 111), and is placed over the frames of the lower hive. The perforations are a scant $\frac{5}{32}$ of an inch in width; this space allowing the workers to pass through, but restraining the queen. Mr. D. A. Jones, of Canada, who raises his comb-honey at the side of the brood-chamber, places a sheet of this perforated zinc between the brood-frames and sections. They have been extensively used in England, and a number of bee-keepers in America have lately tried them, but I do not think they are likely to come much into favour. I believe that any advantages gained by their use will be more than counterbalanced by some great disadvantages, such as creating a tendency to swarm by confinement of the brood-chamber, and the extra time taken by the workers getting into the supers. It appears to me that it would be more profitable to run the risk of having brood in the surplus boxes occasionally, rather than that of having extra swarms.

TAKING SURPLUS HONEY.

When storing in the surplus boxes is in full swing, and honey coming in rapidly, it should be taken away as soon as ready. Sections should be removed as soon as finished (but not until every cell has been capped), and empty ones put in their places. Any propolis or wax that may be about the edges of the boxes should be scraped off, and the boxes placed for a day or two on shelves in the honey room, which must be dry and well ventilated, after which they can be crated for market. Any unfinished sections at the end of the season should have the honey extracted from them, and be put away for next season's use.

Frames of honey for extracting may be removed as soon as about one-third capped. Some prefer to leave the combs in the hives until the whole of the cells are sealed, while others, again, remove them as soon as they are full, whether sealed or not, and ripen the honey after it is extracted. There is certainly a great saving of labour by the latter method, little or no uncapping to be done, and the honey can be thrown from the combs with much less trouble. I prefer to see combs partly capped, but I think there is no necessity to wait till they are sealed all over, if proper precautions are afterwards taken to finish the ripening of the honey before putting it up for market.

<h3 style="text-align:center">RIPENING HONEY.</h3>

In California, where they have no rain during the height of the honey harvest, some of the bee-keepers ripen their honey in what is termed a " sun evaporator." This is a

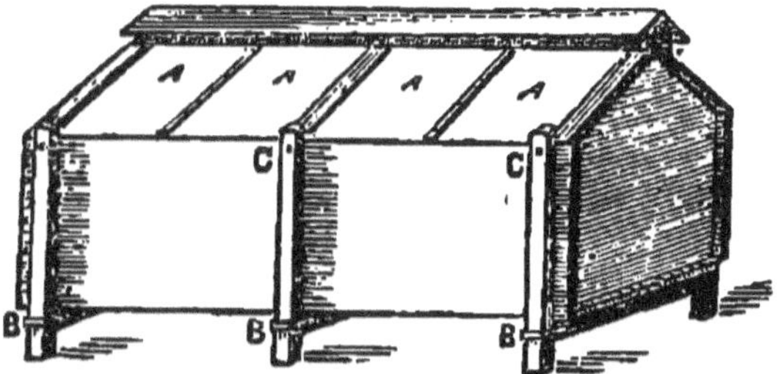

Fig. 112.—BLISS' SUN EVAPORATOR.

wooden tank lined with tin, one side of the cover being fitted with sheets of glass (A A, Fig. 112); and to allow surplus moisture to evaporate, there is a space two inches wide, the length of the tank immediately under the ridge, covered with wire-cloth. The tank is exposed to the sun, and a pipe is led into it from the extractor. When full, the honey is allowed to remain for two or three days to ripen before being tinned. However, if the arrangement of honey-house and tanks given in Chap. VII., and the methods of management therein detailed be adopted, I believe they will be found to be as good as any, if not the best yet recommended.

COMB BASKETS.

When removing surplus honey from the hives it is necessary to have something to place the frames or sections in, both for convenience of carrying and to keep the honey secure from the bees until it is safe in the honey or extracting house. I find tin comb-baskets, similar to the one shown below, answer

Fig. 113.—COMB BASKET.

the purpose admirably. They are light, strong, clean, and handy. They should be made so that the frames will hang in them the same as in a hive, and should have a space of at least two inches below the bottoms of the frames, to hold any honey that may drip from the combs after they have been uncapped. To hold six frames conveniently they should be $7\frac{1}{2}$in. wide inside.

MARKETING HONEY.

The first aim of the bee-keeper should be to produce a first-class article, and the next, to place it upon the market in the most suitable sized packages, prepared in a neat and attractive manner. Until within the last few years, most of the honey in well-got-up packages seen upon the Australasian markets was of foreign production, the locally produced article having usually been put into all sorts of odd things, from a pickle bottle to a kerosene tin. To create a demand for our produce, we must adopt a form of package for placing it upon the market that will not add much to the cost of the honey, and at the same time be capable of being made attractive to the eye of the general public.

EXTRACTED HONEY.

When put up in small packages, I believe the most suitable for the Australasian markets are 2lb., 10lb., and 20lb. tins. Although I find the two larger sizes sell fairly well, I would advise putting the bulk in 2lb. tins, and only a limited quantity in the others. Some bee-keepers prefer selling their honey wholesale in barrels of any size, rather than take the trouble of tinning it. This may be a convenient plan in some cases, for instance, if there should be an opportunity of selling direct to manufacturers who require to use the honey in the preparation of other articles, otherwise it is scarcely to be recommended as a means of establishing a character for the honey, or of securing the best price in the market. A great deal of mischief has been done to the industry in England and America by dealers buying up honey in bulk and adulterating it before putting it on the market in retail packages. This is much less likely to be practised in future, and besides, it is true that after the apiarist has disposed of his honey in the public market, whatever way he may have made it up, he cannot be accountable for what may be done with it afterwards. But he can and should always hold himself responsible for the quality of his honey in the *original packages*, and this he may do by tinning it before it leaves the apiary.

The following are the sizes of the different tins mentioned :

2lb. tins, round	$3\frac{3}{8}$in. diameter, by $4\frac{3}{8}$in. high
10lb. ,,	6in. ,, ,, 7in. ,,
20lb. ,,	7in. ,, ,, 10in. ,,
20lb. tins, square	6in. by 6in. ,, $10\frac{3}{4}$in. ,,
60lb. ,,	$9\frac{1}{2}$in. by 9in. ,, $13\frac{1}{2}$in. ,,

I find tins with a top the same as made for fruit tins the handiest for filling and packing ; a little practice will enable any handy person to solder the discs satisfactorily.

The look of the tins depends much on the label, and this ought to be neatly lithographed in colours, bearing some kind of design, with a trade mark and the name of the apiary and proprietor. Our Matamata labels also contain the following :— "Notice.—This honey is separated from the combs by the 'Honey Extractor,' thereby retaining its original aromatic flavour," on one end, and on the other :—"Directions.—Should this honey granulate, which may be taken as one of the surest tests of its purity, it may be liquidated again by immersing the tin in hot water." Our labels are made to go round and cover the whole circumference of the two-pound tins, and the same labels answer very well for the larger ones. The tops and bottoms of the two-pound, and the whole of the large tins, are japanned before the labels are put on. Black japan can be bought at most colourmen's, and it only requires thinning down with an equal or greater quantity of turpentine to bring it to the required shade ; a broad camel-hair brush is the best to put it on with. Our packing cases are made to hold thirty 2lb. tins, six 10lbs., or four 20lbs.—each case being stencilled with the name of the apiary and the nature of its contents.

The 60lb. tins are a very convenient size for making up honey to be sold in bulk or for export. Two of these tins in one packing-case are heavy enough, though not too heavy for handling by one man, and will pack together conveniently for any mode of transport.

Honey while in a clear liquid state looks very well in glass. I should not be disposed to put much up in that way when it is liable to granulate quickly, nor in any case in bottles holding more than about two pounds. In this form, if the bottles are well made, of clear white glass, and neatly stoppered and labelled, they generally find favour for retail sale, although necessarily something dearer than if made up in tins.

COMB-HONEY.

This, as I have already observed, is more readily sold in one-pound sections than in any other form. It is not advisable to raise more comb-honey in any district than can be sold in it, or at some market near to it. The demand for it is more limited than that for extracted honey, and it runs great risk of getting damaged when sent to distant markets. To preserve the combs from flies and dust when exposed for sale the sections should be packed in crates having a sheet of glass in front, similar to the one shown in the engraving. The cover should be tight fitting, and the whole should be neatly painted. It is better to affix a label to each section rather than to put anything on the crate, as it might be made use of to retail other honey from. A price can be put on the crates when selling the honey, and made returnable. I find crates holding

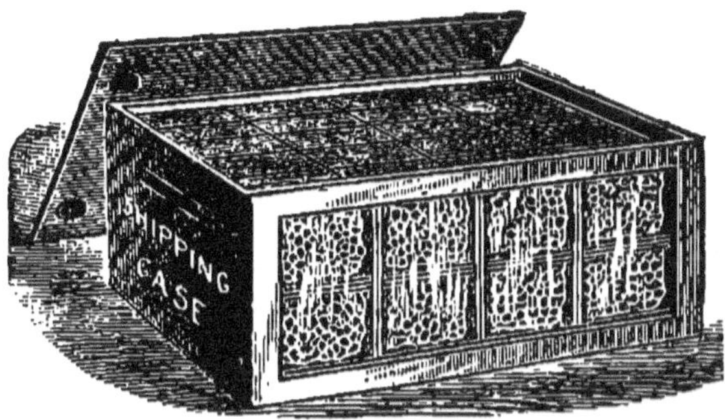

Fig. 114.—SHIPPING CRATE AND SHOW CASE FOR COMB-HONEY.

twenty-four 1lb. sections a handy size. Care should be taken that the sections next the glass are a fair sample of the rest, and every section should be full and the combs completely capped, or it is likely there will be leakage.

BENEFITS OF CO-OPERATION.

The whole question of the most suitable and effective manner of placing the honey crop of any particular district upon the market, as well as the more general one of promoting by judicious means a more extensive consumption of honey, is one which peculiarly calls for the co-operation of the bee-keepers concerned. There are few matters in which a local association can make itself more useful than in deciding such points according to the local circumstances; obtaining the best information for its members, with samples of the sorts of packages used in different places; and then, by a joint action, obtaining the most suitable kinds for all who require them, on the most favourable terms.

CHAPTER XIV.

WINTERING.—UNITING.

A STRIKING feature in the practical working of apiaries in the Australasian colonies (and indeed in all the habitable regions of the Southern Hemisphere, with the exception, perhaps, of the most southern extremity of South America), is the facility for carrying the stocks safely through the winter, as compared with the difficulties experienced in that respect in the most important bee-keeping districts in the Northern Hemisphere.

DIFFERENCE IN CHARACTER OF WINTER SEASONS.

In no part of New Zealand or Australia is there a lower mean winter temperature than 40° Fahr., which is as mild as that of Northern Italy, and in the greater portion of the Northern Island of New Zealand, in New South Wales, Victoria, and South Australia, it is up to or above 50°, with a perfect freedom, in the level country, from anything deserving the name of heavy frost or falls of snow. There are certainly many countries in the northern part of the world similarly circumstanced, some of which, like the countries of the Mediterranean, have been the earliest seats of bee-culture, and others, like the Southern and Western States of the American Union, are now amongst the most advanced in that industry; but in consequence of scientific apiculture having grown up in Germany, England, and the Northern States of America, all of which countries have to contend with severe and protracted winter seasons, it happens that all the works which have been published on the subject of bee-culture naturally dwell much upon the difficulties to be contended with in wintering bees, and the modes of overcoming those difficulties; and nearly all those works treat the subject as if it were a difficulty inherent in the practice of bee-keeping, and not merely dependent upon peculiarities of climate. At least, there is not one which has as yet treated on the subject of "wintering" in semi-tropical climates ; yet that is a matter which, although not attended with the same difficulties as in colder climates, nor indeed with much difficulty of any sort, cannot be passed over lightly by Australasian bee-keepers.

AUSTRALASIAN WINTERS.

Nowhere in these colonies can there be any question of housing or cellaring the hives during the winter months, or of confining the bees to their hives. There is nothing to prevent the bees being left on their summer stands, and upon all fine days the bees will fly as freely, and collect honey and pollen (if to be had) as regularly as in summer-time. At no time is the day temperature so low as to check this tendency ; there is no snow lying on the ground to entrap unwary bees that may be induced by a gleam of sunshine to venture away from home ; it is only on days of continued rain or of very high winds that flight would be likely to be injurious to bees, and on such occasions they may be safely trusted to remain comfortably at home.

PRECAUTIONS FOR WINTERING.

The precautions necessary to insure safe wintering may be confined to the following points :—1. Reducing the interior room in the hive to the size suitable to the strength of the stock. 2. Providing a fair supply of food to be drawn upon when the bees are prevented from working. 3. Securing protection from the effects of night frosts by sufficient mats or cushions between the frames and the hive covers. 4. Reducing the entrance so as to effectually exclude mice, and to enable the bees easily to keep out other enemies. 5. Making sure that all covers are perfectly water-tight, and not liable to be blown off by high winds. 6. Last, but not least, providing winter forage in the neighbourhood of the apiary.

REDUCING INTERIOR ROOM IN HIVE.

Usually the season for taking surplus honey closes a month or six weeks before the time when it is necessary to prepare for wintering. In such cases the spare supers or surplus boxes should be removed, and the bees confined to as few frames as they can cover.   If the stocks are strong enough at the end of autumn to fairly cover all the ten frames of the Langstroth hive, no furthur reduction of space will be necessary, and it will be very desirable to keep as many stocks as possible up to that strength, by uniting weak colonies.   If, however, the stock to be wintered does not cover more than six or seven frames, the space should be contracted by the use of division boards, which, whether made of solid wood or of frames covered with cloth and stuffed with chaff, should fit close to the ends and bottom of the hive, as explained in Chapter XIII.   It is not, as a rule, desirable to winter a stock covering less than six frames ; in such cases unite two to form one good stock (see page 250).

PROVIDING FOOD.

As the matter of providing winter food has already been fully dealt with in Chapter IX., under the head of "Feeding for Winter," I must refer the reader to that portion of the book for information on the subject.

INNER COVERING OF FRAMES.

The nature of the covering over the frames inside of the hive cover should depend upon the degree of sharp night frosts to be apprehended during the winter nights.   With a mean winter temperature over 50°, and where the night temperature seldom falls below 32°, and then only for a few hours at a time, nothing more is required than the ordinary mat which is used all through the year.   Where severer night frosts are likely to occur, a thin chaff cushion may be laid over the ordinary mat. The use of a sheet of ordinary cotton wadding has been recommended by Mr. Beloe, who has used it with advantage in the Waikato district.   In more southern and colder districts it may be advisable to leave an additional half-story box over the main hive, filled with a loose-fitting sack or cushion of chaff.

REDUCING ENTRANCE.

One of the great advantages of the form of Langstroth hive, as herein recommended, is the facility with which the size of the entrance can be enlarged or diminished by simply sliding the hive forwards or backwards on the floor-board.   In winter the back end of the hive should be within an inch or less of the back of the floor-board ; the entrance will then only admit of one or two bees going in or out at the same time, which will enable the sentinels to guard it effectually, and will render it impossible for the smallest of mice to force their way in.

SECURING COVERS.

It is equally important, in summer as in winter, to see that all hive covers are impervious to wet, and not liable to be displaced by sudden gusts of wind ; it is only more necessary to look well to these points at the beginning of winter than at other times, because the hives will not require to be so constantly under the eye of the bee-keeper during the few months which are to follow, and indeed, if everything has been done that ought to be done, the less the hives are meddled with during that time the better. Not that I would recommend leaving the apiary to take care of itself in winter, any more than at the busy season of the year ; on the contrary, a prudent bee-keeper will take care that a general outward inspection of the hives is made every morning, if possible, to make sure that nothing unusual has occurred, and occasionally to brush away spiders and their webs. The security of the covers from being displaced by wind will of course depend very much upon the judicious shelter of the apiary. In some climates, however, sudden violent gusts of wind, sometimes coming in the form of small passing *whirlwinds*, cannot be well guarded against, and in such cases it may be prudent to secure the covers to the boxes by means of clasps or hooks and eyes.

PROVIDING WINTER FORAGE.

The necessity of attending to this point in a country where the bees fly and work all through the year, and the facilities for doing so in these colonies by the planting of such evergreens — especially acacias and eucalypti — as are peculiarly suited to the climate, and which afford honey and pollen all though the winter and early spring months, will be found fully dwelt upon in the chapter upon " Bee Forage," to which it is only necessary in this place to direct special attention.

VENTILATION.

A certain amount of ventilation is required during the winter months to prevent moisture condensing within the hive. Mould showing on any of the combs is a sign of insufficient ventilation. With a hole bored in each end of the cover, and such mats as I have recommended, together with a contracted space suitable to the strength of the colony, there need be no fear of bad results under ordinary circumstances.

CHAFF-HIVES.

Hives made with double walls, the space between (about 2in. wide) being filled with chaff, cork-dust, or some such material, are frequently used in England, the Northern States of America, and in some of the colder parts of the European continent. Some of our New Zealand bee-keepers recommend their use in the southern parts of the colony; but I doubt very much that there is any part of the Australasian colonies so cold as to require the use of chaff-hives for the safe wintering of bees. Mr. T. G. Brickell, of Otago, has had some in use, and speaks very favourably of them. In response to a request to furnish me with a description of his way of making them, he very kindly sent me the following and a photograph of one, from which the engraving has been made.

"TO MAKE A CHAFF-HIVE.

" Both outside and inside walls are made of half-inch stuff, planed on one side, and mine is made to take the Langstroth frame, to which the following dimensions apply : For the inside end walls two pieces 15½in. long, 8⅝in. wide are checked, ⅛in. deep, 14¼in. between the shoulders ; on bottom edge of one piece, an equal distance from each end, cut out a piece 8in. long, ⅜in. deep for entrance. Cut two pieces, 18¾ by 8⅝in. for the sides. Nail together, and fix on the bottom.
" For outside end walls cut two pieces, 19¼ by 10½in., and two pieces, 24½ by 10½in. for sides. Nail the sides into the ends, and cut in centre of one end a slot for entrance 8¼in. from top, 8in. wide, ⅜in. deep to and correspond with that cut in the inner end. Two battens, 2 by 1½in. 19¼in. long, are nailed across the bottom, projecting 2in. on either side ; slip the outside case over the inside one, so that the bottom edge is half an inch below the battens, and level with the second and lower bottom when screwed on, taking care that the space between the two boxes is equal all round. Cut pieces 4in. wide, planed to the necessary bevel, and fix them fair on the upper edge of the cases—mitred at corners. Two pieces of tin, 14¼ by 1½in., turned up ¼in. on one edge,

are tacked on to form metal supports, and four pieces of tin, 2in. wide, bent neatly over the mitred corner, ensure its being waterproof. A fillet rebated to fit the super in use, fixed round the top, ⅛in. deep from outside shoulder, ensures ⅜in. between the frames. Four pieces, 10¼in. long and 1½in. square, are rebated 1in. each way, and fixed up to the angles ; the two slots for entrance are connected, and an alighting board screwed on. Turn it upside down to fill it with chaff, and put

Fig. 115.—BRICKELL'S CHAFF-HIVE.

a piece of tarred felt over the chaff before the bottom is screwed on ; this prevents the damp from turning the chaff musty. The advantages of chaff hives made in this way are that they can be manipulated as easily as the ordinary two-story hive in the summer, and yet afford the necessary protection in winter."

PROVIDING SPACE ABOVE FRAMES IN WINTER.

This is another expedient which many bee-keepers in cold climates have recourse to for the better wintering of their bees. The object in the first place was to provide a winter passage from comb to comb in the warmest part of the hive, and for this purpose small sticks were placed across the frames to raise the mats and chaff cushion a little ; but some four years ago a Mr. Hill conceived the idea of providing a space large enough to allow of the bees clustering in it, and it was found to answer so well that many have adopted the plan since. The contrivance for placing on top of the frames to raise the mats or chaff cushion, shown in the engraving below, and which has been named after Mr. Hill, is made of small pieces of wood sawn or bent to form part of a circle of about 11 inches in diameter; the two centre pieces are 9 inches in length, and the two outside ones 8 inches. A thin piece of hoop iron 12 inches long is tacked on the back of the pieces to

Fig. 116.—HILL'S DEVICE.

keep them in position. I have tried them on small colonies and find them to answer very well indeed. The bees cluster on frosty nights right up in the space thus formed. Care must be taken that the side frames are well covered with mats.

UNITING WEAK AND QUEENLESS COLONIES.

The advisableness of keeping none but strong colonies, and of uniting together two or more of any found to be below a certain standard of strength, has been more than once pointed out in previous chapters ; but to impress it upon the mind of the reader I would here remark that special attention should be paid to the matter of uniting weak colonies in the early spring months and when preparing the hives for winter (see pages 232 and 246). Colonies that become queenless at a time when no queens are available should also be united as soon as discovered to others possessing queens. Queenless colonies are frequently to be found at the latter part of winter and in early spring.

When the hives containing the colonies to be united are located some distance apart, move one a few feet every day till it is alongside the other. The queen may now be removed from one of the colonies, and in the evening place the frames on which the bees that still have a queen are clustered, to one side of the queenless hive, and the queenless bees on the other, with a vacant comb or two in the centre ; do this as quietly as possible, and close the hive. If this is done on a cool evening, without much disturbance, the bees will remain quietly in their respective quarters during the night and unite peaceably the next day, when their combs may be placed together.

Another and very good plan is to place a frame with bees from each colony alternately in the hive, and so mix them up thoroughly ; and I have found that a little syrup sprinkled over the bees when practising this method tends to prevent them interfering with each other. I have also united colonies by simply placing one hive over the other, and after a day or two have lifted off the upper one and put all the bees into the lower hive. In any case united colonies should be watched for a while, and if fighting should commence use the smoker freely ; in most cases two or three strong applications of smoke will put a stop to it.

UNITING SWARMS.

When swarms issue late in the season it is often advisable —especially when increase is not desired—to hive two or more together in order that there may be a larger working force brought into union, and so be better able to provide themselves with ample winter food during the remainder of the season. Swarms that issue on the same day may be hived together without the least trouble, and a swarm may be united with one that was hived a day or two previous, by placing the hive containing the new swarm as a super on top of the other, watching the bees for a while, as already advised. Swarms can be united in the same way at any time.

AXIOM.

" THE BEE-KEEPER WILL ORDINARILY DERIVE ALL HIS PROFITS FROM STOCKS, STRONG AND HEALTHY, IN EARLY SPRING."

Langstroth.

CHAPTER XV.

ROBBER BEES.

ALTHOUGH bees have so many commendable qualities, as we all know, still their most ardent admirers cannot deny that they possess one which does not redound much to their credit, according to our ideas of morality, namely, a propensity to rob their neighbours in order to increase their own store. They have this faculty very strongly developed, although it must certainly be admitted, by way of qualification, that they very seldom steal while honey is to be easily obtained in a legitimate way. When they once enter upon this bad course, however, they are almost quite incorrigible; and if we adopt the theory of the "survival of the fittest," we should expect to find the honey-bees of to-day much superior to their ancestors, since they never display the slightest compunction about robbing, starving, and killing their weaker sisters of other colonies. It may, however, be objected to this conclusion, that the worker bees do not propagate their own race, and the charge of robbery can be brought home to them only, and not to the queens or drones.

CAUSES OF ROBBING.

Immediately after the close of the regular honey season, as soon as their ordinary sources of supply begin to fail, the bees not only become more jealous of any interference with their own stores, but they are at once on the alert to pick up honey or any kind of saccharine matter wherever it is to be found. At such times they will visit grocery stores, breweries, jam factories, or in fact any place where saccharine matter can be obtained. On one occasion, when transferring some bees for Captain Moore, of s.s. *Vivid*, Thames, I noticed that the honey cells were of a very peculiar colour. I examined the contents, and on tasting found it to be *raspberry syrup*, which, no doubt, had been obtained from a neighbouring cordial factory. Captain Moore informed me that Mr. Koefoed, of tomato sauce fame, told him he lost on one occasion several pounds of tomato sauce that he had forgotten to cover up, thanks to the robber bees.

" Robbers " will at this time be found visiting all the hives in the neighbourhood, flying round the covers and joints, trying, like burglars, to effect an entrance without being seen by the inmates. If, in these expeditions, they should come across a weak hive, they are bold enough to go in by the entrance door. If one bee should be fortunate enough to get away with a pilfered load, he will soon be back with some more of his hive companions ; and this will go on systematically until in the end the hive is emptied of its honey, leaving the inmates in a starving condition. In most cases these villains will commit murder ; they will find the queen and kill her, and while the colony is, in consequence, in a demoralised state, " do a good business " in their nefarious practices. They also exert an evil influence over the inmates of the robbed hive ; for these will sometimes join with the robbers, and help to steal their own stores, leaving their old home to take up their quarters with the plunderers.

HOW TO KNOW ROBBER BEES.

It is rather difficult at first for the inexperienced to detect robber bees approaching a hive, but by careful observation one may soon become familiar with them. A robber bee, instead of alighting at once at the entrance, will fly and " dodge " about, making now and then a feint to settle ; but should one of the sentinel bees at the entrance approach her, she at once starts back, as it were, out of the sentinel's range. If she finds the entrance too well guarded, she will try to find some other place to enter, hovering about the sides and back of the hives, and examining the joints of the floor-board, super, and cover, which the proper inhabitants of the hive would not be likely to do. Should there be reason to suspect that robbing has been started in any particular hive—and sometimes it is not easy to detect it at first—watch the bees as they come *out* of the hive ; and if they appear to be loaded (which can be seen by the increased size of their abdomen), and find it difficult to rise and fly away, you may be certain that they are robbers. To find out which hive—if in your own apiary—the robbers have come from, dust the suspected bees with flour as they leave, and watch the other hives to see which they enter with their loads. A sure sign that robbing has been going on in a hive is the appearance of small pieces of wax—cappings of cells—scattered about near the entrance.

PRECAUTIONS TO BE OBSERVED.

More can be done in the way of precaution to guard against the starting of robbing than to put a stop to it when once started. A very slight cause may give rise to the first attempts; and if the bees once get infected with the desire, it may materially affect the prosperity of the apiary for the time being. No honey, pieces of comb, or sweets of any sort should be left lying about where the bees can get at them. No hives should be opened more than is absolutely necessary when robbers are about. If much manipulation should be necessary at such times, a bee-tent, such as described below, should be used to enclose the hive to be operated on. Robbing is likely to occur both in autumn and spring, but the first four or five weeks after the close of the honey season I have always found to be the worst in that respect. Entrances should be contracted at this time, to give the bees a better opportunity to defend them; and if all the colonies are kept strong, there will be less danger of any of them being attacked.

BEE TENTS.

A most useful appliance is a small tent (Fig. 117), which can be placed over a hive when necessary. The artist has made the tent in the engraving rather large in proportion to the hive inside. It may be made of a light framework of wood, covered with mosquito netting, and should be large enough to cover the hive and an empty one beside it, or the super when taken off, and to give the manipulator room to work comfortably. It should be made to fold when not required for use, so that it may be stowed away conveniently. It will answer for transferring combs under its shelter, manipulating bees in at the close of the season, and in other ways for keeping off the attacks of robbers.

HOW TO STOP ROBBING.

I have adopted the following methods to stop or check robbing, when it has taken place. I take a watering-pot, fitted with a fine rose, containing water, and, with a cloth laid across the entrance of the robbed hive, pour the water on to the cloth and over the bees that are flying about the front. In a short time I remove the cloth and let any bees that wish to do so come out, without letting those outside get in. As soon as they are out, I put the cloth back, and again wet it. This process I repeat two or three times. This appears to frighten

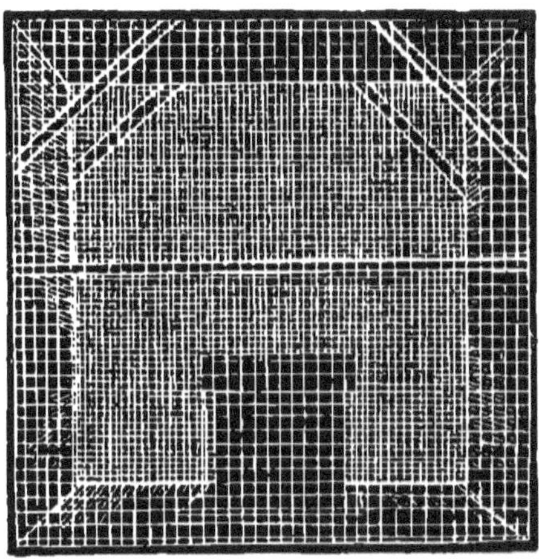

Fig. 117.—BEE TENT.

the robbers, and in most cases I have found it effectual; but in extreme cases I have tried the following plan, which has had a satisfactory result: Found out where the robbers have come from, and changed the places of the robbed and robber hives. There are other methods; but having found the two above described successful, I think it is not worth while to go into them and perhaps confuse the beginner.

It has been already noticed that when a colony becomes hopelessly queenless, that is, without the possibility of the bees rearing a new queen for themselves, they lose the energy for working and for defending their stores, and become an easy prey to robbers. In a well-kept apiary such cases will not be allowed to occur. Keeping all colonies supplied with good queens, and consequently "strong," is the sovereign remedy against robbing, as well as against many other evils in the apiary.

CHAPTER XVI.

DISEASES OF BEES.

UNTIL quite recently it had been generally supposed that bees were subject to two diseases only, viz., dysentery, and what was termed "foul brood." The scientific investigations of Mr. Frank Cheshire, however,—which are still being carried out—point to the existence of other diseases, and there is good reason to believe that when he shall have finished his researches, many things connected with apiculture which are now inexplicable will be made clear.

The former of the two diseases mentioned is a very common and often a dangerous one in cold climates, but in Australasia it is not very troublesome and rarely dangerous.   The latter, foul brood (now generally known by the scientific name of *bacillus alvei*), is of a most dangerous nature wherever it breaks out, and cannot be too carefully guarded against, as will be seen further on.

DYSENTERY AND ITS PREVENTION.

The presence of this disease may be known by the bottom-board within and around the entrance to the hive being spotted with a dirty, yellowish, disagreeable-looking excrement, which has an intolerably offensive smell.   The abdomens of many of the bees become so distended that they can no longer fly, but may be seen dragging themselves outside the hive to die.   The chief causes of the evil are—cold, damp hives, unwholesome food, and long confinement during severe winters.   It usually makes its appearance in the latter part of winter and early spring.   Ineffective ventilation with a low temperature is productive of dampness by causing the condensation of the watery vapour given off by the bees, which, settling on the combs, dilutes and sours the honey, and consequently makes it unwholesome as food. The ventilation of hives, both in summer and winter, should receive close attention. Owing to the disease usually appearing in the northern part of New Zealand about the time when peach trees commence to blossom, it has sometimes been wrongly attributed to the honey gathered from that source. There is, however, reason to believe that it is sometimes caused by the honey gathered from the wharangi shrub, which blossoms about the same time, and which will be found more particularly mentioned in Chapter XVIII.

The best preventive measure that can be taken is to provide the bees with clean, warm, tight, well-ventilated hives, and the surest cure is a spell of warm, genial weather.

BACILLUS ALVEI (FOUL BROOD).

This disease is the most fatal to bees, and if prompt action is not taken to stamp it out on its first appearance it may very rapidly lead to the loss of the whole apiary. The disease appears to have taken a pretty firm hold in a few districts in both Australia and New Zealand, and the difficulty of getting rid of it in these places is increased by the carelessness and wilfulness of box-hive bee-keepers, who, in the utter disregard of the advice given them by more careful men, will persist in leaving the old boxes with their combs lying about in which diseased colonies have died for other bees to enter, and occasionally hiving stray swarms in them, only to propagate the disease and finally perish as the others have done before them. A gentleman who went to much trouble to assist some box-hive bee-keepers in his neighbourhood to get rid of the disease, and supplied them with a remedy free of cost, wrote me some short time since, that as they would not follow on with the work he had commenced, he had given the whole thing up in disgust, and that he had no hopes of the district ever being clear of the disease while such men remained in it. Fortunately, I have had no experience of the disease beyond inspecting pieces of diseased combs that have been sent to me.

SYMPTOMS OF BACILLUS ALVEI.

Dr. Dzierzon, who once lost 500 colonies by it, describes the symptoms as follows :—

"An infallible symptom of the presence of foul brood (*bacillus alvei*) is the discovery of dead, dried-up, shrivelled larvæ or nymphs in separate cells amongst healthy brood. These dead larvæ have passed into a pap-like or tough mass, and later on into a greyish-brown or quite black crust on the floor or the lower surface of the cells. If the majority of the cells are in that condition, the infection took place some time ago, and the evil has already become very great. Because a stock with foul brood generally ventilates considerably, the evil may be recognised in hives with immovable combs by an unpleasant smell proceeding from the entrances; the smell is similar to that of putrid glue or meat. As the bees take the trouble to bring out separate larvæ that have not yet entirely rotted, such will be found sometimes on the floor of the hives affected. The bees take the trouble partially to remove to the outside the blackish-brown crust forming finally from the rotten matter. There are therefore found on the floor a dark-coloured dust and entire skins torn off, which, when rubbed down between the fingers, give off the same unpleasant smell. In spring, when other stocks are already diligently building, the foul-broody do not generally make any preparation for it; at most they will only do so when they are still fairly strong and unusually good pasture sets in. If the combs are examined, the sealed brood is never found *en masse*, but standing in isolated irregular patches. To be thoroughly satisfied, a piece of brood-comb must be cut or torn out; and if it shows cells with the matter described above, foul brood is certainly present."

He says that there are two kinds of foul brood,

"One kind that is mild and curable, and another kind malignant and incurable; both kinds are, however, contagious;" and that "the curable kind may occur of itself, under certain conditions of ingathering, ... and sometimes disappear again of itself when the conditions have changed."

It has long been the opinion of many bee-keepers, that bees occasionally contract this disease while working on certain plants; Dzierzon mentions two of this class, "bilberries and pines;" while on the other hand it has been said that it never appears in apiaries situated in the neighbourhood of certain other plants. As the name, foul brood, implies, it was thought —until Mr. Cheshire proved otherwise—to be a disease of the larvæ, and that the old bees and queen were not affected by it. This and other errors, however, of earlier investigators have been brought to light by the above-mentioned gentleman.

Beyond the fact that it was a germ disease, nothing reliable was known about foul brood up to a very recent period. The fungoid growth was supposed to belong to the sort known to microscopists as *micrococci;* it was supposed to originate in the decaying bodies of the larvæ. A treatment with salicylic acid and borax, used in solution, to spray over the affected combs

and to mix with syrup as food for the bees was recommended,
and has in some cases proved to be efficient as a cure ; but the
successful application of this treatment was so uncertain, that
many very experienced bee-keepers in America still held to
the opinion that the only safe and effectual course to be followed•
was to stamp out the plague and prevent its ruinous spreading
by *burning* everything supposed to be infected—bees, combs,
larvæ, and hives. New and very important discoveries have,
however, been made, and the bee-keeping world is now relieved
from much of the dread caused by the apprehension of this
disease, owing to the

INVESTIGATIONS OF MR. FRANK CHESHIRE.

This distinguished English scientist, in the month of July, 1884,
read a paper before the International Conference of Bee-keepers
at the International Health Exhibition at Kensington, giving
the results of his long-continued investigations into the nature
of this disease, "the means of its propagation, and the method
of its cure "—results which bid fair to solve all the difficulties
of the case, and to lay all apiculturists under a deep debt of
gratitude to the investigator. It would be out of place here
to give anything like a *résumé* of the paper referred to ; it has
been published at length in the Bee Journals ; but it is necessary to state as shortly as possible the conclusions arrived at
by Mr. Cheshire. After mentioning the fact that " science has
recently shown that all putrefactive changes, fermentations,
and very many diseases are brought about entirely by minute
organisms, which are, in fact, rudimentary vegetables," and to
which the general name of *Schizomycetes* is given, divided into
four genera, *micrococcus, bacterium, bacillus,* and *spirillum*, he
proceeds to point out the chief characteristic differences between
bacilli and *micrococci* in the following words :—

" Under certain conditions the *bacilli* produce spores, or seeds (Fig.
119), which the *micrococci* never do ; while in addition *bacilli*, unlike
micrococci, are provided at their extremities with wondrously delicate
filaments, called flagella, with which they strike the fluid containing
them, and so swim much as a fish does by the use of its fins ; so that
shape and the power of locomotion sharply divide one from the other."

It must be remembered that although the powerful micro-
scopes now used enable the observer to discern these peculiar-

ities of form, and approximately to measure the size of these organisms, still they are so minute, that, as calculated by Mr. Cheshire, *one thousand millions* may be contained in the larva of a bee, and of the spores or seeds, as many as *one hundred millions* in the bee's egg, which is itself only one-fourteenth of an inch in length, and one-seventeenth of an inch in diameter.

<table>
<tr><td>Fig. 118.—BACILLI
(greatly magnified).</td><td>Fig. 119.—SPORES OF BACILLI
(greatly magnified).</td></tr>
</table>

Having premised so much, it may be stated that Mr. Cheshire has established, as the first result of his investigations, that the fungoids present in cases of foul brood are *not micrococci,* but always *bacilli,* that they are not the result of the disease generated by the decaying larvæ, but the *cause* of it, generated in the fluids of the bees themselves, and possibly transmitted even by the eggs of an infected queen. He says :

"Foul brood, then, is a *bacillus* disease ; and in these days, when the 'germ theory' is the question of questions amongst pathologists and physiologists, it is extremely interesting for us to note that science has lately shown that different species of *bacilli* also cause consumption, cholera, typhoid, leprosy, and many other diseases afflicting the human family ; whilst amongst animals glanders, splenic fever, sepitcæmia, etc., arise from a similar cause."

The particular species found in the hive, and which causes the disease, has been named by Mr. Cheshire and the late lamented secretary of the British Bee-keepers' Association, *bacillus alvei,* by which name the disease is now more correctly designated than by its old one.

BACILLUS ALVEI UNDER THE MICROSCOPE.

Mr. Cheshire's description (illustrated) of the disease, as seen by him under a very powerful microscope, is very interesting ; he says :

"Taking a small quantity of the juices of a healthy grub, and spreading it out under a thin glass under the microscope, one is presented with such an appearance as is seen in Fig. 120 ; fat globules are numerous, while blood-discs abound, and everywhere may be noticed tiny particles which are constantly slowly dancing with what are called Brownian movements.   But if a speck of coffee-coloured, foul broody matter be similarly treated, we find neither fat globules, blood discs, nor molecular base, but observe the field crowded with very small ovoid bodies, as we have them represented in Fig. 121.   These are the micrococci of Schonfield ; but if this substance be stained according to the modern plan of Weigert and Koch, and then carefully examined, in all probability we shall discover, associated with the ovoid bodies, a

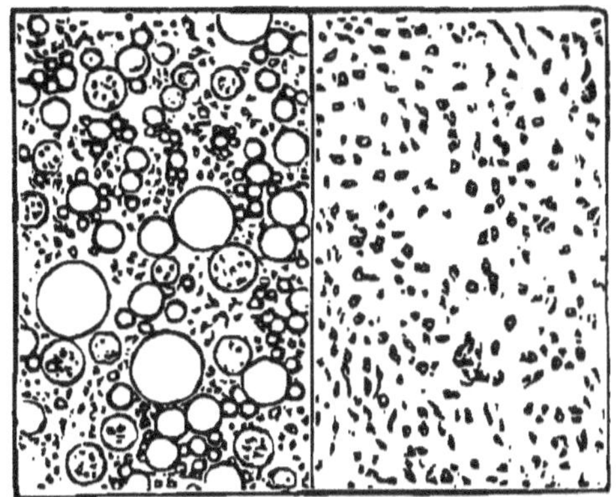

Fig. 120.—HEALTHY JUICES Fig. 121.—BACILLUS ALVEI
OF LARVA. (last stage).

very few other organisms, longer and rod-shaped, while we notice that the so-called micrococci are neither round nor dumb-bell like, but oval or boat-shaped.   This led me at once to suspect an error, and further searching showed me if, instead of coffee-coloured matter, such as that usually sent for microscopic examination, the body of a grub, dead, but in a fresher condition, were taken, the number of the rod-like bodies very considerably increased, while that of the ovoid ones diminished, as seen in Fig. 122.   My own inoculated stock—inoculated for experimental purposes—was cured, and gave me no material ; but soon I obtained a comb from a suffering hive, and then had the opportunity of expressing the juices from a death-stricken larva.   These, when examined under a power of 600 diameters, and carefully illuminated, were seen, to my great delight, to be full of active rods, swimming backwards and forwards, and worming their way between the degenerate blood discs and fat globules, as represented in Fig. 122, while here and there were long strings of them, the lepothrix form previously referred to."

As to the means by which the disease is propagated, Mr. Cheshire has no hesitation in saying—

" That the popular idea that honey is the means by which it is carried from hive to hive, and that mainly through robbing, is so far an error, that only occasionally and casually can honey convey it from stock to stock."

And after giving his reasons at full length, he concludes this part of his subject by saying :

"Although I would not dogmatise, my strong opinion is, that commonly neither honey nor pollen carries the disease, but that the feet and antennæ of the bees usually do. I also think it probable that, occasionally at least, nurse-bees infected bring the disease germs to

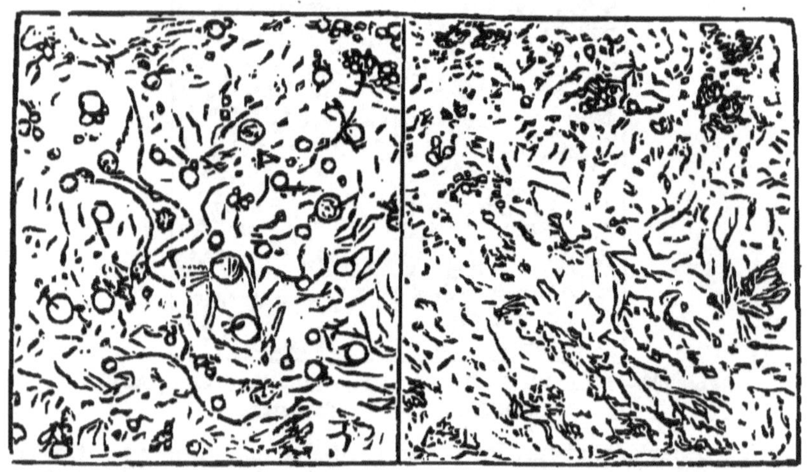

Fig. 122.—B. ALVEI (early stage).   Fig. 123.—B. ALVEI (late stage).

the mouth in feeding the larvæ, and then turning foragers, leave a germ or germs in the nectary of a flower, which, visited by another bee, becomes the means of infection to it ; the malady is thus carried by adult bees into other and perhaps somewhat distant apiaries. Balancing all the probabilities, it would appear that most generally the adult bee takes the disease, and then carries it directly or indirectly to the brood."

THE CHESHIRE CURE.

As to the method of cure, Mr. Cheshire is entirely opposed to the use of salicylic acid and of borax. He considers the spraying does more harm than good ; that any beneficial results from the acid can only be obtained by its use mixed with food ; that even in that way it requires to be used with great discrimination as to the dose, because salicylic acid is likely to act as a cumulative poison, and that borax, itself a drug nauseous to the bees, is only of use as a solvent for the salicylic acid, if used for spraying; and when mixed with it in the food, only reduces its curative effect, and renders the treatment somewhat dangerous.

Mr. Cheshire's principle is, that as the disease originates with the adult bee, and is conveyed by it to the larvæ, so the fungicide intended to cure it must be administered in food to the adult bee, and be transmitted in the same way to the larvæ. On the suggestion, as he informs us, of Mr. Robert Sproule, he tried experiments with phenol as a fungicide, and has come to the conclusion that when used in the proper manner, which he has taken great pains to arrive at, it is a thoroughly effective cure for *bacillus alvei*. Phenol is generally known as "pure carbolic acid," but care must be taken to obtain the really pure article—absolute phenol, sold as Calvert's phenol No. 1. Mr. Cheshire warns us to be cautious about this. "Carbolic acid is an impure phenol, and is useless. It contains creosote and creosols, and bees abhor it."

Having obtained the pure article, the next point is to know how to administer it. Mixed with syrup or honey, in the proportion of one part in two hundred, it will be refused by the bees altogether. In the proportion of 1 to 400 it might be administered to a sound stock without any injurious consequences; but Mr. Cheshire found that "1 to 500 dissipated foul brood quickly, even while honey was coming in," and that "1 in 750 appeared enough when it was not being gathered."

These, then, are established as the correct quantities under the circumstances mentioned. If honey be coming in, it will be useless to place the medicated syrup in a food bottle in the hive, as the bees will not touch it; but he recommends to

"Take out the brood combs, and to pour from a bottle having a dropping-tube loosely placed in its neck the medicated syrup into those cells immediately around and over the brood, and the bees will use a curative quantity of phenol. The syrup is best poured in by holding the comb at an inclination of ordinary writing, not by placing it on its side."

By pursuing this course of treatment for some time, the adult bees should become all sound, the diseased larvæ be removed by them from the cells, and all the new brood be brought out in a healthy condition.　Mr. Cheshire says at the conclusion of his paper :

"I could take an apiary at the beginning of March with every stock diseased, and by May 1st, with but very little labour, deliver it up clean and strong, as strong as though the disease had never appeared."

After the reading of the paper the late editor of the *British Bee Journal*, in company with many prominent bee-keepers, examined a stock that had been treated by Mr. Cheshire for the cure of bacillus alvei, and which had been placed in the Health Exhibition.　This stock had been sent to Mr. Cheshire five weeks before, with seven of its combs " affected with foul brood in its most virulent form, being a mass of corruption." These seven combs, with two others, when examined, were perfectly clean and had

" literally not one single cell affected.　Whole sheets of brood in all stages were to be seen quite healthy ; young bees hatching out and eggs being laid in the vacated cells.　This wonderful change had been effected by the bees alone, aided simply by the administration of the medicated food."

The simple manner of preparing a syrup of the strength 1 to 500 is to take one ounce of the phenol with three ounces of warm water, to thoroughly dissolve it, and then mix with thirty-one pounds of sugar syrup or of diluted honey.

Writing later in the year, Mr. Cheshire announced that he found the plan of pouring the medicated syrup into the combs to answer admirably in spring and summer ; but if the cure has to be effected in autumn, he recommends to give the phenol in a cake made of sugar and pea-flour, placed on the top of the frames, under the mat, as the bees are not disposed at that season to clean out the combs, as they would in spring and summer, and the use of the liquid syrup is likely to start robbing.　The cakes he prepares "in the usual way ; but after removal from the fire, during the stirring and cooling process, painstakingly mix with it one-fifth ounce of phenol to each seven pounds of sugar."

THE SALICYLIC ACID REMEDY.

Notwithstanding what has been said against the use of salicylic acid in cases of *bacillus alvei*, many reports of cures having been effected by it have from time to time reached the bee journals. Mr. A. E. Bonney, who has been very successful in eradicating the disease from his apiary by Muth's method, described the plan in a very able paper read by him before the South Australian Bee Keepers' Association, on January 5th, 1885, in the following words :—

"Remove from its stand the hive containing the diseased colony, and put in its place a clean hive with starters of foundation in all the frames. Brush the bees into the clean hive, and feed them with honey or sugar syrup, adding to every quart of food an ounce of the following mixture, namely, sixteen grains salicylic acid, sixteen grains soda borax, one ounce water. This feeding should be kept up for about ten days. The diseased combs should be cut from the frames and burned up, and the hive and frames scraped and well scrubbed with carbolic soap and water. Calvert's medical soap, containing 20 per cent. of acid, is most effective for this purpose. That is the whole operation; but when there are more hives than one, certain precautions must be taken, or else the bee-keeper will discover that in curing this one he has spread the disease into others. The best time for this work is the early morning, and everything should be prepared the previous evening. The entrances to all adjacent hives should be closed with perforated zinc, which must not be removed until after the operation is completed. Before removing the zinc, the alighting-boards and fronts of the hives should be washed over with a solution of 1 of phenol in 200 of water. While the disease exists in the apiary the apiarist should make a practice of always washing his hands, smoker, etc., with the above solution before going from one hive to another. A small piece of sponge is convenient for this purpose. After treating a bad case I have my clothes washed before wearing them again amongst healthy hives. Carbolic acid No. 5, well mixed with water, in the proportion of one ounce of acid to two quarts of water, should be sprinkled on the ground where the diseased hive stood. This will destroy the germs in any foul broody matter which may have been carried out of the hive. All honey extracted from diseased combs should be thoroughly boiled with one-fourth of its quantity of water before being fed back to the bees. . . . . Thirteen of my hives have been cured by Mr. Muth's method, and it is to me a source of much pleasure to go from hive to hive and see large sheets of healthy brood in all stages, knowing that not long ago some of these colonies were dying out in rottenness. . . . . Mr. Stevens, of Goodwood, by the persistent use of this cure got rid of foul brood last year, and has not since been troubled with it."

Writing in June, 1885, he says :

"It is quite possible that in many places in Australia, where the honey supply is intermittent, or in the more rigorous climate of New Zealand, the Cheshire cure may be the best to adopt, but it certainly is not in this locality (Adelaide)."

Mr. R. Harding, of Hawke's Bay, New Zealand, also reported favourably of Muth's remedy, and it seems quite possible that it may be the most suitable one for districts like that around Adelaide, where more or less honey is gathered throughout the greater part of the year.

I believe it to be a good plan to give all spare hives, frames, and bottom-boards a brush over with a solution of carbolic acid at the end of the season, before putting them away for winter ; it can do no harm, but may do a great deal of good.

BACILLUS GAYTONI.

In the further pursuance of his investigations, Mr. Cheshire has discovered that there are other species of *bacilli* which affect bees with diseases different altogether in their symptoms from those of foul brood. Through the close observation of Miss Gayton, a well-known successful bee-keeper, who forwarded to him for examination some bees of the glossy black and hairless appearance which has heretofore been very generally supposed to indicate " old robbers," he found them filled with a *bacillus* altogether different from the *bacillus alvei*, and to which he has given the name of *bacillus Gaytoni*. He says " it is a very mild offender beside the *bacillus alvei*, but it will be very interesting to note whether it succumbs to the same treatment."

OTHER DISEASES OF BEES.

It is probable, as I previously intimated, that we may be still further indebted to Mr. Cheshire for a knowledge of other diseases which have heretofore puzzled the bee-keeper. There is one sort of symptom which has attracted attention of late in America and here, generally alluded to as that of " trembling bees," about which we are still in the dark. Mr. Cheshire says, in a paper which appeared in the *British Bee Journal* of September, 1884 :

" During the last two months I have been able to make out no less than five, or possibly six, distinct disorders arising from that number of specifically different germs, all of which will require prolonged attention, if anything very definite is to be arrived at respecting them. In addition I suspect strongly that true dysentery will also turn out to be an infectious disorder ; but since specimens fail me, the question must remain, as far as I am concerned at least, till another season." Then, after describing the *bacillus Gaytoni*, he adds : " With regard to the other germs found, my knowledge is at present so slender, that I must advance nothing beyond the discovery of an enormously large *bacillus* which takes what is called the *zooglea* form—two, or possibly three, very minute kinds of bacilli and a micrococcus. The micrococcus will most probably turn out to be a putrefactive kind accidentally present."

ARRENOTOKIA.

This name is given to a certain defective condition of queens· Mr. Cheshire, in carrying out his investigations, required a number of queens, which have been furnished him by different bee-keepers. Amongst them he discovered two drone-breeders, each with its spermatheca " furnished completely with spermatozoa." In explanation he says :

"The name 'arrenotokia,' applied by Leuckart in 1857 to a case similar to the one we are considering, indicates that the queen, as distinguished from a normal drone breeder, is fully furnished with spermatozoa, and is yet incapable of fertilising her eggs. The possible causes are various, since the mechanism, so wondrously delicate and complex, which pays out the spermatozoa as they may be required, and which I explained a few months since, may fail in its muscles or nerves, or even the spermatozoa themselves may be defective, as actually appears to be the case in this instance."

In speculating upon the probable cause of defective spermatozoa, he asks :

"Can the lateness of the season at which this queen was hatched in any way explain the matter ? Drones, at the date given (October), are normally gone ; but the progeny of fertile workers are then discoverable in the prime of youth, as well as old drones permitted to live in queenless stocks. Speculation is easy, and the possibility suggests itself, that the defective spermatozoa owe their faults to the fact that old or abnormal drones yielded them." The first case he examined he thinks " was probably due to paralysis of some of the muscles attached to the spermathecal valve ;" and further says : " This production of drones only has been artificially produced by pinching the extremity of the abdomen, so that the last ganglion is injured."

I have myself known young queens—to all appearance perfectly healthy—after laying worker eggs for a time, suddenly turn to drone-breeders in some unaccountable manner. It certainly would be interesting to know the cause of such a change ; —injury to the abdomen would be almost certain to cause it.

SPRAY DIFFUSER.

A very handy appliance is shown in the following engraving; it is useful for spraying combs, when necessary, and is some-

Fig. 124.—SPRAY DIFFUSER.

times used for sprinkling bees with scented syrup to prevent fighting when uniting two or more colonies. There are some of a different construction to that shown, any of which can be purchased for a few shillings at most chemists.

RULE.

PROFITABLE BEE-KEEPING DEPENDS UPON STRONG COLONIES. No BEES NO HONEY.

CHAPTER XVII.

ENEMIES OF BEES.

THE Australasian colonies are certainly favoured with an exemption from many of the natural enemies of the bee, which are very troublesome in some of the older bee-keeping countries.

AUSTRALASIAN EXEMPTIONS.

There is here a total absence of wasps, hornets, and toads; and I can answer for it that in New Zealand at least the ants are harmless, and the wax-moth not of a very formidable character. The principal enemies here are spiders, mice, the bee-hawk (*libellula*), and the bee or wax moth. There are also other insects which occasionally enter weak hives, viz., wood-lice, earwigs, ants, and beetles; but it is thought these latter only enter the hives for shelter rather than plunder. I am not aware that there are any birds here that attack bees, although I have been keeping a sharp look-out, and I have come to the conclusion that we have little to fear from that quarter.

SPIDERS.

These insects, if opportunity offers, spread their webs in front and around the hives to capture unwary bees; therefore the fewer corners or angles about them the better. Porches to hives, so much admired by some amateur bee-keepers, are not only useless, but make a very convenient place for spiders to carry on their work of destruction. The same may be said of bee-sheds and other unnecessary fixings.*

* In the *American Bee Journal* for August, 1885, there is an interesting communication from the Rev. L. Langstroth, showing that in his experience he has found "spiders one of the bee-keeper's best friends, to preserve empty combs from the ravages of the bee-moth." He even recommends *rearing spiders;* that is, to place the so-called "spider bags," or webs, full of eggs, in the boxes in which the empty combs are kept over winter. This

MICE.

These little pests are sometimes as great a nuisance in the apiary as they are in domestic dwellings. They destroy the combs, create disgusting smells, and when they gain a footing in a hive it usually ends in the destruction of the colony. It is a good plan to keep a cat or two about the apiary, especially if it should be some distance from the dwelling-house. If the hives, however, are made in accordance with the instructions given in Chapter VI., it will be almost impossible for these little wretches to enter, as the entrance will not be large enough to admit them.

ANTS.

In some countries ants are the cause of much annoyance to bees and bee-keepers. In the warmer parts of the Australian colonies some of the larger species are numerous, and there they become very troublesome. In South Australia Mr. Bonney says, "In some districts the hives must be protected from them (the ants), otherwise they would destroy the bees. I find Italians protect themselves against ants much better than the blacks." In New South Wales I am informed that "ants are sometimes troublesome;" and Mr. David Gloss, writing from Victoria, says he never found or heard of ants being troublesome in the cool districts (Kyneton and Ballarat), but they may be in the warm districts. The smaller species, such as I have seen in New Zealand, do but little damage; and as it is a rare occurrence to find them in hives that are properly managed, they are scarcely worth taking into consideration. But where the larger kinds exist, I would advise the bee-keeper to keep a sharp look-out for their nests, and destroy them. When in the hive, they usually congregate above the mat in the upper part, where they may be swept off and destroyed.

Some writers recommend placing poisoned saccharine matter or meat in vessels well protected from the bees by a covering of wire-cloth, and putting these near the hives where the ants are numerous. Large numbers may be trapped in this way. A narrow strip of fur tacked completely round the under edge

hint, like everything coming from the same source, is worthy of all attention. But though in this way we may indeed convert the spider into a friend or ally of the *bee-keeper*, we must continue to class it amongst the "enemies of bees, and to banish it from the vicinity of working hives.

of the bottom-board—hairy side down—affords a good protection against these insects entering the hive. If this practice is to be followed, the cross-pieces or stands below the bottom-board would need to be cut shorter to allow room for the strip of fur to pass them, and nothing in the shape of grass or weeds must be allowed to touch the hives, or the fur would be useless.

BEE-HAWK (LIBELLULA).

This is a handsome four-winged insect (Fig. 125), commonly known as the dragon-fly. It captures the bees while on the wing, swooping down on them with great swiftness, and no

Fig. 125.—BEE-HAWK (*Libellula*).

doubt it is from this circumstance it derives its name. I once caught a splendid specimen, measuring fully three inches in length ; it had just caught a bee in its formidable jaws and settled close to where I was standing. It is easily scared or frightened away, and may be killed on the wing by using a whip. It is found in greatest numbers near still water, in which it lays its eggs and the larvæ are reared.

THE BEE OR WAX MOTH.

I believe there will be found to be many separate species or varieties amongst the moths which infest beehives in different countries. Langstroth, following Linnæus and Réaumur, speaks of the *tinea cereana* and *tinea mellonella;* Cook, following Fabricius, calls the bee-moth *galleria cereana*, and says it belongs to the family of snout-moths, *Pyralidæ*, and that "its members are very readily recognised by their usually long palpi, the so-called snouts."

The moth has been found a serious evil in some of the Australian colonies, at least previous to the introduction of the Italian bee. Mr. Fullwood states that the race of black bees was nearly exterminated in Queensland by the moths, but that when Ligurians were imported they soon defended themselves, and obtained the mastery; and Mr. E. Palmer, of New South Wales, says, "The bee-moth is the great scourge of the wild and cultivated bees, and the only serious obstacle to successful bee farming of which I have, during a series of years, had any experience." I believe the Australian moth to be identical with the American one. In New Zealand, the moth found in hives is of a smaller species, and is not likely to give much trouble in well-kept hives.

TINEA CEREANA.

The following description of this moth in America is taken from Dr. Harris's report on the insects of Massachusetts, as quoted in Langstroth's work :—

Fig. 126.—BEE-MOTH (*Tineæ cereana*).

"Very few of the *tineæ* exceed or even equal it in size. In its adult state it is a winged moth, or miller, measuring, from the head to the tip of the closed wing, from five-eighths to three-quarters of an inch in length, and its wings expand from an inch and one-tenth to one inch and four-tenths. The four wings shut together flatly on the top of the back, slope steeply downwards at the sides, and are turned up at the ends somewhat like the tail of a fowl. The female is much larger than the male, and dark coloured. There are two broods of this insect in the course of a year. Some winged moths of the first brood begin to appear towards the end of April or early in May—earlier or later, according to climate and season. Those of the second brood are most abundant in August; but some may be found between these periods, and even much later."

The bee-moth I have seen here in New Zealand does not exceed a quarter of an inch in length, and, as far as my experience teaches me, is identical with the clothes moth, and will be easily recognised by most of my readers.

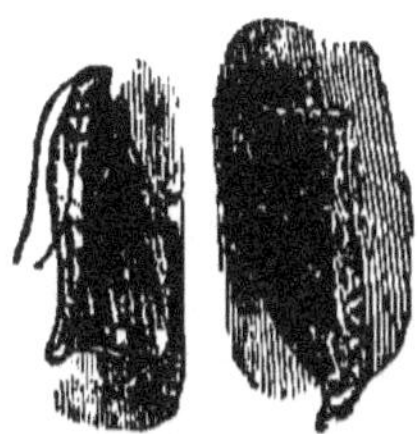

Fig. 127.—MALE AND FEMALE BEE-MOTH.

The engravings shown in this chapter represent the bee-moth and its ravages described by Langstroth; but as its habits are the same as those in Australasia, they convey all the information necessary to put the apiarist on his guard.

DAMAGE TO COMBS.

In warm evenings the female moth may be often seen about the hives, seeking for a place to deposit eggs, which she usually does in any cracks or crevices about it. These eggs are white, round, and very small. In a short time they hatch into dirty grey-looking caterpillars, and it is in this stage they commit their ravages by destroying the combs.

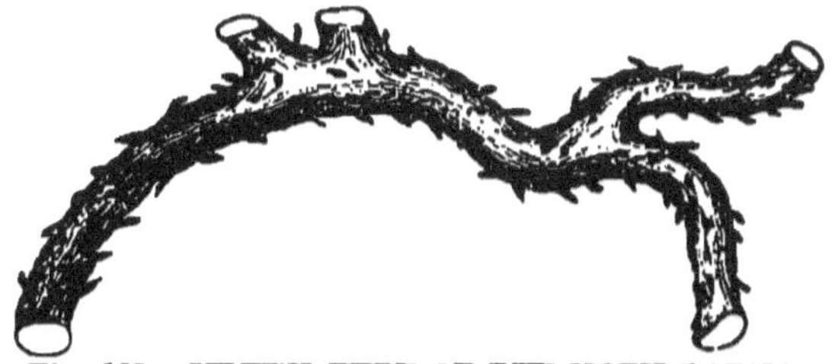

Fig. 128.—SILKEN TUBE OF BEE-MOTH LARVA.

As a defence from the attacks of the bees, each of the larvæ envelopes itself in a silken tube (Fig. 128), which they extend

T

through the combs as they advance, often detaching them and and causing them to fall together. These tubes will, in time, extend through the whole of the combs, killing the larvæ of the bees, and ultimately destroy the whole colony.

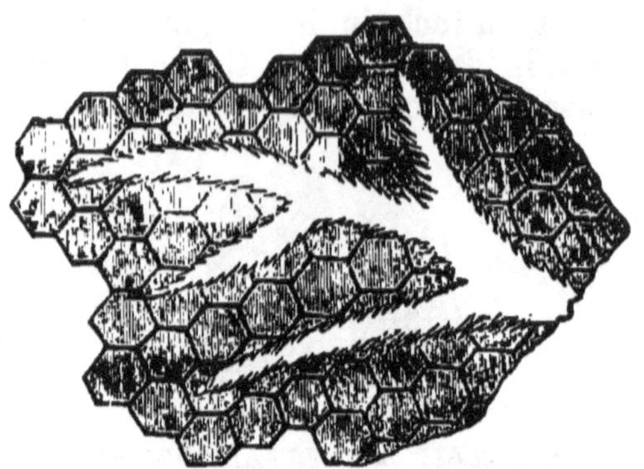

Fig. 129.—SILKEN TUBE IN COMB.

Cook says :

"In three or four weeks the larvæ are full grown (Fig. 130). . . . . . They now spin their cocoons, either in some crevice about the hive, or, if very numerous, singly (*a*, Fig. 131), or in clusters (*b*, Fig. 131), or on

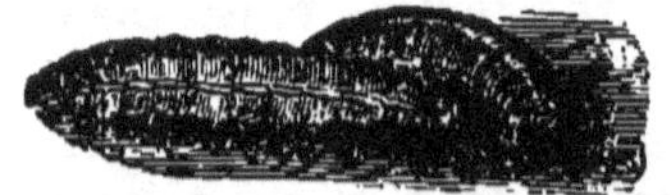

Fig. 130.—LARVÆ OF BEE-MOTH

the comb, or even in the drone cells (*c*, Fig. 131), in which they become pupæ ; and in two weeks—even less sometimes—during the extreme heat of summer, the moths again appear. In winter they may remain as pupæ for months."

And Bevan remarks :

"Wax moths are remarkably active in their movements. 'They are,' says Réaumur, 'the most nimble-footed creatures I know.' And if the approach to the apiary be observed on a moonlight evening, the moths will be found flying or running round the hives, watching an opportunity to enter; whilst the bees that have to guard the entrances against their intrusion will be seen acting as vigilant sentinels, performing continual rounds near this important post, extending their antennæ to the utmost, and moving them to the right and to the left alternately. Woe to the unfortunate moth that comes within their reach! 'It is curious,' says Huber, 'to observe how artfully the moth knows how to profit by the disadvantage of the bees (who require

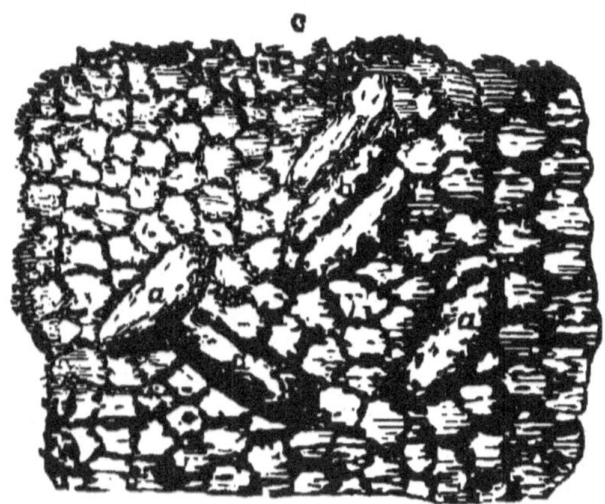

Fig. 131.—COMB DESTROYED BY BEE-MOTH LARVÆ.

much light for seeing objects), and the precautions taken by the latter in reconnoitring and expelling so dangerous an enemy.'"

REMEDIES.

The most effectual remedies against all attacks of insects, including the moth, is to *keep all the colonies strong*, and by using movable frame hives an enemy can be easily dislodged. The bottom-board of a suspected hive should be occasionally examined; for if the bee-moth has gained a footing in the hive, the eggs and larvæ may be often found upon it, when of course they must be at once removed and destroyed. If, however, any colony should have become, through neglect, hopelessly injured by these moths, then the bees should be transferred to another hive, and the old one, with its combs, fumigated with sulphur.

It is now an established fact, that wherever Italian bees are introduced, there this terrible scourge ceases to exist, and therefore this one feature alone is enough to justify the introduction of Italian bees in the place of blacks.

BEE-MITE.

My attention was called last May (1885), by Mr. A. Bow, of Hokianga, to an insect he had found in his hives in great numbers, which, he said, resembled cheese-mites. In a foot-note to his letter, which was published in the late *Journal*, I suggested the probability of their having been bred in the pollen, but could not say what insects they were. Since then I have examined under a microscope several lots of pollen taken from combs that had been stowed away for winter, and in each case I discovered these mites. On describing their appearance to Mr. T. F. Cheeseman, curator of the Auckland Museum, he said he had no doubt they were mites, but whether parasitical or not he could not say. Professor Cook says : " The bee-mite is very small, hardly more than five milimeters (one-fiftieth of an inch) long. The female is slightly larger than the male, and somewhat transparent. The colour is black, though the legs and more transparent areas of the females appear yellowish." Those I examined were of a greyish colour, and semitransparent ; the full-grown insects possessing eight legs, and some having long bristly-looking hairs scattered over their bodies.

I have no doubt that where very numerous they might become very troublesome, and do considerable injury to the bees, though I doubt much whether they would be able to gain a footing in strong colonies. I would suggest as a remedy making an examination of the pollen in the hives where they are found, and removing all that is infested by them ; also to shift the bees into clean hives, and wash the bottom-boards with a solution of carbolic acid. Probably spraying the infested combs with a solution of phenol would rid them of the pest.

A CAUTION TO IMPORTERS OF BEES.

I have it from good authority that the bee-moth mentioned by Mr. Fullwood as having worked such sad havoc amongst the black bees in Queensland was accidentally introduced into that country in the following manner :—In one of the earlier attempts to introduce Italian bees (previous to the one made by Mr. Fullwood in 1880) an imported hive with bees had been landed at the apiary of a bee-keeper in the vicinity of Brisbane. This hive, it appears, not only contained bees, but also their greatest enemy, the wax moth peculiar to the country from whence the bees had come, in all stages of growth, from the larva to the perfect insect. The bee-keeper, in blissful ignorance of the fact, opened the hive, and before he could prevent it some of the moths took flight, and thus it was that this terrible enemy came to be introduced into Queensland. It is easy to understand that in so favourable a climate it would take but a very short time for the country to be over-run with the pest, and it is quite likely that ere this the moth has spread over the greater part of the warm districts of Australia.

Such a circumstance nearly occurred to myself. Some of the boxes containing queens imported direct from Italy, that came to Matamata in the latter part of 1884, were crowded with eggs and larvæ of an enormous bee-moth, as well as the perfect insect. Some of the larvæ measured over an inch in length. As soon as I made the discovery that the boxes contained such unwelcome intruders, I opened them very cautiously, shifted the queens and bees to clean boxes, and burnt the others, with their combs, without delay. The boxes were opened in a room, so that if any moths had flown, they could have been easily captured.

I would therefore warn all who may hereafter import bees to be very cautious in opening hives or boxes, and to see that no insects escape, and to examine the combs, or what is better, put the bees into a clean hive, on clean combs, and make a fire with those they came in.

FUMIGATING COMBS.

I have previously referred to fumigating combs to destroy the bee-moth larvæ, or any other insects that may have harboured in them. The following is the method :—Hang them in a small close room (or if only a few, a large close box will do as well), burn one pound of sulphur to every one hundred cubic feet contained in the room or box. To burn it, get an iron pot, put some ashes in the bottom, with hot embers, and pour on the sulphur ; shut the door of the room, or close up the box, for two or three days.

When combs are stowed away during the winter months, they should be looked over occasionally, and if necessary fumigated again as above described. Combs containing honey stowed away should also be fumigated, if eggs or worms of bee-moth are detected in them.

CHAPTER XVIII.

BEE FORAGE.

THE vegetable origin of both honey and pollen being understood, and the facts being borne in mind, that bees can only collect the substances they require from plants and trees that are in blossom, and that the quality, colour, and flavour of the honey vary very considerably according to the source from which it is obtained, it will be easily seen how necessary it is for the bee-keeper to make himself acquainted with the resources within his reach ; and not only with the names of the different plants and trees, but with the character of the honey or pollen afforded by each, and with the usual periods of their coming into and remaining in blossom. These circumstances, as well as their bearing upon the practical operations of the apiary, vary very importantly in different countries and climates. In high latitudes, where the winters are severe, and the bees confined to their hives for four or five months of the year, it is clear that even if there were plants which blossomed in winter, they would be useless to the bees, and the bees equally useless to them. In temperate climates, however, where the bees can work more or less freely all through the winter, though only producing surplus honey in summer, it becomes a matter of great importance that there shall be a variety of nectar and pollen-bearing plants coming into blossom at different periods of the year, even in the depths of winter, in order to afford forage at all times, and keep up the health and strength of the bees without artificial feeding. And in more tropical climates, where the bees can not only work, but also breed, swarm, and store surplus honey, more or less, all the year round, such a variety in the habits of the plants and trees as will afford a succession of bloom at all seasons becomes essential to the prosperity of the apiary. An intelligent apiarist will therefore carefully study the peculiar circumstances of his immediate neighbourhood, see what he has to expect from the natural or existing flora of his district, and consider whether he can improve his position by raising any varieties of plants or trees suited to supply the wants of his bees at times when the existing resources would be likely to fail.

ORDINARY SOURCES.

Some of the best and most generally known sorts of bee forage are of a character which plainly invites the establishment of apiaries in their neighbourhoods, wherever they are found extensively, such as clover, heather, wild thyme or wild sage, large orchards with their abundance of fruit blossoms, or forest trees of some particular descriptions. The heaths of Scotland and Ireland, the wild thyme of Greece and the wild sage of California, are not to be sought for in New Zealand or Australia; but, on the other hand, the clovers and the fruit trees of all kinds flourish so well in all parts of these countries as to hold out the strongest inducements for their cultivation, and the great majority of the bush or forest trees are peculiarly valuable, not only on account of their honey-bearing qualities, but also because they vary so much in their times of blossoming, that some of them are available at almost every season of the year.

NATIVE FLORA OF NEW ZEALAND.

Most of the forest trees indigenous to New Zealand are honey-bearing, some of them remarkably so. The following are the native, English, and botanical names of the principal ones :—

1. Rewa-rewa .. Honeysuckle, *Knightia excelsa.*
2. Pohutukawa ... Christmas tree, *Metrosideros tomentosa.*
3. Rata ... ... Oak-elm, *Metrosideros robusta.*
4. Hinau ... ... *Elaeocarpus hinau.*
5. Kahikatea ... White pine, *Podocarpus dacrydioides.*
6. Matai ... ... Black pine, *Podocarpus spicata.*
7. Miro ... ... *Podocarpus ferruginea.*
8. Puriri .. ... New Zealand oak, or teak, *Vitex littoralis.*
9. Kohekohe ... Cedar, *Hartighsea spectabilis.*
10. Tawai ... ... *Leiospermum racemosum.*
11. Kowai ... ... *Edwardsia microphylla.*
12. Nikau ... ... Native palm, *Areca sapida.*

In the bush are also generally found the native fuchsia, wild clematis, rata creeper (white and red), karaka (laurel, *corinocarpus*), koromiko (*veronica*), and many other flowering shrubs or creepers ; in the open, both on hillsides and swampy places, cabbage trees (*Cordyline Australis* and *Dracœna Australis*), and the New Zealand flax (*Phormium tenax*), and in the fern-lands generally, manuka (tea-tree, *Leptospermum scoparium*), rawiri, (*Leptospermum ericoides*), and tutu (*Coriaria sarmentosa*).

Most of the forest trees, especially the first five in the foregoing list, afford, in seasons when they blossom freely, not only excellent quality, but also a great quantity of honey. They do not, however, blossom equally well every year ; some of them even do not blossom at all in some seasons. The same may be said of the flax, which, however, in favourable seasons, exudes such quantities of nectar, or honey-dew, that it may be collected by hand. Dieffenbach observed, when examining the Taranaki district in 1840, large patches of land covered with *Phormium tenax* of great size. "The leaves in many instances were twelve, and the flower-stalks twenty, feet long ; their flowers contain a kind of sweet liquid in considerable quantities, the extraction of which forms a favourite occupation among the New Zealand children." The honey obtained from it is generally so thick as to be difficult to extract. The periods of blossoming of all these trees and plants varies considerably with their geographical situation ; and it is very desirable that the particulars of their habits in that respect, as well as the character of the honey afforded by each, should be carefully noted by bee-keepers in all the different parts of the country. Mr. J. Blair, of the Great Barrier Island, about fifty miles north of Auckland, reports as follows with regard to the native flora of his district :—

"Here the cabbage tree blooms in October—November; flax, November—December. Tea-tree, it is possible to get a specimen of bloom all the year, but for practical purposes it blooms from the last week in March to the end of December, and the bees work on it all the time. My bees have been working on it now a fortnight, but it only gives honey in quantity from the beginning of October to the end of December, and during that time they gather honey only. From March to October they gather both honey and pollen. From October to December any one can both taste and see the honey in the blossom. Rewa rewa blossoms from September 20th to December 20th : any one can lick the thick honey off it with the tongue. When the bees get properly started on these, they don't take notice of any honey lying about.

"In the middle of November the tree rata and pohutukawa begins, and lasts to the middle of February. The white rata (creeper) begins in December, and lasts till April; red rata (creeper) begins in January, and lasts till May; koromiko begins in January, and lasts till June; nikau begins in February; puriri begins in March, and lasts till November, although it is easy to get a specimen all the year; hohere (it has a thick stringy bark), kohekohe cedar, mangeo, and another I do not know the name of, all bloom in April, May, and June. These are the principal trees we get honey from here. There are plenty of other trees that bees work upon, but they either give honey in small quantities, or we have not got them in sufficient number for the bees to store honey from."

In the Thames district, about as far south of Auckland as the Barrier Island is north of it, my own experience is as follows : Kowai commences to blossom in September, as also tauro and mahoi—the first lasts four weeks, the second sixteen, and the last eight ; hinau in November, lasting four weeks ; rata and flax commence in December, and blossom four or five weeks; cabbage-palm, beginning of December, lasting four weeks. Besides those enumerated, there are other native trees and shrubs which blossom between October and March ; amongst the best for honey are pohutukawa, kahikatea, puriri, matai, tawai, tariri, miro, karaka, native fuchsia, and nikau. Tea-tree blossoms in September, but yields no honey in the Thames district.

The honey from most of the native flora of New Zealand is of a first-class quality, though not equal to that obtained from white clover. Some of it granulates very slowly ; I have kept samples nearly all through the winter without granulating.

NATIVE FLORA OF AUSTRALIA.

All the species of those two great families of trees so peculiar to Australia, EUCALYPTUS and ACACIA, are good for bee forage, yielding both nectar and pollen in abundance, and, what is of especial importance in a climate like that of most of the Australasian colonies, where the bees can gather surplus honey nearly all the year round, they seem as if specially designed to supplement each other; the eucalypti blooming, as a rule, in the summer half, and the acacias in the winter half of the year.

NEW SOUTH WALES.

Mr. Thomas E. Willis, to whose kindness I am indebted for much information with reference to bee culture in New South Wales, reports on the bee forage of that colony as follows :—

"The honey season on the coast takes place in spring, whilst inland the summer is the best season ; this is owing in a great measure to the vast forests of gum trees being then in full bloom, whilst on the coast there is a profusion of winter and spring flowers and golden wattles, the latter being a favourite with the bees. The principal bush trees about Sydney are, spotted gum,* black butt,† iron bark,‡ woolly butt, and honeysuckle : from all these the bees derive sustenance   I have also noticed them on the large flowers of the dogwood, a kind of dwarf gum. In Baron Müller's Botanic Teachings, he mentions the honey eucalyptus (*E. melliodora*) as a favourite with bees, as its blossoms exude much nectar, and also the Cape honeysuckle, or *Protea mellifera*. The *Wigandea caracasana*, which keeps in flower from August till November, is an especial favourite with the bees ; it has large leaves, about eighteen inches by twelve, and bears numerous spikes of blue flowers ; it is an ornamental shrub, and thrives well in gardens about Sydney. It is a native of America, tropical and sub-tropical."

SOUTH AUSTRALIA.

With reference to the district immediately around Adelaide, Mr. A. E. Bonney informs me that the chief sources of the honey crop are the blossoms of the Eucalypti :—

"These trees, of which there is a great variety, flower on the plains from December to the end of March, and on the hills from October to March. The quality of the honey is very fine. Perhaps the most valuable is *E. rostrata* (red gum), which produces a copious flow of honey during January and February. In spring the dandelions, or Cape marigold, yield a large harvest, but the honey is of poor quality. In favoured localities, such as Mount Barker, the honey season lasts all the year round ; but as a rule, from January 1st to February 28th may be called the honey season. At Mount Barker, Mr. Justice Boucant has been getting surplus honey during the past eighteen months; also one of my colonies has been filling section boxes for eleven months, only stopping for four weeks whilst I was treating them for foul-brood."

Acacias, or wattles, are also plentiful here as elsewhere in Australia. The black wattle (*A. decurrens*) and the coast wattle are described as furnishing very good bee forage on the south-east coast. Mr. Bonney, in writing to the *New Zealand and*

* *E. goniocalyx.*　　† *E. pilularis.*　　‡ *E. leucoxylon.*

Australian Bee Journal, about one species known here in New Zealand as common wattle, says :

"The botanical name of the acacia referred to by T. J. M., in 'Notes about Bee Forage' (Part I.), is *Acacia lophantha.* I can fully endorse all he says about the usefulness of this tree as a honey producer; but in South Australia it is objectionable, because of its rank growth and very unpleasant smell. At the end of the second part of his ' Notes on Bee Forage,' T. J. M. remarks that none of the red or blue gums in his neighbourhood are more than seven or eight years old, and scarcely any have as yet commenced to bear blossoms. This is an objection to most of the eucalypti; it is so long before they flower, and when they do, little honey appears to be secreted until the trees attain a large size. *Eucalyptus carophylla* (the red gum of Western Australia) is an exception, and is a tree that I would like to bring under the notice of all bee-keepers throughout Australia and New Zealand. According to Müller, this tree, in its native land, grows to a height of 150 feet, with stems occasionally ten feet in diameter. It is the most ornamental gum I have seen; the foliage is denser and more horizontal than that of any other species we know of, and the tree is readily distinguished by its large seed-pods. Around Adelaide it does very well, comes into flower a few years after planting, and offers a fine pasturage for bees. In some instances the foliage is almost hidden by the large masses of beautiful white blossom. The tree remains in bloom several months, and during that time it is always crowded with bees. I have never known a season when this gum did not flower."

VICTORIA.

As regards this colony, Mr. David Gloss informs me that the chief sources of honey supply in the cool districts are white clover and thistles, and in the hot districts eucalypti, or gum trees; and that in several hot districts, which are also treeless, bees have generally died out when introduced. Another correspondent praises the acacias, black wattle and coast wattle, as invaluable for bee forage.

QUEENSLAND.

Mr. Chas. Fullwood says, with regard to Queensland :

" Our principal source of honey is the various species of eucalypti, which afford supplies the greater portion of the year, as they bloom during all the months from August to April and May, and, in fact, some bloom during the winter months, when the bees gather, through the warmer hours of the day, a small amount of honey. What is most popularly known as 'tea tree' here is just now (June) going out of blossom; this supplies a large amount of honey of rather high colour and rank sweet flavour, which it to a large extent loses while ripening, but which is easily detected when first gathered.

"There are many tropical and European trees, shrubs, plants, etc., cultivated here, that yield good supplies of excellent honey, blooming at various periods during the warm months, yet, as a rule, not cultivated in sufficiently large numbers in any localities to supply any number of stocks with a particular kind of honey. Clover is not much cultivated ; so we do not obtain 'clover honey.' Lucerne is largely grown in localities near town ; and where it is allowed to bloom and mature, I believe the bees work on it well. In the neighbourhood of orange orchards bees gather very fine honey from thence ; still our principal source is eucalypti, which affords a very sweet article, ranging in colour from amber to dark orange ; wild flowers and grasses, too, yield nectar. I am not able to state if any of the latter grow in sufficient numbers to grade honey as from such sources, or detect by the flavour from whence it is secured.

"As the industry progresses, and keener observation is awakened, we shall be able to glean much more reliable information, of which we may avail ourselves in the future."

TASMANIA.

This island, as the native place of the *E. globulus*, or Tasmanian blue gum, is well supplied with that and other varieties, as well as with acacias. Mr. Hood informs me that amongst the native trees, "the different varieties of the eucalyptus, box, and lightwood are splendid honey producers. When the box is in bloom, a friend tells me, 'you would think there must be a swarm of bees in each tree.' Immense quantities of honey are got from the bush. One friend tells me he felled a tree in which there were three large colonies of bees, and secured 630 lb. of honey." There are also large quantities of a native heath which affords good forage for bees. Its splendid native flora, fruit orchards, and beautiful climate must altogether make Tasmania a grand country for apiculture.

EUCALYPTI AND ACACIAS IN NEW ZEALAND.

Von Hochstetter, when comparing the native flora of New Zealand with that of Australia, mentions as a very remarkable fact, "that under the families of *Myrtaceœ* and *Leguminosœ*, exactly those genera which are most numerously represented in Australia, the eucalypti and acacias, are entirely wanting in New Zealand, although when introduced there they flourish with extraordinary luxuriance." The latter observation is so true, that the different varieties of gum trees and wattles which have been already introduced here are likely, in course of time, to become as general as in Australia, if we except the great natural forests, and confine ourselves to the plantations in settled districts. The eucalypti heretofore generally planted in New Zealand are *E. globulus* (blue gum) and *E. rostrata* (red gum). Of the acacias a greater variety have been already cultivated; amongst the rest, the following :—

1. *A. lophantha*, common wattle.
2. *A. decurrens*, black wattle.
3. *A. dealbata*, silver wattle.
4. *A. pycnantha*, golden or broad leaf wattle.
5. *A. longifolia*, long-leaf wattle.
6. *A. melanoxylon*, lightwood or blackwood.
7. *A. undulata*, Kangaroo Island prickly acacia.

Of these, the most valuable as bee forage in this country is the first (notwithstanding the objections referred to by Mr. Bonney), as it blooms luxuriantly from May to August, both months inclusive, and sometimes also in April and in September, and during all that time affords ample stores of both nectar and pollen; so that where these trees are convenient, bees cannot know what want is during the late autumn and the whole of winter. Nos. 2, 3, and 4 are the sorts valuable as producing tanners' bark; they are all great nectar and pollen producers in July and August. No. 5 is an evergreen shrub, with long lancet-shaped leaves, which blossoms luxuriantly from July to September, but which also exudes nectar from a pore in the upper edge of the leaf, near the stalk, and furnishes food for the bees in that way for a couple of months before the blossoming time. It is thus available for bee forage (and is a great favourite with the bees) from early in May until the end of September. No. 6 is somewhat similar to the last as regards its leaves and blossoms, but is a tree, not a shrub, and one of the largest of the acacias. No. 7 is used as a hedge plant, and flowers from the beginning of August to the middle of October; it is therefore available as bee forage when all the other acacias are out of bloom, but at a time when there is generally abundance of other spring forage.

EUROPEAN PLANTS AND TREES.

The climate of New Zealand and of most parts of the Australasian colonies is so favourable to the growth of many European plants, that we already enjoy the advantages of most of the old-world bee forage. The clovers (and especially white clover), which are the sources of probably the finest quality of honey, not only grow well when sown in pasture or meadow lands, but spread rapidly over the still uncultivated lands in the neighbourhood. Dandelion, or capeweed (something different from the British dandelion, but equally good as bee forage), is very abundant, and thistles, which furnish a very good honey, spring up everywhere, for a time, in newly laid down lands, and spread themselves along the roadsides and into the waste or fern lands, wherever these have been for a while used as cattle runs; they thus become a very considerable source of the honey supply in some parts of the country. The white thorn, sweet-briar, and furze, or gorse, are all more or less used in different districts as hedge plants; and whatever objections there may be to the two latter sorts on the score of their spreading over the lands and being troublesome to eradicate, wherever they do occur they can only be welcome to the bee-keeper. Fruit blossoms of all sorts are valuable in spring, and the willow, which is here a widely spread tree, is one of the earliest sorts of spring bee forage.

It may be interesting here, before proceeding to notice some particular sorts of honey-bearing plants and flowers, to copy an alphabetical list given in the work of Thomas Nutt, an English apiarist, who wrote in 1832, as a catalogue of British "trees, plants, and flowers most frequented by bees," and all of which either are already or may easily be cultivated here.

"Alder, almond, *Althea fontex*, alyssum, amaranthus, apple, apricot, arbutus (alpine), ash, asparagus, aspin, balm, bean, beech, betony, blackberry, black currant, borage, box, bramble, broom, bugloss (viper's), buckwheat, burnet, cabbage, cauliflower, celery, cherry, chestnut, chickweed, clover, cole (or coleseed), coltsfoot, coriander, crocus, crowfoot, crown imperial, cucumber, currants, cypress, daffodil, dandelion, dogberry, elder, elm, endive, fennel, furze, golden-rod, gooseberry, gourd, hawthorn, hazel, heath, holly, hollyhock (trumpet), honeysuckle, honey-wort (*cerinthe*), hyacinth, hyssop, ivy, jonquil, kidney-bean, laurel, laurustinus, lavender, leek, lemon, lily (water), lily (white), lime, liquid amber, liriodendron, lucerne, mallow (marsh), marigold (French), marigold (single), maple, marjoram (sweet), millelot, melon, mezereon, mignonette, mustard, nasturtium, nectarine, nettle (white), oak, onion, orange, ozier, parsley, parsnip, pea, peach, pear, peppermint, plane, plum, poplar, poppy, primrose, privet, radish, ragweed, raspberry, rosemary (wild), roses (single), rudebchiæ, saffron, sage, saintfoin, St. John's-wort, savory (winter), snowdrop, snowberry, stock (single), strawberry, sunflower, sycamore, tamarisk, tansy (wild), tare, teazel, thistle (common), thistle (sow), thyme (lemon), thyme (wild), trefoil, turnip, vetch, violet (single), wallflower (single), woad, willow-herb, willow tree, yellow weasel-snout."

In many parts of continental Europe the chief honey harvest is derived from the linden (*Tilia Europœa*) and the acacia (*robinia* or *pseudacacia*), trees which are easily grown in these countries also. Buckwheat, which is also much grown in America, affords a late harvest of honey of dark colour and inferior quality.

AMERICAN PLANTS AND TREES.

The American species of linden, the basswood (*Tilia Americana*) stands at the head of the honey producers in the Northern and Eastern States; the honey locust, named by Quinby *robinia* or *pseudacacia*, but to which Professor Cook gives the botanical name of *Glidischia triacanthus*, the latter being the correct one, according to Johnson's "Gardener's Dictionary." The liriodendron, or tulip-tree, the Judas-tree, or red-bud, the magnolia, and the sugar maples, are all prized, in various parts of the States, as bee forage.

Speaking of the basswood (Fig. 132) in his "A B C of Bee Culture," A. I. Root says of it: "With perhaps the single exception of white clover, the basswood, or linden, as it is often called, furnishes more honey than any other one plant or tree known." Comparing it with clover, he says: "The best yield of honey we have ever had from a single hive in one day was from the basswood bloom: the amount was 43 lb. in three days. The best we ever recorded from clover in one day was 10 lb." Fully ripened basswood honey is noted for its delicious flavour. It blossoms in July in the United States, and in many parts it is the chief dependence of the apiarists. It does well in New Zealand.

The sages (*salvia*) of California are perhaps amongst the most wonderful honey plants known at the present time. There are different varieties growing wild on the mountains and in the cañons of this State, but I believe most of the honey is gathered from what are commonly known as the white, black, and button sages. The estimated crop of honey last season (1884), for California, was about 4,500 tons. This plant would thrive well north of Auckland, and in the warmer parts of Australia. I grew some from seed sent me by Mr. Wilkin, of San Buenaventura, when I was living at the Thames, and they flourished splendidly; but on taking the plants to

Matamata, I lost them the first winter, as they could not withstand the frosts there.

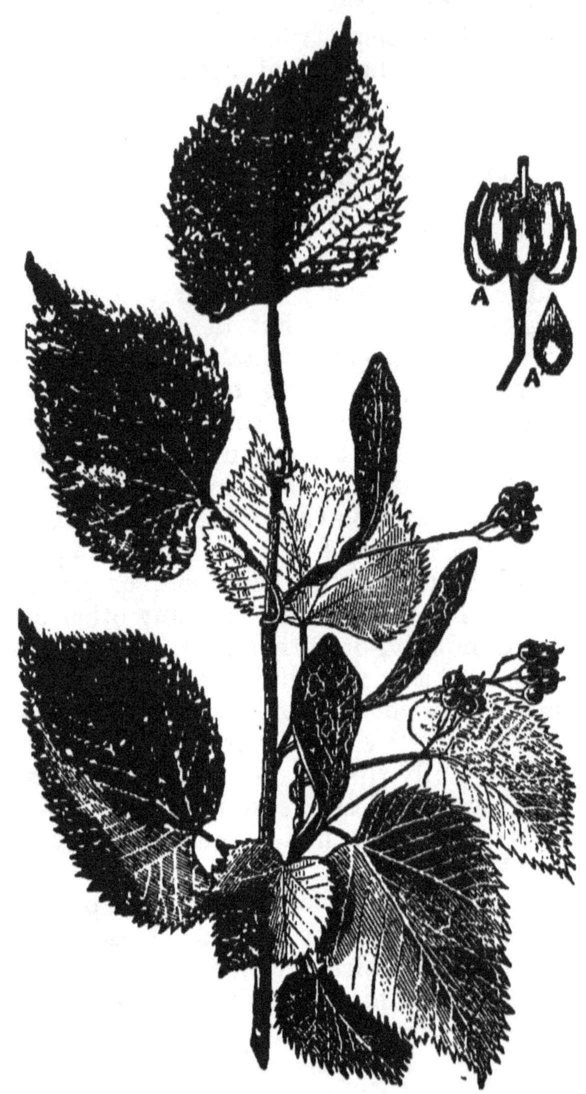

Fig. 132.—AMERICAN LINDEN, or BASSWOOD (*Tilia Americana*).

The horse-mint (Fig. 133) is another plant (allied to the
salvias) from which enormous crops of honey have been taken
in the State of Texas, though it does not appear to yield honey

in quantity every season. Mr. B. F. Carroll, of Dresden, Texas, reported a yield of nearly 1000 lb. from one colony, gathered principally from this shrub. I have now some young plants growing, which have withstood the frosts remarkably well. They would do well in most parts of Australasia.

The United States can boast of a number of what may be termed fine honey-bearing weeds, among which figwort (Fig. 134) stands high. I have grown this plant myself for several years, and can speak well for its honey-yielding qualities. As soon as the first little pitcher-like flowers open, the bees are constantly visiting the plants till they are out of blossom. They

Fig. 133.—HORSE-MINT OF TEXAS (*Mouarda punctata*).

flower for about two months, commencing at the end of November. Being tuberous-rooted plants, and dying down every winter, frost does not affect them. The honey, as far as I can judge of it, is very fine.

The golden-rods (Fig. 135) are members of a very numerous family to be found growing nearly all over the United States. They must be very valuable honey plants; for A. I. Root states that "in some localities they furnish the bulk of the great yield of fall honey." These and the asters (Fig. 136),

which are also autumn-blooming plants, and allied to each
other, would be of great value on that account in these colonies, more especially in New Zealand, where there is rather a

Fig. 134.—FIGWORT (*Scrofularia nodosa*).

scarcity of honey in the fall of the year.   With regard to the
quality of the honey from these plants, Mr. Root, when speaking of the golden-rods, says : "The honey is very thick, and of
a rich golden colour, much like the blossoms.   When first

gathered, it has, like the honey of most other fall flowers, a rather rank, weedy taste; but after it has thoroughly ripened it is very rich.

Spider plant (*Cleome pungens*) (Fig. 137) is another common weed in some parts of the United States, which yields a good quantity of nectar; it is allied to, and very much like, the Rocky Mountain bee plant (*Cleome integrifolia*). The peculiarity about this plant is that it only yields nectar at night, commencing late in the evening, at which time and very early morning the bees

Fig. 135.—THREE VARIETIES OF GOLDEN-RODS (*Solidago*).

visit it. I have had some very good reports of this plant from those I have supplied with seeds. It lasts a long time in blossom, and is rather handsome when in flower. Mr. Root speaks very highly of both the plant and the quality of the nectar it yields. The seeds of the foregoing plants—that is, figwort, horse-mint, golden-rods, asters, and spider plant—together with any of the herbs, such as horehound, catnip, etc., might be scattered about in waste places and odd corners, without doing any harm, and with profit to bee-keepers.

The next two plants, mellilot clover and giant mignonette, are not natives of America. I obtained some seeds of them from that country, have grown them both, and am therefore acquainted with their honey-yielding qualities, and believe, where there is ground to spare, they will pay to grow for honey, though I do not at present advocate growing special crops for honey, unless these crops are profitable in some other direction.

Mellilot, or sweet clover (Fig. 138), is a great favourite with bees, though I think of little use for anything else than its

Fig. 136.—ASTER (*Starwort*).

honey, of which it yields a large quantity. It has a delightful perfume when in flower, and scents the air for a long distance. It blooms for nearly four months, lasting right up to near the commencement of winter. It will grow almost anywhere, but does best on fairly good land. It is a biennial, and blossoms the second year; but it can be made to perpetuate itself by sowing the same land two years in succession. I have had it growing to the height of seven feet.

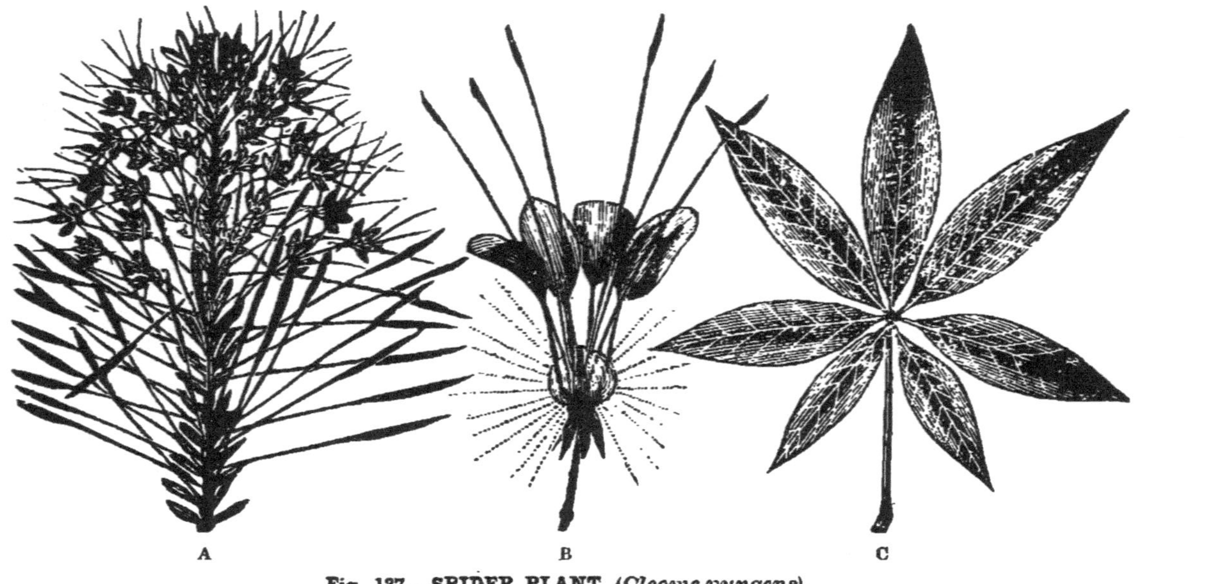

Fig. 137.—SPIDER PLANT (*Cleome pungens*).

EXPLANATION OF FIGURE.—A is a flower stalk, of which their are a dozen or more to each plant ; B, floweret with honey sparkling at base of petals ; C, leaf of plant.

The spider plant is an annual, and grows here to the height of about five feet, sending out a number of branches, on the ends of which are the flower-stalks. The honey is secreted in the flowerets at night, and the bees may be seen on them at the first peep of day. The seeds can be sown in beds, and the young plants pricked out in rows when large enough to handle. They should be planted three feet apart each way. It continues to blossom for two or three months, and is well worthy of cultivation by all bee-keepers.

Giant mignonette (Fig. 139) is another grand bee plant. All the mignonettes are good for bees, and handsome as garden plants. The giant variety has not the sweet scent of the smaller and common kind, but the flowers are exactly similar in appearance. I have had it in blossom for seven successive months, and during the whole of this time it was continually visited by bees, even when clover was at its best.

DURATION OF THE HONEY SEASON.

From what has been already said under the heads of "Climate" and of "Native Flora," both in this and in the

Fig. 138.—MELLILOT CLOVER (*Melilotus*).

first chapter, it will readily be seen that within the range of New Zealand and Australia there must be considerable differences in the time of commencement, and in the duration of the so-called honey season, or that in which the apiarist expects to take his crop of surplus honey. This is one of the points deserving of the most particular consideration, as the judicious bee-keeper will have to regulate many of his operations in accordance with the peculiar circumstances of his district in this respect, in order to obtain the best general results, and to grade his honey according to the sources of supply which predominate during different portions of the honey season. I shall here add only a few observations with reference to my own experience in the Thames and Waikato districts, some thirty miles south of Auckland.

Here the dawn of the honey season opens with the willows and early flowering peaches. These blossom about the beginning of September, and enable the bees very quickly to replenish the stores of honey, at this time nearly exhausted.

Then follow other fruit trees, dandelions, daisies, hawthorn, wild rose, some native trees, &c., and the honey season may be said to have fully commenced in October, when clover begins to blossom, and lasts without cessation, in a fairly good season,

Fig. 139.—GIANT MIGNONETTE.(*Reseda gigantea*).

till near the latter end of February, when the hot dry weather causes a break, which lasts till the rains commence again (usually in the latter part of March); when, if the weather is showery, a considerable amount of honey will be gathered from different sources through the latter part of March, April, and a part of May.

The time of commencement and duration of the autumn crop entirely depends upon the rains; if these should be late it will of course affect the crop in the same manner, and also shorten the honey season.

The autumn gathering is very valuable; it enables the bees to fill up empty brood combs with honey, thus putting the colony in good condition for wintering, and renders feeding unnecessary, a very important matter, especially to beginners.

In this district it is seldom that any surplus honey can be obtained from the autumn crop; but I have no doubt in more favourable situations a considerable amount might sometimes be taken.

FLIGHT OF BEES.

It is generally understood that the usual range of a worker bee's flight is from $1\frac{1}{2}$ to 2 miles in all directions from the apiary, although bees are known to go much further when pasturage is scarce within that distance.

Of course the greater the flight the less honey is stored; so that the apiarist will understand how necessary it is, where practicable, to have good and abundant pasturage near the apiary. It is Mr. Langstroth's opinion that "although bees will fly in search of food over three miles still if it is not within a circle of two in every direction from the apiary they will be able to store but little surplus honey."

OVER-STOCKING.

On the question of over-stocking many different opinions exist, some thinking it almost impossible to over-stock a good district, while others, again, believe it may be done. The former argue in support of their opinion that each day's secretion of honey in the flowers will, if not gathered the same day, dry up or be wasted before evening. This being so it would take an enormous number of bees to visit all the honey-producing flowers every day, and until this is done, there can be no fear of the district being over-stocked. This appears to me to be a very reasonable way of looking at the matter.

My own idea is, that in most districts of New Zealand, and the settled districts of Australia, a very large number of colonies might be profitably supported in every square mile, and the possibility of such districts ever being over-stocked is very remote.

CHAPTER XIX.

APICULTURE IN RELATION TO AGRICULTURE.*

THE benefits derived by both agriculturists and horticulturists from the labours of the bee are now very generally understood and acknowledged ; but still cases do sometimes occur, though rarely, of farmers objecting to the vicinity of an apiary, and complaining of bees as " trespassers," instead of welcoming them as benefactors.

ARE BEES TRESPASSERS ?

It is not, perhaps, surprising that at first a man should imagine he was being injured in consequence of bees gathering honey on *his land*, to be stored up elsewhere, and for the use of other parties ; he might argue that the honey belonged by right to him, and even jump at the conclusion that there was so much of the *substance of the soil* taken away every year, and that his land must therefore become impoverished. It is true that if he possessed such an amount of knowledge as might be expected to belong to an intelligent agriculturist, working upon

A paper, from which the matter of this chapter is abridged, appeared in the three numbers of the *New Zealand and Australian Bee Journal* for the months of August, September, and October, 1884, and has since been partially reprinted in more than one of the American bee papers. Since it was first written, the subject of which it treats has been brought prominently forward in consequence of the action taken by a farmer in one of the United States to claim damages from a neighbouring bee-keeper for supposed injury done to his grazing sheep by trespassing (?) bees. Just now, whilst these sheets are passing through the press, the American Bee Journals are full of communications from bee-keepers, pointing out the absurdity of such claims, and calling for united action in opposing all such attempts that may be made to check the progress of bee culture. However unfounded and unreasonable such claims really are and must appear to those who understand the nature and habits of both bees and grazing animals, the mere fact of their being seriously advanced is sufficient to show the necessity of bee-keepers adducing such facts and arguments as are calculated to *prove satisfactorily* the groundlessness of all assertions to the effect that bees occasion *any injury* to the farmer, either as regards the fertility of his soil, the condition of his crops, or the safety and comfort of his grazing stock. [Since the above was in type the lawsuit—Sheep *v.* Bees—has been dismissed by the Judges.]

rational principles, he should be able, upon reflection, to see that such ideas were entirely groundless. Nevertheless the complaint is sometimes made, in a more or less vague manner, by persons who ought to know better ; and even bee-keepers appear occasionally to adopt an apologetic tone, arguing that "bees do more good than harm," instead of taking the much higher and only true stand by asserting that bees, while conferring great benefits on agriculture, *do no harm whatever*, and that the presence of an apiary on or close to his land *can be nothing but an advantage* to the agriculturist.

BENEFICIAL INFLUENCE OF BEES ON AGRICULTURE.

We have already, in Chapter III., dwelt upon the value of the intervention of bees in the cross-fertilisation of plants, and can here only refer the reader for further information to the works of Sir J. Lubbock and of Darwin. The latter, in his work on "Cross and Self-Fertilisation of Plants," gives the strongest evidence as to the beneficial influence of bees upon clover crops. At page 169, when speaking of the natural order of leguminous plants to which the clovers belong, he says, "The cross-seedlings have an enormous advantage over the self fertilised ones, when grown together in close competition ;" and in Chapter X., page 361, he gives the following details of some experiments, which show the importance of the part played by bees in the process of cross-fertilisation :—

"*Trifolium repens* (white clover).—Several plants were protected from insects, and the seeds from ten flower-heads on these plants and from ten heads on other plants growing outside the net (which I saw visited by bees) were counted, and the seeds from the latter plants were very nearly ten times as numerous as those from the protected plants. The experiment was repeated in the following year, and twenty protected heads now yielded only a single abortive seed, whilst twenty heads on the plants outside the net (which I saw visited by bees) yielded 2,290 seeds, as calculated by weighing all the seeds and counting the number in a weight of two grains. *Trifolium pratense* (purple clover).—One hundred flower-heads on plants protected by a net did not produce a single seed, whilst one hundred on plants growing outside, which were visited by bees, yielded sixty-eight grains weight of seeds ; and as eighty seeds weighed two grains, the hundred heads must have yielded 2,720 seeds."

Here we have satisfactory proof that the effect of cross-fertilization, brought about by bees, upon the clovers and other plants growing in meadows and pasture lands, is the certain production of a large number of vigorous seeds, as compared with the chance only of a few and weak seeds if self-fertilization were to be depended upon. In the case of meadow cultivation it enables the farmer to raise seed for his own use or for sale instead of having to purchase it; while, at the same time, the nutritious quality of the hay is, as we shall see further on, improved during the process of ripening the seed. In the case of pasture lands, such of those vigorous seeds as are allowed to come to maturity and to fall in the field will send up plants of a stronger growth to take the place of others that may have died out, or to fill up hitherto unoccupied spaces, thus tending to cause a constant renewal and strengthening of the pasture. The agriculturist himself should be the best judge of the value of such effects.

The beneficial effect of the bees' visits to fruit trees has been well illustrated by Mr. Cheshire, in the pages of the *British Bee Journal*, and by Prof. Cook, in his article upon " Honey Bees and Horticulture," in the *American Apiculturist*. In fact, even those who complain of bees cannot deny the services they render; what they contest is the assertion that *bees do no harm.*

CAN BEES HARM THE SOIL, OR THE CROPS ?

Is then the question to be considered. The agriculturist may say, " Granting that the visits of bees may be serviceable to me in the fertilisation of my fruit or my clover, how will you prove that I am not obliged to pay too high a price for such services ?" For the answer to such a question one must fall back upon the researches of the agricultural chemist, which will furnish satisfactory evidence to establish the two following facts :—First, that saccharine matter, even when assimilated and retained within the body of a plant, is not one of the secretions of vegetable life which can in any way tend to exhaust the soil, being made up of constituents which are furnished everywhere in superabundance by the atmosphere and rain water, and not containing any of the mineral or organic substances supplied by the soil or by the manures used in agriculture; and secondly, that in the form in which it is appropriated by bees, either from the nectaries of flowers or as honeydew from the leaves, it no longer constitutes a part of the plant, but is in fact an excrement, thrown off as superfluous, which, if not collected by the bee and by its means made available for the use of man, would either be devoured by other insects, which do not store honey, or be resolved into its original elements and dissipated in the air.

The foregoing statements can be supported by reference to authorities which can leave no doubt as to their correctness, namely, Sir Humphrey Davy in his "Elements of Agricultural Chemistry," written more than fifty years ago, and Professor Liebig in his "Chemistry in its Application to Agriculture and Physiology," written some ten years later, and the English version of which is edited by Dr. Lyon Playfair and Professor Gregory.   These works, which may be said to form the foundation of a rational system of agriculture, were written with that object alone in view, and the passages about to be quoted were not intended to support any theory in favour of bee culture or otherwise; they deal simply with scientific truths which the layman can safely follow and accept as true upon such undeniable authority, although he may be incapable himself of following up the processes which have led to their discovery or which prove their correctness.

SACCHARINE MATTER OF PLANTS NOT DERIVED FROM THE SOIL.

Liebig, when describing the chemical processes connected with the nutrition of plants, informs us (at page 4*) that—

"There are two great classes into which all vegetable products may be arranged. The first of these contain nitrogen; in the last this element is absent. The compounds destitute of nitrogen may be divided into those in which oxygen form a constituent (starch, lignine, &c.), and those into which it does not enter (oils of turpentine and lemon, &c.)."

And at page 141 that

"Sugar and starch do not contain nitrogen; they exist in the plants in a free state, and are never combined with salts or with alkaline bases They are compounds formed from the carbon of the carbonic acid and the elements of water (oxygen and hydrogen)."

* The edition to which reference is made is the fourth, published 1847.

Sir Humphrey Davy had already stated that, " according to the latest experiments of Guy Lussac and Thenard, sugar consists of 42·47 per cent. of carbon and 57·23 per cent. of water and its constituents."

Now Liebig, in several parts of his work, shows that the carbon in sugar and all vegetable products is obtained from carbonic acid in the atmosphere ; and that " plants do not exhaust the carbon of the soil in the normal condition of their growth ; on the contrary, they add to its quantity."

DERIVED FROM THE ATMOSPHERE AND RAIN WATER.

The same authority shows (as we have already stated in Chap. IV.) that the oxygen and hydrogen in these products are derived from the atmosphere and from rain water ; and that it is only the products containing nitrogen (such as gluten or albumen in the seeds or grains), and those containing mineral matter (silex, lime, aluminium, &c.), which take away from the soil those substances that are required to be returned to it in the shape of manures. The saccharine matter, once it is secreted by the plant and separated from it, is even useless as a manure. Liebig says on this head, page 21 :—

"The most important function in the life of plants, or, in other words in their assimilation of carbon, is the separation, we might almost say the generation of oxygen. No matter can be considered as nutritious, or as necessary to the growth of plants, which possesses a composition either similar to or identical with theirs ; because the assimilation of such a substance could be effected without the exercise of this function. The reverse is the case in the nutrition of animals. Hence such substances as sugar, starch, and gum, themselves the products of plants, cannot be adapted for assimilation ; and this is rendered certain by the experiments of vegetable physiologists, who have shown that aqueous solutions of these bodies are imbibed by the roots of plants, and carried to all parts of their structure, but are not assimilated ; they cannot, therefore, be employed in their nutrition."

NECTAR OF PLANTS INTENDED TO ATTRACT INSECTS.

The secretion of saccharine matter in the nectaries of flowers is shown to be one of the normal functions of the plant, taking place at the season when it is desirable to attract the visits of insects for the purposes of its fertilisation. It may then be fairly asserted, that the insect, when it carries off the honey from any blossom it has visited, is merely taking with it the fee or reward provided by nature for that special service.

SOMETIMES THROWN OFF AS SUPERFLUOUS.

There are, however, occasions when considerable quantities of such matter are thrown off, or exuded by the leaves, which effect is taken to indicate an abnormal or unhealthy condition of the plant. At pages 106 and 107 of Liebig's book (speaking of an experiment made to induce the rising sap of a maple tree to dissolve raw sugar applied through a hole cut in the bark) he shows (in a passage already quoted at page 86) that,

"When a sufficient quantity of nitrogen is not present to aid in the assimilation of the substances destitute of it, these substances will be separated as excrements from the bark, roots, leaves, and branches."

In a note to this last paragraph we are told that

"Langlois has lately observed, during the dry summer of 1842, that the leaves of the linden tree became covered with a thick and sweet· liquid in such quantities that for several hours of the day it ran off the leaves like drops of rain. Many kilogrammes might have been collected from a moderate-sized linden tree."

And further on, at page 141, he says :—

"In a hot summer, when the deficiency of moisture prevents the absorption of alkalies, we observe the leaves of the lime tree, and of other trees, covered with a thick liquid containing a large quantity of sugar; the carbon of the sugar must, without doubt, be obtained from the carbonic acid of the air. The generation of the sugar takes place in the leaves; and all the constituents of the leaves, including the alkalies and alkaline earths, must participate in effecting its formation. Sugar does not exude from the leaves in moist seasons; and this leads us to conjecture that the carbon which appeared as sugar in the former case would have been applied in the formation of other constituents of the tree, in the event of its having had a free and unimpeded circulation."

These quotations will probably be considered sufficient to justify the assertion that the gathering of the honey from plants can in no possible way tend to exhaust the soil, or affect its fertility. There is no difference of opinion amongst scientific men as to the sources from which the saccharine matter of plants is derived. Since Liebig first put forward his views on that subject, as well as with regard to the sources from which the plants derive their nitrogen, the principles of agricultural chemistry have been studied by the most eminent chemists, some of whom combatted the views of Liebig on this latter point (the source of nitrogen and its compounds), and Liebig himself seems to have modified his views on that point; but there has been no difference of opinion about the saccharine matter, as to which Liebig's doctrine will be found given unaltered in the latest colonial work on the subject, MacIvor's "Chemistry of Agriculture," published at Melbourne a few years ago.

SUPERFLUOUS NECTAR EVAPORATED IF NOT TAKEN BY INSECTS.

That the nutritive quality of the plants in any growing crop is not diminished by the abstraction of honey from their blossoms would appear to be evident from the fact already referred to, that those plants have actually thrown off the honey from the *superfluity* of their saccharine juices, as a matter which they could no longer assimilate. There would appear, on the other hand, to be good reason to believe that the plants themselves become daily *more* nutritive during the period of their giving off honey, that is, from the time of flowering to that of ripening their seeds. This is a point upon which, I believe, all agricultural chemists are not quite agreed, but the testimony of Sir H. Davy is very strong in favour of it. In the appendix to his work already quoted, he gives the results of experiments made conjointly by himself and Mr Sinclair, the gardener to the Duke of Bedford, upon nearly one hundred different varieties of grasses and clovers. These were grown carefully in small plots of ground as nearly as possible equal in size and quality; equal weights of the dried produce of each, cut at different periods, especially at the time of flowering and at that of ripened seeds, were "acted upon by hot water till all their soluble parts were dissolved; the solution was then evaporated to dryness by a gentle heat in a proper stove, and the matter obtained carefully weighed, and the dry extract, supposed to contain the nutritive matter of the plants, were sent for chemical analysis." Sir H. Davy adds his opinion that this "mode of determining the nutritive power of grasses, is sufficiently accurate for all the purposes of agricultural investigation." Further on he reports, " In comparing the compositions of the soluble products afforded by different crops from the same grass, I found in all the trials I made, the largest quantity of truly nutritive matter in the crop cut when the seed was ripe, and the least bitter extract and saline matter, and the most saccharine matter, in proportion to the other ingredients, in the crop cut at the time of flowering." In the instance which he then gives, as an example, the crop cut when the seed had ripened showed nine per cent. *less* of sugar, but *eighteen per cent.* more of mucilage and what he terms " truly nutritive matter " than the crop cut at the time of flowering. From this it would follow, that during the time a plant is in blossom and throwing off a superfluity of saccharine matter in the shape of honey, the assimilation of true nutritive matter in the plant itself is progressing most favourably. In any case it is clear that the honey, being once exuded, may be taken away by bees or any other insects (as it is evidently intended to be taken) without any injury to the plant, by which it certainly cannot be again taken up, but must be evaporated if left exposed to the sun's heat.

QUESTION AS TO GRAZING STOCK.

There is, however, a plea put in by the agriculturist on behalf of his grazing stock, and one which he generally seems to consider unanswerable. He says, " Even if it be admitted that the removal of the honey from my farm is neither exhausting to the soil nor injurious to the plants of the standing crops, still it is so much fattening matter, which might be consumed by my stock, if it had not been pilfered by the bees."

Now it may at once be admitted that honey consists, to a great extent, of fattening matter, though it may be allowable to doubt whether, in that particular form, it is exactly suitable as food for grazing cattle. Although it is quite true that the saccharine matter assimilated in the body of a plant tends to the formation of fat in the animal which eats and digests that plant, still one may question the propriety of feeding the same animal on pure honey or sugar. We may, however, waive that view of the subject, as we shall shortly see that it is only a question of such homoeopathically small doses as would not be likely to interfere with the digestion of the most delicate grazing animal, any more than they would considerably increase its weight. Admitting, therefore, that every pound of honey of which the grazing stock are deprived by bees is a loss to the farmer, and therefore to be looked upon as a set-off, to that extent, against the benefits conferred by the bees in other ways, it will be necessary to consider to what extent it is possible that such loss may be occasioned.

QUANTITY OF HONEY FURNISHED BY PASTURE LANDS.

In the first place it must be recollected that a large proportion—in some cases the great bulk—of the honey gathered by bees is obtained from trees, as, for instance, the linden in Europe, the basswood and maple in America, and in this country the forest trees, nearly all of which supply rich forage for the bee, and everywhere from fruit trees in orchards. A large quantity is gathered from flowers and flowering shrubs reared in gardens—from clover and other plants grown for hay, and not for pasture; and even in the field there are many shrubs and flowering plants which yield honey, but which are never eaten by cattle. Pastures therefore form but a small part of the sources from which honey is obtained; and in dealing with this grazing question we have to confine our inquiries to clovers and other flowering plants grown in open pastures, and such as constitute the ordinary food of grazing stock. In order to meet the question in the most direct manner, however, let us assume the extreme case of a large apiary being placed in a district where there is nothing else but such open pastures, and growing only such flowering plants as are generally eaten by stock. Now, the ordinary working range of the bee may be taken at a mile and a half from the apiary on all sides, which gives an area of about 4,500 acres for the supply of the apiary; and if the latter consists of a hundred hives, producing an average of a hundred pounds of honey, there would be a little more than two pounds of honey collected off each acre in the year; or if we suppose so many as two hundred hives to be kept at one place, and to produce so much as ten tons of honey in the season, the quantity collected from each acre would be four to five pounds.

PROPORTION POSSIBLY CONSUMED BY STOCK.

Let us next consider what proportion of those few pounds of honey could have found its way into the stomachs of the grazing stock if it had not been for the bees.   It is known that during the whole time the clover or other plants remain in blossom, if the weather be favourable, there is a daily secretion of fresh honey, which, if not taken at the proper time by bees or other insects, is evaporated during the mid-day heat of the sun.   It has been calculated that a head of clover consists of 50 or 60 separate flowers, each of which contains a quantity not exceeding 1-500th part of a grain in weight, so that the whole head may be taken to contain *one-tenth* of a grain of honey at any one time.   If this head of clover is allowed to stand until the seeds are ripened it may be visited on ten or even twenty different days by bees, and they may gather on the whole one, or even two grains of honey from the same head, whereas it is plain that the grazing animal can only eat the head once, and consequently can only eat one-tenth of a grain of honey with it.   Whether he gets that one-tenth grain or not depends simply on the fact, whether or not the bees have exhausted that particular head on *the same day just before it was eaten*   Now, cattle and sheep graze during the night and early morning, long before the bees make their appearance some time after sunrise ; all the flowering plants they happen to eat during that time will contain the honey secreted in the evening and night time ; during some hours of the afternoon the flowers will contain no honey, whether they may have been visited by bees or not ; and even during the forenoon, when the bees are most busy, it is by no means certain that they will forestall the stock in visiting any particular flower.   If a field were so overstocked that every head of clover should be devoured as soon as it blossomed, then, of course, there would be nothing left for the bees, but if, on the other hand, as is generally the case, there are always blossoms left standing in the pasture, some of them even till they wither and shed their seeds, then it must often happen that after bees shall have visited such blossoms ten or even twenty times, and thus collected one or even two grains of honey from one head, the grazing animal may, after all, eat that particular plant and enjoy his one-tenth of a grain of honey just as well as if there had never been any bees in the field.   If all these chances be taken into account, it will be evident that out of the four or five pounds of honey assumed to be collected by bees from one acre of pasturage, probably not one-tenth, and possibly not even one-twentieth, part could, under any circumstances, have been consumed by the grazing animals—so that it becomes a question of *a few ounces* of fattening matter, more or less, for all the stock fed upon an acre during the whole season ; a matter so ridiculously trivial in itself, and so out of all proportion to the services rendered to the pasture by the bees, that it may safely be left out of consideration altogether.

BEE-KEEPING AS A BRANCH OF FARMING.

There is still one point which may possibly be raised by the agriculturist or land-owner : " If the working of bees is so beneficial to my crops, and if such a large quantity of valuable matter may be taken, in addition to the ordinary crops, without impoverishing my land, why should I not take it instead of another person who has by right no interest in my crop or my land ?" The answer to this is obvious. It is, of course, quite open to the agriculturist to keep any number of bees he may think fit ; only he must consider well in how far it will pay him to add the care of an apiary to his other duties. No doubt every one farming land may, with advantage, keep a few stands of hives to supply his own wants in honey ; the care of them will not take up too much of his time, or interfere much with his other labours ; but if he starts a large apiary with the expectation that it shall pay for itself, he must either give up the greater portion of his own time to it, or employ skilled labour for that special purpose ; and he must recollect that the profits of bee-keeping are not generally so large as to afford more than a fair remuneration for the capital, skill, and time required to be devoted to the pursuit. In any case, he cannot confine the bees to work exclusively on his own property, unless the latter is very extensive. When such is the case, he may find it greatly to his advantage to establish one or more apiaries to be worked under proper management, as a separate branch of his undertaking ; but in every case, whether he may incur or share the risks of profit and loss in working an apiary or not, the thing itself can only be a source of unmixed advantage to his agricultural operations, and consequently, if he does not ocucpy the ground in that way himself, he should only be glad to see it done by any other person.

CHAPTER XX.

USES OF HONEY IN FOOD—DRINKS—MANUFAC-. TURES—MEDICINE.

IN Chapter I. the great antiquity and universality of the use of honey has been dwelt upon with a view to impress upon the reader its importance as one of those gifts of nature so evidently intended by a bountiful Providence for the use and benefit of man ; in this chapter it is proposed to glance at the variety of uses to which it is, or may be, applied at the present day, with the object of turning the attention of the bee-keeper to the practical question of developing a market for his produce.

HONEY AS FOOD.

Of the ordinary use of honey as a condiment for the break-fast or tea table it is unnecessary to say much ; every one knows, and has known since his childhood, the enticing ap-pearance of the luscious sweet in its delicate white comb—the aroma and the delicious flavour, so different from most other sweets ; and of late years people are becoming accustomed to appreciate also the pure honey, separated from its comb (with-out any unappetising process of squeezing through cloths), and served either in its clear liquid state or in a thick granulated condition which is now gradually gaining favour with honey eaters. It is probable that the greater portion of all the honey now produced is consumed in this way—either as comb honey or in its extracted state, but still it has not reached the tables of the hundredth part of those who could afford to use it as a cheap luxury, if it were as generally produced as it might easily be, and brought within easy reach of all consumers at reasonable prices. It is capable of being made not only a cheap luxury, but a really economical article of food, for those to whom strict economy is a necessity. Under proper arrangements it can be placed on the table, as extracted honey, at half or less than half the price, weight for weight, of ordinary butter, and might with decided advantage be used as a substitute for some portions of that admittedly desirable and wholesome article of consumption. Dr. W. G Phelps, writing lately in an American paper (the *Practical Farmer*) says on this point :—

"Honey can really no longer be considered one of the mere luxuries of life. For the poor it has become a cheap and wholesome substitute for the too frequently impure butter. Millions of pounds are to-day consumed by rich and poor alike, where ten or fifteen years since but a few thousands were used. The severe stab which the manufacturer of the miserable glucose has received is due to a great extent to this production of extracted honey. This being the pure article, and produced even at a profit for ten cents (five pence) per pound, has virtually gained the mastery in competition with the above ' falsely so-named cheap sweet.' "

FOR DOMESTIC COOKERY.

But in addition to this well-known use of honey it can be used with the greatest advantage in a hundred different ways in domestic cookery. Mr. Newman, in his " Bees and Honey," remarks : " The use of honey instead of sugar for almost every kind of cooking is as pleasant for the palate as it is healthy for the stomach." And again : " In fact, honey may replace sugar as an ingredient in the cooking of almost any article of food, and at the same time greatly add to its relish." At the end of this chapter will be found a few of the recipes usually given for honey-cakes of different kinds.

FOR PRESERVING FRUIT.

Honey is an excellent medium for stewing peaches or other fresh fruits for table use, or for fruit pies ; and it is also strongly recommended for preserving fruits of all kinds. Dr. Phelps, already quoted above, says :—

"Used instead of sugar for preserving raspberries and other fruits, I know of nothing its equal, as to many such compounds it imparts a peculiarly delicious flavour."

And Mr. Newman observes :—

"Well purified honey has the quality of preserving, for a long time in a fresh state, anything that may be laid in it or mixed with it, and preventing its corrupting in a far superior manner to sugar ; thus many species of fruit may be preserved by being laid in honey, and by this means will obtain a pleasant taste, and give to the stomach a healthy tone.   One who has once tried it will not use sugar for preserving fruit."

MODERATION IN USE.

There is no doubt honey, like everything else, may be abused as well as used. Dr. Phelps remarks :—

"Honey in its purity is a God-given sweet, and in its proper use is conducive to health and strength.   Indulged in immoderately, and only then at rare intervals, it may, like many other excellent articles of food, provoke an attack of colic or indigestion.   Used however frequently, and in connection with other food, it has a tendency to produce pure blood and give tone to the human system."

This is no more than might be expected, and something of the same sort has been said a long time ago.   Solomon, though describing honey as something which is " sweet to the soul, and health to the bones," does not omit to warn against its excessive use, as in Proverbs xxv. 16, " Hast thou found honey, eat so much as is sufficient for thee, lest thou be filled therewith and vomit it."   And again in verse 27, " It is not good to eat much honey, etc."   There are besides some people of exceptional constitutions who cannot with impunity eat even in moderation any of the purest honey, but such cases are very rare.   Referring to them, Dr. Phelps says : " To the rare individual for whom the temperate use of honey may produce functional disorders, I would say try heating honey before using it, and see if all such trouble is not remedied."   Perhaps however it would be wiser for such persons to abstain altogether from the use of honey.

DELETERIOUS HONEY.

As to some particular kinds of honey which are found to be deleterious, if not absolutely poisonous, and which have been referred to in Chapter IV., it is of course the duty of every beekeeper to see that no such article finds its way into the market from his apiary. His brand should be a guarantee for purity in this respect as well as in all others.   The cases are indeed exceedingly rare in which it can be truly asserted that the pure honey itself is in any degree deleterious.   When disagreeable results sometimes arise from partaking of such honey, it is much more likely to be caused by some portion of pollen or bee bread, or even propolis taken perhaps in old parts of the comb, and it is quite possible that even in such cases, if the honey could be cleanly extracted, it would be found by itself to be quite harmless. The only instance at present known of honey gathered in New Zealand being suspected of injuring either bee or man, is that obtained from the Wharangi shrub —and even in that case it is by no means proved that the evil lies in the honey itself—and as that shrub only blossoms in a few places where it is well known, and only for a short time, and not in the season of the regular honey harvest, there is very little chance of its becoming even mixed with the surplus honey of an apiary. At all events it can scarcely be stored without attracting the attention of the beekeeper, as its first effects are shown upon the bees themselves, who are to be seen lying stupefied, as it were, about their hive, though they seem mostly to recover soon from these attacks.

Heretofore I have only referred to what may be termed the domestic consumption of honey. It is used besides in the preparation of certain articles of human consumption, upon so large a scale as to be more properly noticed further on under the head of manufactures.

FERMENTED DRINKS.

Mead and metheglin are two names meaning nearly the same thing, and derived, no doubt, from the same root, which may be traced through a great number of the most ancient languages, thus showing the antiquity and the general use of the article designated, which is simply a fermented drink, made chiefly from honey. *Methu* is wine in the Greek, as is *medo* in the Zend, or ancient Persian, and *madia* in Sanscrit. When we come to the less ancient nations of the north and west of Europe, who did not enjoy originally the juice of the grape, but made their first fermented drink from honey, we find the Teutons called that liquor *meth;* the Saxons, *medo* or *medu;* the Gaels in Wales, *mez ;* and in Ireland and Scotland, *miodh* or *meadh.* These words were evidently all intended to mean a "wine," or strong drink, made from honey. The Russians seem to have adopted the name for honey itself, which with them is *med* or *meda*. The name metheglin seems to be peculiarly Welsh ; *mezyglin*, derived, according to Owen, "from *mezyg*, a physician, and *llyn*, water,—a medicinal liquor." But we must recollect that mead in the same language is *Bez*. Mr. Harris, the author of "The Honey Bee," who seems to have turned his attention to the history of honey drinks in Britain, says, "Properly speaking, the word *Betheglin* was applied to the superior sorts of mead, the two beverages being related much in the same way as effervescing cider and the ordinary draught cider." He tells us that the manufacture of mead was considered of such importance, that the brewer of that beverage for former princes of Wales was the physician of the household, and ranked eleventh in point of dignity; that Athelstan, when king of Kent, is recorded to have expressed his satisfaction that "there was no stint of mead" when he paid a visit to his relative Aethelfleda ; and that, "according to an antique rule of the Welsh court, there were 'three things which must be communicated to the king before they were imparted to any other person : first, every sentence of the judge ; second, every new song ; and third, every cask of mead.'" Mr. Harris also gives us the recipe according to which the mead was made every year for Queen Elizabeth, who was, it appears, very fond of that beverage. It will be found amongst the recipes at the end of this chapter.

There appears to be no sufficient reason why the making of mead should have been given up, except that the great increase of population, without a corresponding increase in the production of honey, rendered it impossible to supply the requisite quantity of such drinks, and brought into fashion the use of beer and ale, which can be manufactured in any quantities from malted grain. It is nevertheless indisputable that mead continued in great favour, as Mr. Harris remarks, long after the introduction of malt liquors, and it is probable that it only ultimately gave way to foreign wines and to more potent, but less wholesome, distilled spirits.

Considering the increased supply and reduced price of extracted honey at the present day, there is no reason why the taste should not be revived, and some portion at least of the beverages now in common use be replaced, probably with advantage in a sanitary point of view, by drinks prepared from honey. For home use in country places where bees are kept, or where honey can be cheaply obtained in the immediate neighbourhood, such drinks are sure to come into favour. Annexed will be found several recipes for the preparation of a variety of beverages, varying in strength from a light summer drink, like ginger-beer, to the potent miodomel of the monks of Tokal.

MANUFACTURES.

The chief purposes for which honey is used in such large quantities as may be considered to bring the operations under the head of manufacturing industry, are beer brewing, liqueur and vinegar making, confectionery, biscuit making, and fancy soap making.

As regards the use of honey in beer breweries, an American writer, Mr. G. W. House, says :—

"Grape sugar is now largely used in the manufacture of beer. In a conversation with a prominent brewer of Brooklyn, who is now using dark extracted honey in place of grape-sugar, he said 'that the honey could be used at eleven cents (fivepence-halfpenny) per pound, and be cheaper than grape-sugar, besides making a beer that could not be excelled in purity and healthfulness.' If bee-keepers will go to a little trouble, they can establish a trade in this line that would demand more honey than is now produced."

There can be little doubt that brewers in all parts of the world, where they find they can get a sufficient supply of honey at a reasonable rate, will be glad to follow the advice of their American colleague, or at least to give the system of manufacture a fair trial.

Great quantities of honey are required for the manufacture of liqueurs, vinegar, and for confections of various sorts. Some fancy confectioners, or sugar bakers' establishments, use up as much as ten to twenty tons of honey each in a year.

A large demand has lately arisen for fine honey to be employed in the manufacture of biscuits. In May, 1884, it was announced in the *British Bee Journal* that Messrs. Huntley and Palmer, the well-known firm of biscuit manufacturers, of Reading, were prepared to contract for a supply of two tons of honey per week, or one hundred tons at the end of the season, for the manufacture of honey biscuits. There are other extensive biscuit makers who may be expected to follow this example, though, probable, not upon so large a scale. Honey is also used in the preparation of honey soap, and probably in some other branches of manufactures.

HONEY AS MEDICINE.

Gibbon, the historian, remarks of the ancient Greeks, and especially the Athenians, "They taught that health might be preserved, and life prolonged, by the external use of oil, and internal use of honey." Physicians have continued to use honey, more or less, in all ages up to the present time, both as a medicine and as a medium for administering some drugs. Mr. Newman, who has collected much information with regard to the uses of honey in his work, "Bees and Honey," gives the following quotation from a pamphlet by Herr Karl Gatter, a German teacher, and editor of the *Bienenvater*, at Vienna, who considers that his own life was saved by the use of honey for the cure of diseased throat and lungs :—

"In medicine, and especially in the healing of wounds, was honey, already in early times, used as a universal remedy ; it yet constitutes the principal ingredient of several medical preparations, is used with the best results in many internal and external diseases, serves as a means for taking powders, for the preparation of salves, and the sweetening of medicine.

"Honey mollifies ; promotes festering ; causes gentle purging, divides and dissolves, warms, nourishes, stops pains, strengthens the tone of the stomach, carries away all superfluous moisture, aids digestion, thins and purifies the blood, and animates and strengthens the breast, nerves and lungs. Honey is therefore to be used when suffering from *cough, hoarseness, stoppage of the lungs, shortness of breath, and especially with the best results in all affections of the chest.*

"Many persons afflicted with various species of consumption, thank the use of good honey, either for their entire restoration to health, or for the mitigation of their often painful condition of body and mind.

"Honey is also an excellent remedy for the occasional inactivity of the abdominal organs, and a means of strengthening weak nerves. For severe coughing, barley-water mixed with honey and the juice of lemons, drank warm, is a pleasant relief. It appeases and mitigates fevers, and owing to its taste and its soothing qualities, it is used as a gargle.

"Honey can also be used with advantage in asthma, in constipation, in sore throat ; promotes perspiration, lessens phlegm, and is very healing to the chest, sore from coughing.

"With old persons the use of honey is very useful, since it produces warmth and a certain activity of the skin. For persons leading a sedentary life, and suffering from costiveness, and especially from piles, pure unadulterated honey, either mixed in their drink, used alone, or on bread, is the best and healthiest means of relief.

"Honey has also great value as a medicine for children, and is readily partaken of by them as a choice dainty dish. It is especially useful to children afflicted with scrofula or rickets. In difficult teething, rub the gums with a mixture of honey and an emulsion of quinces. For the removing of worms, honey has often been beneficially used, and it is often used in diseases of the mouth and throat.

"Honey mixed with flour and spread on linen or leather is a simple remedy for bringing to head, or to maturity, boils, &c. Also, honey mixed with flour or fried onions, serves an excellent purpose as a covering for any hard swelling or callosity or abscess; and for ulcers it is often mixed with turpentine, tar, and tincture of myrrh. A plaster made of unslaked lime and honey has sometimes relieved most obstinate sciatica.

"If good honey is applied to inflamed wounds or boils, it lessens the drawing, quiets the pain and produces a good festering or suppuration. Undoubtedly, for all wounds, pustulous inflammations, bruises and bad festerings, honey is the best and most reliable remedy, and affords quicker and safer help than all other known plasters; all that is needed is to spread it rather thick on a piece of linen, place it upon the fresh wound, bind it fast, and renew the plaster every four or five hours. Of course, if bones are broken, surgical aid must be had.

"Honey-dough—*arto mele*—a plaster made out of honey and rye flour or rye bread, into which henbane or other narcotic substances is mixed, is an excellent means of irritation; which should be used in festering and bringing the sore to a head, and assuage the drawing and pain. It should be warmed, spread on a piece of linen and placed upon the sore part.

"For persons who are weakened through fast living, honey is, of all helps, the best nourishment, since it not only removes the poisons in the system, but also through its virtues strengthens the system; hence it has made itself so necessary to the inhabitants of the Orient."

RECIPES.

HONEY CAKES.

In Germany "Honig Kuchen" or honey cake, is quite an institution. The booths exhibiting this article in all its varieties form one of the most characteristic features of the German Kirmesses or Fairs, which are held periodically in most of the towns, as well as at the Schutsen Fests and all similar public festivities. Some of the following recipes taken from Mr. Newman's book no doubt represent the articles exposed for sale on such occasions :—

HAMBURG HONEY CAKE.—The flour intended for this cake should be well dried and sifted before being weighed; then take twelve pounds of flour and twelve pounds of honey; bring the honey to a boiling heat, pour it in the flour, and mix thoroughly. Dissolve two and a half ounces of pearlash in two gills of rose-water, the evening before ; take one pound of butter or lard, two tablespoonsful of West India rum, the grated rind of two lemons, the candied or sugar-coated rind of two oranges, and a very small quantity of pounded cloves. The solution of pearlash is to be added when the dough has become cool, and the mass must be thoroughly kneaded. The dough may be prepared several days in advance of the baking.

HONEY BROWN CAKE.—To four pounds of flour take four pounds of honey, one half pound of pulverized loaf or lump sugar, one-half ounce of Canella, three ounces of lard, a small quantity of cloves, one ounce of pearlash, one gill of rose-water, and two spoonsful of rum or French brandy. The honey and lard are to be incorporated by boiling, and when again cooled off, add the pearlash previously dissolved in the rose water. Knead the mass well, let it stand for several days, and then work it over again very thoroughly. Some persons prefer to omit the cloves, and substitute for them pounded cardamon seeds, grated lemon peel, or sugar-coated orange peel.

HONEY APPLE CAKES.—Soak three cups of dried apples over night ; chop slightly, and simmer in two coffee cups of honey for two hours, then add one-and-a-half coffee cups of honey, one-half coffee cup of sugar, one coffee cup of melted butter, three eggs, two teaspoonsful of saleratus ; cloves, cinnamon, powdered lemon or orange peel, and ginger syrup, if you have it. Mix all together, add the apples and then flour enough for a stiff batter. Bake in a slow oven. This will make two good-sized cakes.

HONEY FRUIT CAKE.—Four eggs, 5 cups of flour, 2 cups of honey, 1 teacupful butter, 1 cup sweet milk, 2 teaspoonsful cream of tartar, 1 teaspoonful soda, 1lb. raisins, 1lb. currants, ½lb. citron, 1 teaspoonful each cloves, cinnamon and nutmeg ; bake in a large loaf in a slow oven. This will be nice months after baking as well as when fresh.

MEAD, METHEGLIN, MIODOMEL.

MEAD—QUEEN ELIZABETH'S RECIPE.—Take of sweetbriar leaves and thyme each one bushel, rosemary half a bushel, bay leaves one peck. Seethe these ingredients in a furnace full of water (containing probably not less than 120 gallons), boil for half an hour ; pour the whole into a vat, and when cooled to a proper temperature (about 75° Fahr.), strain. Add to every six gallons of the strained liquor a gallon of fine honey, and work the mixture together for half an hour. Repeat the stirring occasionally for two days ; then boil the liquor afresh, skim it till it becomes clear, and return it to the vat to cool. When reduced again to a proper temperature (about 80° Fahr.) pour it into a vessel from which fresh ale or beer has just been emptied ; let it work for three days, and then barrel it. When fit (after fermentation) to be stopped down, tie up a bag of beaten cloves and mace (half an ounce of each), and suspend it in the liquor from the bunghole. When it has stood for half a year, it will be fit for use.—*Harris's Honey Bee.*

METHEGLIN.—Mix honey and water strong enough to carry an egg; let it stand three or four weeks in a warm place to ferment; then strain through a cloth and add some spices to suit the taste. [NOTE. In working on this recipe we find 3lb to 4lb of honey to the gallon of water is sufficient. About a teaspoonful of powdered ginger and half as much allspice to each gallon. The bottles require to be well corked as the liquor is as effervescent as champagne.]

SACK MEAD.—To each gallon of water add 6lbs. of honey, and also the white of an egg, and the shell broken up. Boil this mixture until the scum has all been cleared off, then add one ounce of hops to the gallon, and boil slowly for one hour. Strain away the hops, and when new-milk warm add a small quantity of yeast on a toast; let it stand a couple of days, and then put it into a barrel, which should only have been used previously, if at all, for white wine. Skim off any yeast that rises before being put into the barrel. Let the mead stand two years before bottling, and then when bottled it will keep for any length of time, and the colour will deepen with age.

MIODOMEL.—The following recipe is from the prior of the celebrated stronghold of Tokal, situated on the banks of the Bug, built in the thirteenth century, as a security against the invasion of the Tartars and Muscovites. This monastery enjoys a widely spread fame, through its miraculous Holy Virgin, and still more through its excellent miodomel. To twenty-four gallons of water put twelve gallons of honey, and 12lbs. hops; boil them together over a *very slow* fire, till the whole is reduced one-third. Care must be taken that the fire be not too strong, yet the heat must increase gradually; from a sudden and excessive heat, a burnt taste will be communicated to it. From the boiler empty it into a large tub or barrel, which must be deposited in a warm place during eight days to undergo fermentation, afterward filter through a wooden filter into a barrel, and place in a cellar for use; the older it is the better and stronger it becomes. After twelve months it may be bottled and kept for years. The peasantry generally keep it in barrels, where it is preserved as well as in bottles. Half a pint of good old miodomel taken every second night before going to bed, improves—and even restores to the stomach—the power of digestion; but if the miodomel be very old, say from ten years upwards, a wine-glassful is quite sufficient and more effective. It is also good in cases of gout and rheumatism.

FRUIT WINE WITH HONEY.—Take ten pints and a half of ripe fruit, which may be either gooseberries, currants, raspberries, blackberries, peaches, cherries, plums, or sloes; pound them in twenty-one pints of water, let them steep for four days, and then pour off the liquid. Press the skins between the hands and add twenty-one pints of water; let it stand six hours, squeeze the skins hard, pour off the liquid and throw away the refuse. Mix the two liquids together; add 9lbs. of honey, mix well, put into a cask, which must be entirely filled, and place it where the temperature is from 59° to 68° Fahr. The liquid will soon ferment and work out of the bunghole, which should be left open. The cask should be constantly filled up with some of the diluted juice reserved for that purpose, and when fermentation has nearly ceased the cask may be securely bunged up.   It has a particular aroma, becoming better as it grows older, and is more valuable as a drink than most wines sold by merchants.

WINE MEAD.—To make mead, not inferior to the best foreign wines, put 3lbs. of the finest honey to two gallons of water, two lemon peels to each gallon ; boil it one half hour, and skim well.   Put in the lemon peel while boiling.   Work this mixture with yeast, and then put it in a vessel to stand five or six months; then bottle for use.   If you choose to keep it several years, add 4lbs of honey to a gallon of water,

LIGHT BEVERAGES.

CHEAP HARVEST DRINK. — To those engaged in harvesting and other occupations tending to create thirst, we recommend the following preparation, which makes a very palatable· and healthful drink in hot weather :—Take 12 gallons of water, 20lbs. of honey, and 6 eggs, using the whites only.   Let these boil one hour ; then add cinnamon, ginger, cloves, mace, and a little rosemary.   When cold, add one spoonful of yeast from the brewer.   Stir it well, and in 24 hours it will be good.

HONEY MEAD.—Take three gallons of water of blood warmth, three half-pints of honey, two-thirds of a tablespoonful of ginger ; one third of a tablespoonful of allspice, and mix well together with a gill of yeast ; let it stand over night, and bottle next morning.   It will be in good condition to drink in 24 hours.

HYDROMEL.—This is a very nice drink and easily made.   For 11lbs. of honey take from 26 to 52 pints of water, according to the strength you wish to give the drink ; boil in a copper saucepan for an hour or two on a moderate fire ; take off the scum as soon as it forms.   Remove from the fire, let it cool, and pour it into a clean barrel, which must be quite filled, and place it with the bung-hole open in a dry, wholesome place, having a temperature from 60° to 66° Fahr.   At the end of two or three days fermentation takes place.   If long in fermenting add a little yeast ; it will be active enough in a few days.   Take care to fill the cask out of a bottle previously filled for that purpose.   In a month or six weeks the cask may be closed and put in a cellar.   The liquid clears and is soon fit to drink.

MEDICINAL.

HONEY GARGLE (Consumptive Hospital recipe for sore throat).— Borax, 1 drachm ; honey, 2 drachms ; water, 4 ounces.   Mix.

HONEY PASTE (*pate au miel*), for chaps, etc.—Clarified honey and cold cream, equal parts, rubbed smooth together.

HONEY AND BORAX, for sores in children's mouths.—Dissolve 1 ounce of borax in 1 ounce of glycerine, and then add 6 ounces clarified honey.

To CLARIFY HONEY.—Melt in a water bath (*i.e.*, place a vessel containing honey in a saucepan of water and heat), and strain while hot through flannel previously moistened with warm water.

For Asthma honey is an excellent remedy. Mix 1 ounce of castor oil with 4 ounces of honey. Take one tablespoonful night and morning. A simple and beneficial remedy.

One of the best uses that honey can be put to medicinally is in cases of sore and inflamed eyes. There are numerous reports of late in the bee journals from people who, after suffering for some time with these complaints, and trying many remedies without getting relief, have been cured by using honey wash.

Honey Eye Wash, for sore or inflamed eyes.—One part of honey to five parts of water. Mix, and bathe the lids, putting a few drops into the eye two or three times a day until well.

Honey also makes excellent vinegar. The washings of the extractor, cappings, or vessels that are used for honey, that otherwise would be wasted, will do for this purpose. About 3 or 4 pounds are sufficient for 5 gallons. Put in a cask and stand in the sun. Makes splendid pickling vinegar.

These recipes have been taken from various sources and may be relied upon, many of them having been tried by myself.

Proverb.

" Pleasant words are as an honey-comb, sweet to the soul, and health to the bones."

Prov. xvi. 24.

CHAPTER XXI.

CALENDAR AND BEE KEEPERS' AXIOMS.

VARIABILITY OF SEASONS.

No *invariable* rules can be laid down for the work to be done in an apiary each month, which can be strictly followed in *every* place, nor even in the same place in *every* year. The whole plan of operations must be suited in the first place to the normal climate of the district in which the apiary is situated, the nature of the bee forage available both in the spring and the honey season, and to the natural habits of the bees as influenced by their local peculiarities. If these circumstances be properly taken into account, a set of general rules may be established suitable to the average of seasons ; but even these must be liable to modifications at the judgment of the apiarist, according to the variations, or the more or less abnormal features, of different seasons.

USE OF METEOROLOGICAL OBSERVATIONS.

To any one at all acquainted with the subject it will be apparent how much depends upon the manner in which the bees get through the winter months, upon their condition in early spring, upon the period at which they commence to rear brood extensively and to prepare for swarming, and upon the time when those trees or flowers which are to furnish the chief crop of honey come into bloom ; and it must be equally apparent that all these things depend mainly upon the meteorological character of the winter, the spring, and the summer, especially upon the rates of temperature and of moisture which vary a good deal occasionally in each season, even in the best climates. We are all accustomed to speak in a vague manner about mild or severe winters, early or late springs, and wet or dry seasons, but without attaching any very exact value to these different terms. To the bee-keeper it may be strongly recommended to avail himself constantly of the records of the nearest meteorological observatory—to make himself acquainted with the normal mean temperature and rainfall of each month and each season of the year—that is, with the average results of observations continued over a large number of years, and then to watch, note, and compare with this normal standard, the mean temperature and the rainfall of each passing month. He will thus, at least, impress upon his mind some definite idea of how far the seasons of the current year correspond with or differs from the average of years, and this will considerably assist him in anticipating the probable course of events in the approaching honey season, and will suggest to him the most suitable mode of treatment for getting his bees into a condition to make the best use of the harvest season when it arrives.

The system of recording such observations as those alluded to above in a *graphic* form upon a diagram prepared for the purpose is now so universally used in similar cases that it is only necessary here to suggest its adoption to any one who wishes to keep before his eyes a clear picture of the few simple facts bearing upon the character of the months and seasons.

CALENDAR.

IN arranging the apiary work for the different months I have chosen the dates of the bee-seasons for the latitude of Auckland, New Zealand, as being nearest to those in the majority of places in Australasia. Where they differ much the dates can be altered to suit by carefully noting the time when willows and early flowering peaches blossom, which in the latitude of Auckland is at the latter end of August and beginning of September.

JANUARY.

In average seasons a good quantity of honey is gathered during this month, and in late seasons the bulk of the crop is often secured. Late swarms should be expected, but unless increase is required all precautions should be taken to prevent swarming at this time. Remove sections as soon as completely sealed, and place them in the honey house for a few days to ripen before packing them for market; take care that there are no bee moths in the house. Extract as often as necessary the surplus honey in frames.

FEBRUARY.

This is sometimes a hot dry month with a scarcity of bee forage toward the latter part. Beware of robber bees as soon as honey gathering slackens. The first month or six weeks after the close of the honey season is the worst time for robbing. If necessary to open a hive while robbers are about use a bee tent to cover it. Keep down weeds and tall grass around the hives.

MARCH.

Breeding will now begin to diminish, and a good look-out should be kept for queenless stocks. All poor queens should be superseded. With occasional showers the autumn flowers now coming into blossom will yield some nectar which will provide winter stores. Timber should now be stacked under cover to season for making into hives and frames during the winter. Look out for robbers; contract entrances if necessary. Rape and mustard seeds may now be sown for early spring forage.

APRIL.

Any colonies likely to be short of food for winter stores should be supplied with sufficient towards the end of the month. There should be from 25lbs. to 35lbs. in each hive at the commencement of winter. All implements in the honey-house and apiary, for which there is no further use, should be cleaned and stowed snugly away till required again.

MAY.

Now will be the time to prepare the hives for winter by removing all boxes not occupied by bees. Unite all weak and queenless colonies, also queen-rearing nuclei, and see that all are well supplied with food. Contract entrances and cover the frames with an extra mat as the weather becomes cooler.

Stow all spare combs in the fumigating room, and give the spare hives and boxes a brush over with a solution of carbolic acid before putting them away for the winter.

JUNE.

If the instructions given for last month have been attended to, the bees will need but little attention this month, and the less they are meddled with during cold weather the better. The planting of evergreens or other plants for shelter or bee-forage may now be commenced. This is the best time for shifting evergreens. The timber for hives and frames will now be ready to make up as opportunities offer. See that the hives are sufficiently ventilated to prevent dampness.

JULY.

Look out for leaky covers, and remove damp mats, replacing them with dry ones. Contract hives with division boards where necessary, and confine the bees to as few frames as possible. Select a warm day when the sun is shining to examine the hives, and get through the work quickly. Examine the combs in the fumigating room, and if moths or their larvæ are seen fumigate with sulphur.

AUGUST.

All hives should be well overhauled on the first fine days this month and the condition of each noted. See to the food supply, and feed where short, as a larger quantity will now be required for feeding the brood. Clean the bottom boards, and put in division boards where required. As breeding will have commenced, care should be taken to keep the interior of the hives warm. Remove any combs that are mouldy. Place the hive on a stand alongside while cleaning the bottom board. Unite weak and queenless colonies, and stimulate those required for queen-rearing purposes. Make up hives, frames, etc., and send orders to the manufacturers for material required for the coming season. Willows and early-flowering peach-trees blossom at the end of the month. Sow seeds of honey plants as soon as frosts are passed.

SEPTEMBER.

Where spring forage is scarce, the bees should be fed sparingly to stimulate brood rearing. As more room is required in the brood nest, shift the division boards and place a clean empty comb in the centre. Early in the month place a clean drone comb in the centre of the brood chamber of the colony chosen for raising drones. About the middle of the month, if the season promises to be favourable, start cell-building for queen rearing. Finish uniting, and beware of robbing. Give frames of emerging brood from the strongest to the weaker colonies. Sow seeds of honey plants for successional blossoms.

OCTOBER.

Hives for the expected increase should now be ready to set out before the swarming season commences, which will begin in ordinary seasons about the third week in this month. Cut out queen cells as soon as ready, and form nuclei. Transfer in the early part of the month. If increase is not desired, put on surplus boxes before the bees prepare for swarming. Prevent after-swarming. Enlarge entrances as the weather gets warmer, and keep the apiary clear of weeds and long grass.

NOVEMBER.

As the surplus honey is secured, keep each kind as far as possible by itself, more especially white clover honey. Extract when necessary, and remove sections when filled and sealed. Provide plenty of room in the brood chamber for breeding purposes by extracting the honey from the combs carefully, if necessary. Keep some spare queens on hand in nucleus hives. Shade all newly-hived swarms, and give plenty of ventilation. Buckwheat and mustard may now be sown for autumn flowering.

DECEMBER.

The instructions given for last month will apply to this. Keep down swarming by giving plenty of storage and breeding space, and deprive before the hives get too full. Supersede all poor and feeble queens. The main crop of buckwheat should now be sown.

BEE-KEEPERS' AXIOMS.

Mr. Langstroth, in his invaluable work so often alluded to in these pages, has given the following axioms as "a few of the *first principles* in bee-keeping," which ought to be as familiar to the apiarian "as the letters of his alphabet." They are so true, that they are still, and must continue to be, as important to all bee-keepers, whether novices or experts, as they were when first penned. I have already given each a prominent position at the close of some of the preceding chapters, in order the better to impress them on the mind of the reader :—

1. Bees gorged with honey never volunteer an attack.
2. Bees may always be made peacable by inducing them to accept of liquid sweets.
3. Bees, when frightened by smoke or by drumming on their hives, fill themselves with honey, and lose all disposition to sting, unless they are hurt.
4. Bees dislike any *quick* movements about their hives, especially any quick movement which *jars* their combs.
5. Bees dislike the offensive odour of sweaty animals, and will not endure impure air from human lungs.
6. The bee-keeper will ordinarily derive all his profits from stocks strong and healthy in early spring.
7. In districts where forage is abundant only for a short period, the largest yield of honey will be secured by a *very* moderate increase of stocks.
8. A moderate increase of colonies in any one season will, in the long run, prove the easiest, safest, and cheapest mode of managing bees.
9. Queenless colonies, unless supplied with a queen, will inevitably dwindle away, or be destroyed by the bee moth, or by robber bees.
10. The formation of new colonies should ordinarily be confined to the season when bees are *accumulating* honey ; and if this, or any other operation, must be performed when forage is scarce, the greatest precautions should be used to prevent robbing.

The essence of all profitable bee-keeping is contained in Oettl's golden rule—*Keep your stocks strong.* If you cannot succeed in doing this, the more money you invest in bees the heavier will be your losses ; while if your stocks are strong, you will show that you are a *bee-master* as well as a bee-keeper, and may safely calculate on a generous return from your industrious subjects.

FINIS.

GLOSSARY

OF SOME OF THE TECHNICAL AND SCIENTIFIC WORDS MOST COMMONLY USED IN WORKS UPON APICULTURE.

ABNORMAL.—Not according to a general rule.

ABSCONDING SWARM.—A swarm of bees which goes away to some distant place, either before or after first settling in a cluster, or even after being hived.

ANTHER.—In botany, the top of the stamen of a flower, containing the pollen or fecundating dust of the plant.

APIARY.—A place where bees are kept in hives.

APIARIST, APIATOR.—A person who keeps an apiary; a bee-keeper. The second word not much used.

APICULTURE. — Bee-culture; the art of cultivating bees for practical purposes.

APIARIAN, APISTICAL (adjectives).—Relating to an apiary or to apiculture in general.

APIDÆ.—The *family* of insects to which bees belong.

APIS.—Latin for bee; the name of the *genus* of insects of the *family apidæ*, to which all species of bees belong.

APIS MELLIFICA.—The species of the *genus apis*, to which all varieties of the honey-bee belong.

ARRENOTOKIA.—The name applied by Leuchart to denote a certain defective condition of queens.

ARTIFICIAL FECUNDATION, ARTIFICIAL FERTILISATION. — Impregnation of queens under the control of the apiarist, said to have been accomplished, but not generally admitted to be practicable. The former term is most correct.

BACILLUS ALVEI.—The scientific name of the disease known as "foul brood," and also the name of the germ which is the cause of the disease.

BACILLUS GAYTONI.—A germ disease, causing the bees it attacks to become hairless. These hairless bees were formerly thought to be old robbers. (See page 266.)

BALLING A QUEEN.—Bees surrounding a queen in a small compact ball or cluster, usually done for the purpose of injuring or killing her.

BEE-BREAD.—Pollen of flowers prepared by the bees as food for their larvæ. In cold climates wheat, rye, and pea flour are often provided as a substitute for pollen in early spring. (See page 96.)

BEE CULTURE.—(See APICULTURE.)

BEE FORAGE.—Trees, flowers or plants of every description which furnish the materials usually collected and stored by bees. (See Chaps. IV. and XVIII.)

BEE PASTURAGE.—Used in the same sense as bee forage, but not so correctly, as *pasture* infers, strictly speaking, an act of *grazing*.

BLACKS, BLACK BEES.—Terms generally, though incorrectly, applied to the common brown or German bee.

BROOD.—Young bees, in all stages of the change they undergo from the egg until they emerge from the cell.

BROOD CHAMBER, BROOD NEST. — The whole or part of the lower hive occupied by the queen for breeding purposes.

CAPPED BROOD, CAPPED HONEY.—Brood or honey when covered up in the cells of the comb with a *cap* made by the bees of wax, or a mixture of wax and propolis; also termed SEALED.

CAPPINGS.—The caps which are removed from the cells when the honey is to be extracted.

COCOON.—The silky web which the larva of the bee (or other insect) spins round itself previous to its change to the pupa state.

COLONY.—An established stock or collection of bees, consisting of a queen and workers, sometimes with drones, settled in a hive. (See STOCK.)

COMB.—A double set of waxen cells to contain honey, bee bread or brood, built by the bees, one set on each side of a division called the *septum,* which serves as bottom to the cells on both sides—the whole forming a sheet of comb.

COMB-FOUNDATION.—A sheet of wax, so stamped on each side as to form the bases of two sets of cells, constituting an artificial *septum,* which the bees can quickly build into a complete comb

COMB GUIDE.—(See STARTER.)

COMB HONEY.—Honey in the comb, specially raised for table use.

DEPRIVATION. — Removing honey from hives.

DIARRHŒA, DYSENTERY.—For this disease of bees—generally called by the latter name, although the first is perhaps the more suitable—see Chap. XVI.

DRIVING BEES.—Forcing bees to leave one box or hive and to enter another. (See Chap. X.)

DRONE EGG.—An egg that will produce a drone only; an egg unfertilised by the male germ; may be laid by an impregnated queen, a virgin queen, or fertile worker.

DRUMMING.—Rapping on the sides of a hive when driving bees.

DZIERZON (pronounced Tseertsone) THEORY. — The theory of Pastor Dzierzon, formulated into thirteen propositions, and which forms the basis of modern scientific apiculture. (See pages 65 to 67.)

EMBRYO QUEENS.—Queens in their rudimentary or undeveloped state, before arriving at maturity in their cells.

EMERGING BROOD, EMERGING BEES. —Young bees which have undergone all the changes from the egg to the perfect insect and then cut their way through the cappings of the cells.

FARINACEOUS FOOD.—Of a mealy or floury nature, such as the pollen of flowers.

FECUNDATE.—To impregnate. The queen after her " wedding flight " is properly said to be fecundated, though the term fertilised is more generally but not so correctly used.

FERTILISE.—To render fruitful. Applied by botanists correctly to the effect produced by the pollen on the ovules of the flowers; not so correctly applied to the fecundation of the queen bee, as she is, in a certain sense, fertile of herself. (See PARTHENOGENESIS.)

FERTILE WORKER.—A worker bee whose ovaries have been partly developed, and which is able to lay eggs in cases where a colony becomes queenless; but not having been fecundated by a drone, such eggs can only produce drones, like those of an unfecundated queen.

FDN. OR fdn.—An abbreviation of the compound word " comb-foundation " (which see).

FLIGHT (Wedding or Marriage).—The excursion which a young queen usually makes from the hive a few days after she emerges from the cell, for the purpose of meeting the drone in the air and becoming fecundated.

FLIGHT (Cleansing).—The first issue of bees from the hive after a long confinement, for the purpose of voiding their fœces.

FORAGE.—(See BEE FORAGE.)

FOUL BROOD.—A disease of bees, now called bacillus alvei. (See Chap. XVI.)

FOUNDATION.—Sometimes expressed *fdn.* (See COMB FOUNDATION.)

FUNGICIDE.—Any chemical substance which destroys the vitality of fungus sperms, which in a microscopic form are the causes of disease, such as bacillus alvei, etc.

GLUCOSE.—One of the chemical forms of sugar, known also as *grape sugar* and *fruit sugar* ; a cheap and inferior sort of syrup sometimes used to adulterate honey.

GRANULATED HONEY. — Crystallised honey. Nearly all pure liquid honey will granulate and become opaque after a while, unless heated to a high degree and then hermetically sealed while hot. Adulterated honey rarely granulates. On the other hand, there are rare cases of pure honey remaining in a clear liquid state.

GRAPE SUGAR.—(See GLUCOSE.)

HATCHING.—This term is only correctly applied to the production of the larva from the egg of the bee, which happens three days after the egg is laid. The changes from the larva to the complete insect do not properly come under the designation of hatching.

HEXAGON.—A figure having six sides and six angles. If the sides and angles are all equal the figure is a *regular hexagon.* The cells of honeycomb are, as a rule, of this form. (See page 94.)

HEXAPODA. — Six-footed. The subclass of *Insecta,* which includes the order Hymenoptera.

HIVE CRAMP.—A machine for pressing the four parts of a hive firmly together while nailing them.

HONEY BOARD.—A board or framework placed between the frames of the lower hive and super to prevent the bees building comb from the lower to the upper frames and so fastening them together. (See pages 141 to 144.)

HYBRID.—A mule or mongrel produced by parents of different species or varieties. In the latter case, where both parents belong to the same species, the progeny is more correctly termed a cross; but it is the custom of apiarists to call all crosses between varieties of the *Apis mellifica* hybrids.

HYMENOPTERA.—The order of insects with four membranous wings, to which the family Apidæ and the genus Apis belong.

IMAGO.—The last form assumed by insects in their transformation from the larva before emerging as a complete insect.

INSECTA.—Insects. The class of articulated animals which includes the sub-class Hexapoda, the order Hymenoptera, family Apidæ, and the genus Apis.

INTRODUCING A QUEEN. — Giving a strange queen to a colony of bees which has become queenless.

INTRODUCING CAGE.—A small cage in which the queen is placed for protection when introducing her. (See page 223)

JELLY, ROYAL. -The food prepared by bees for the larvæ which are intended to be developed into queens. (See page 73.)

LARVA (pl. LARVÆ).—The maggot or grub hatched from the egg, which is afterwards developed into the pupa or nymph, and ultimately into the imago and perfect insect.

MANIPULATION. — Handling. In a more general sense, the process of treating or dealing with things even where actual manual contact is avoided. Bees and honey may be manipulated without being touched by the hand.

MARRIAGE FLIGHT, OR WEDDING FLIGHT.—(See FLIGHT.)

METAL SUPPORTS. — Uusually, but wrongly, termed tin rabbets; strips of metal (generally tin), about an inch wide, tacked on each end of the hive inside for the purpose of supporting the frames. (See Chap. VI.)

METAMORPHOSIS. — Transformation, as from the larva to the complete insect.

MICROPYLE.—The minute opening in the egg of the bee through which the spermetozoon is introduced. (See page 69.)

NECTAR.—The liquid saccharine matter secreted and exuded by plants in their blossoms or flowers.

NECTARY.—That portion of the flower in which the nectar is exuded for the purpose of attracting insects.

NORMAL. — According to a general rule.

NUCLEUS (pl., NUCLEI).—Literally, the kernel of a nut; figuratively, the source from which something is to be developed. Used by apiarists to signify a small colony of bees intended for queen rearing.

NURSING BEES.—Those which attend to the feeding of the larvæ; generally young bees so employed for about two weeks before leaving the hive to gather honey.

NYMPH (also called Pupa or Chrysalis)·—The second form in the transformation of insects, between the larva and imago stages.

OBSERVATORY HIVE.—A hive constructed so that the operations in progress within it may be observed. (See page 128.)

OVARY.—The organ of a female insect or other animal in which the eggs are formed and developed.

OVIDUCT.—The passage which conveys the egg from the ovary when about to be deposited.

OVULE.— In botany, the "egg" or rudimentary seed.

PABULUM.—Food, or aliment.

PARENT STOCK. — The stock from which a swarm issues.

PARTHENOGENESIS.—The production of young by a virgin. (See page 64.)

PASTURAGE, BEE.—(See BEE FORAGE.)

PISTIL.—In botany, the central organ of a female flower which receives the pollen and contains the ovules from which the seeds are developed.

POLLEN.—The fecundating dust of male plants, obtained from the anthers of flowers, and used by the bees to make bee bread.

PROPOLIS.—A resinous matter used by bees for fastening movable parts of a hive, stopping fissures, and sometimes in covering foreign substances found in their hives. (See page 98.)

PUPA.—(See NYMPH.)

QUEEN CELL.—A large cell of peculiar form, specially built for the purpose of developing a queen from a worker egg or larva. (See page 72.)

QUEENLESS. — Applied to a colony when from any cause it is deprived of its queen.

QUINCUNX.—A term used in gardening with reference to the position of plants or trees—meaning five in a parallelogram, one in each corner, and one in the centre. Applied to the position of hives in an apiary. (See page 103.)

RABBET.—A corruption of the word rebate; a rectangular longitudinal recess made in the edge of anything, as when one part of the edge of a board is planed or cut out lower than the rest.

RACE.—In apiculture this word is used to designate a variety of the single species *A. mellifica* which has acquired some peculiarities of colour or qualities in course of time from natural or climatic causes.

RACK.—A name given to the frame or tray made to hold section boxes for the production of comb honey. (See page 139.)

REVERSIBLE FRAME.—A comb frame so made that the bottom can be turned to the top, and *vice versa*, at the option of the apiarist. (See page 237.)

RIPENING HONEY.—The process by which superfluous moisture is evaporated and the honey rendered safe to keep without fermenting.

ROBBING.—Bees from one hive or colony entering another and taking the honey from it.

ROYAL CELL.—(See QUEEN CELL.)

ROYAL JELLY.—(See JELLY.)

SEALED (Brood or Honey). — (See CAPPED.

SECTION BOX, OR SECTION.—A small frame in which surplus honey is stored and sent to market without being extracted from the comb, usually made to hold from one to two pounds.

SECTION CASE.—A shallow case without frames for holding section boxes on a hive; used as a super.

SECTION, OR BROAD FRAME.—A frame made to hold one or two tiers of section boxes while suspended in a hive.

SECTION RACK.—(See RACK.)

SEPARATOR.—A piece of wood or metal placed between two boxes to confine the bees to build their comb with an even surface. (See page 138.)

SEPTUM.—A partition. In apiculture generally applied to the vertical division between the two series of cells in a comb, which forms the bottoms of all the cells.

SHIPPING CAGE.—A small cage, usually made of wire-cloth and wood, used for sending queens through the post. (See page 227.)

SHIPPING CRATE.—A case used for packing comb honey in to send to market. (See page 243.)

SKEP, SKIP.—Literally, a basket. The name given in Scotland and other places to the old form of straw beehive.

SMOKER.—An implement constructed to burn rags, rotten wood, or other fuel, and furnished with a bellows for blowing the smoke where required. Used to quiet bees when about to be manipulated.

SPECIES.—In natural history, a subdivision of a *genus* or family of animals or plants, which may again be divided into varieties or races possessing some peculiarities but no important structural differences.

SPERMATHECA.—A small vessel attached to the oviduct of the queen bee, and containing, when fecundated, the spermatozoa for impregnating the eggs on their passage from the ovary. (See page 62.)

SPERMATOZOON (pl., ZOA).—A minute spore or germ, of which many millions may be contained in the spermatheca, itself scarcely visible to the naked eye. One of these spermatozoa must be introduced into the egg, through the opening called the micropyle, in order to make it capable of producing a worker bee or a queen.

SPRING DWINDLING.—Colonies which pass through the winter strong may become weak in the spring, in consequence of the old bees dying off before young ones are bred in the same proportion. This result, from whatever cause brought about, is termed spring dwindling.

STAMEN.—In botany, the male organ of fructification in plants, carrying the anther and the pollen.

STARTER.—A narrow strip of comb or of foundation put in a frame or section box to give the bees a "start" in the right direction in building the new comb.

STIGMA.—In botany, the top of the pistil which receives the pollen or fecundating dust of the male plant.

STOCK.—A complete collection of bees, consisting of a queen and workers (with sometimes drones), settled in a hive, and capable of propagating their race. (See COLONY.)

SUPER.—An additional box or hive set over another to increase the space for the bees to work and store surplus honey in.

SUPERING.—Putting on supers.

SWARM.—A portion of the bees of an old stock which leave the hive with either the old queen or a young one just emerged from the cell, to form a new colony. The latter is called an after-swarm.

SWARM BOX. — A box for taking swarms in.

TIERING-UP. — Placing additional stories, one over the other, on a hive, to induce increased production of either frame or section honey.

TIN RABBETS. — (See METAL SUPports.)

TRANSFERRING.—Changing the combs and stock of bees from one box or hive to another. Generally applied to cases of changing comb and bees from a straw or box hive to a movable comb hive.

UNCAPPING. — Removing the wax coverings from cells of capped or sealed honey preparatory to extracting.

UNITING. — Making one colony out of two or more stocks or swarms.

UNRIPE HONEY.—Honey which has not been long enough stored by the bees to get rid of all superfluous moisture and become fit to be capped or sealed.

VARIETY.—As applied to bees, a subdivision of the single species having some peculiarity of colour or qualities. (See RACE and SPECIES.)

VESICLE.—A little bladder or sac.

VIRGIN COMB.—Comb which has only been used once for storing honey and never for brood.

VIRGIN HONEY. — Honey stored in virgin comb only.

Note.—The two latter terms are now obsolete, but were formerly common among box-hive beekeepers.

WAX.—This word, when used without any addition or qualification, is taken to mean bee's wax—the substance secreted by bees, and of which they build their comb. (See pages 87 to 89.)

WAX POCKETS.—The overlapping of the abdominal rings of the worker bee, in which the scales of wax are secreted. (See fig. 27, page 88.)

WINTERING.— Passing colonies of bees safely through the winter months. (See Chap. XIV.)

WIRED FOUNDATION.—Comb foundation strengthened by the introduction of wires in the wax sheet.

WORKER EGG.—An egg laid by an impregnated queen which has been fertilised by receiving the male sperm as it passed the spermatheca on its passage from the ovary; will develop into a queen or worker.

WORKER, FERTILE—WORKER, LAYING.—(See FERTILE WORKER.)

INDEX.

	PAGE
Absconding swarms	200
Acacias in New Zealand	284
Adaptation to women of bee-keeping	20
Adulteration of honey	83
Advantages of hexagonal form of cells	94
Advice to beginners	21
After-swarming, Prevention of	204
After-swarms	202
Agriculture in relation to apiculture	297
Aid (State) to apiculture	22
Air sacs (of honey bee)	49
Alighting board (of hive)	122
Alley's drone and queen trap	210
Introducing cage	223
American linden or basswood	287
Plants and trees	287
Analysis of bee-bread	97
Of wax	89
Anatomy of the honey bee	43
Antennæ of honey bee	54
Ants	270
Antiquity of the use of honey	1
Apiary, Area of ground for	103
Arrangement of hives in	103
General arrangement of	99
House	109
Location of	99
Shade for	100
Stocking	106
Apiculture, as an industry, Importance of	18
In relation to agriculture	297
State aid to	21
Suitability of, for New Zealand and Australia	16
Apis Adansonii	27
Dorsata	37
Fasciata	27
Indica	37
Ligustica	26
Mellifica	25
Sirialis	37
Tregona	38
Unicolor	43
Are bees trespassers?	297
Area of ground for apiary	103
Arrangement of extracting house	155
Of hives in apiary	103
Arrenotokia	267
Artificial pollen	98
Art of bee-keeping, Modern	7
„ „ Origin of	3
Atmosphere and rainwater, sources of saccharine matter	301
Australia, Native flora of	251
Australasian exemptions from bee enemies	269
Automatic basket, Cowan's	151
Axioms, Bee-keepers'	325
Bacillus alvei (foul brood)	257
Salicylic acid remedy	264
Stages of	261, 262
Symptoms of	257
The Cheshire cure for	262
Under the microscope	260
Bacillus Gaytoni	266
Bees, Beneficial influence of, on agriculture	298
Can they harm the soil or the crops?	299
Diseases of	256
Enemies of	269
Flight of	271
Bee-bread	96
Analysis of	97
Bee-hawk (libellula)	271
Bee-keepers' axioms	325
Bee-keeping, adaptation to women	20
As a branch of farming	307
Improved system of	10
In New Zealand and Australia	7
Profits of	19
Bee-mite	279
Bee-moth	271
Larvæ, silken tube of	273
Male and female	273
Bee forage (see Forage).	
Gloves	173
Publications	23
Tents	254
Veils	172
Beginners, Advice to	21
Benefits of co-operation	243
Benton's queen shipping cage	227
Beverages, light, from honey	318
Bingham and Hetherington's uncapping knife	152
Bingham's direct draught smoker	174
Black (or German) bee	28
Bliss' sun evaporator	239
Bottom board (of hive)	121
Brickell's chaff hive	249
Broken-comb basket	154
Calendar, adapted to Auckland, New Zealand	321
Carniolan bee	35
Causes of robbing	252

PAGE

Causes of swarming.. .. 191
Caution to importers of bees .. 276
Cells, advantages of hexagonal
    form 96
Centrifugal extractor, invention
    of 146
Chaff hives 248
    Brickell's 249
Characteristics of structure of the
    honey bee 45
Characteristic structure of honey 83
Cheshire, Mr. Fr., investigations
    of Bacillus alvei 259
    Cure for Bacillus alvei .. 262
Choice of a hive 112
Clarke's cold blast smoker .. 174
Classification of species of the bee 44
Climate, Influence of, on Apicul-
    ture 17
Clipping the queen's wing .. 201
    Process of 202
Comb-foundation 158
    Advantages of its use .. 159
    Board 165
    Gauge for trimming 164
    History of the invention of . 158
    Machines, Root's roller 161
    Mode of fastening in frame .. 165
    Principal points of good .. 159
    Process of manufacture of .. 163
    The Given press 162
    To fasten in sections .. 166
    Wired 167
Comb honey, packing for market 242
    Shipping crate for 243
Comb, how constructed 93
    Baskets 240
    Holder 176
    Stand 216
Combs, Damage to 273
    Destroyed by bee-moth larvæ 275
    Fumigating 112
Co-operation, Benefits of 243
Cover (of hive) 123
Covering for frames (winter) .. 246
Covers, securing 247
Cowan's automatic basket .. 151
Cures for stings 177
Cyprian bee 34

Dadant's uncapping can 153
Dalmatian bees 37
Damage to combs 273
Deleterious honey 310
Development from egg to bee .. 69
Device for securing straight combs 170
Difference in character of winter
    seasons 244
Diseases of bees 256
Distribution (geographical) of the
    honey bee 25
Dividing 206
Division boards 232
Domestic cookery, Use of honey for 309
Drinks made from honey . 311-316

PAGE

Driving 184
    Open 187
Drones, A word concerning .. 209
Drone (bee), The 42
    And queen trap, Alley's .. 210
    Cell foundation 163
Duration of the honey season .. 294
Dysentery 256
Dzierzon, Dr., his discoveries .. 64
    Theory 65
Easterday's wire embedder .. 169
Enemies of bees 269
Entrance guard, Jones' 209
    Reducing 246
Exemptions (from bee enemies)
    Australasian 269
Extracted honey, Manipulation of 145
    Packing for market 241
Extracting house and honey store 104
    Arrangement of 155
Eucalypti in New Zealand .. 284
European plants and trees .. 285
Eyes (of the honey bee) 52
Farming, Bee-keeping as a branch
    of 307
Feeding 178
    For winter.. 179
    Stimulative 179
Feeder, Grey's entrance 182
    Simplicity 181
Fermented drinks from honey 311-316
Fertile workers 75
Fixing combs in frames 185
Flight of bees 296
Flora, native (see Native Flora).
Food, Providing for winter .. 246
Forage, Bee 278
    American linden, or bass-
        wood 288
    Eucalypti and acacias in New
        Zealand 284
    European plants and trees .. 287
    In New South Wales 282
    In Queensland 283
    In South Australia 282
    In Victoria.. .. .. — 283
    In Tasmania 284
    Native flora of Australia as .. 281
    „   „   New Zealand as 279
    Ordinary sources of 278
Forming nuclei 215
    My method 218
Foul-brood (see Bacillus alvei).
Frames, broad or section 131
    Half-story 132
    Narrow, or brood 130
    Number of to a hive 133
Frame form, or gauge 133
Fruit wine, made with honey .. 317
Fuel for smokers 175
Fumigating combs 277
    House 105
Gauge for trimming foundation 164
General arrangement of apiary.. 99

PAGE

Geographical distribution of the honey bee and its varieties 25
German (or black) bee 28
Given press, The 162
Grazing stock, question as to quantity of honey possibly consumed by 305
Grey's entrance feeder 182
Ground for apiary, Area of 103

Handling bees 171
Head (of honey bee) 51
Hearing, senses of, in honey bee 55
Heddon's honey board 142
New practice of transferring 186
Reversible frame 237
Section case 140
Herzegovian bees 37
Hexagonal form of cells, Advantages of 94
Hill's device 250
Hive, An ideal 113
Choice of a 112
How to open a 176
Instructions for making 117
Movable comb 111
Nucleus 127
The Langstroth 114
Various forms now in use 113
With division boards 233
Hives, Arrangement of in apiary 103
Chaff 248
Frames and section boxes for 111
Observatory 127
Hive cramp 125
Hive store 108
Hive making, Timber for 129
Hive painting 129
Hiving swarms 196
Honey, Adulteration of 83
Antiquity of use of 1
Characteristic structure of 82
Deleterious 310
Nature of 80
Honey as food 308
As medicine 314
Cakes 315
For domestic cookery 309
For preserves 309
Moderation in use of 310
Honey bee, Air sacs of 49
Classification of species of 44
General characteristics of structure of 45
Its varieties and distribution 25
Nervous system of 46
Physiology and anatomy of 43
Reproductive organs of queen 62
Respiratory organs of 48
Senses of hearing and smelling of 55
The antenna of 54
The eyes of 52
The head of 51
The honey sac of 59

PAGE

Honey Bee—
The legs of 56
The mouth of 52
The sting of 59
The wings of 55
Honey boards 141
How to construct 142
Honey dew 85
Honey extractor 145
Honey sac (of bee) 59
Honey store 104
Honey, strained, pressed, or melted 145
Honey season, Duration of 294
House apiary 109
How to feed syrup 181
Insert queen cells 216
Know robber bees 253
Open a hive 176
Stop robbing 254
Hybrids—German-Italian 33
Hymettus bees 37
Implements required for transferring 183
Importers of bees, Caution to 276
Improved system of bee-keeping, Introduction of, into New Zealand 10
Importance of apiculture as an industry 18
Increase of stocks 188
Circumstances which affect a decision 188
Modes of attaining the object 189
Preventing 205
What rate is desirable? 188
Industry, Apiculture as an 18
Inmates of the hive, their natural history 40
Inner covering for frames 246
Introducing queens 222
Direct method 224
Virgin queens 225
Introducing cage, Alley's 223
Introduction of bees and bee-culture into Australia and Tasmania 13
German bee into New Zealand 8
Italian bee into New Zealand 9
Investigations of Mr Fr. Cheshire 259
Italian (or Ligurian) bee 29
Markings of 32
Italian bees in N. S. Wales 15
In New Zealand 9
In Queensland 15
In South Australia 16
In Tasmania 16
In Victoria 15
Langstroth hive 114
General description of 115
Instructions for making 117
Larvæ of bee moth 274
Comb destroyed by 275
Legs (of honey bee) 56

PAGE

Libellula (bee-hawk) .. 271
Light beverage, from honey .. 318
Ligurian bee (see Italian).
"Little Wonder" (extractor) .. 149
Location of apiary 99
Management, spring 231
Manipulation of bees 171
    Extracted honey .. 145-154

Manufactures, Use of honey in .. 313
Marketing honey 240
Markings of pure Italian bees .. 32
Mating young queens 218
Mats for covering frames.. .. 134
Mead 316
Medicine, Use of honey as .. 314
Medicinal recipes 318
Meteorological observations, Use
    of 320
Metheglin 316
Method of rendering wax.. .. 91
Mice 270
Miodomel 316
Mite (bee) 275
Mode of securing and marketing
    surplus honey 231
Moderation in use of honey 310
Modern art of bee-keeping 7
Moth (bee or wax) 271
Mouth (of honey bee) 52
Moving hives 107
My method of forming nuclei
    and inserting queen cells .. 218
Native flora of Australasia .. 18
    Of Australia 281
    Of New South Wales 282
    Of New Zealand.. .. 279
    Of Queensland 283
    Of South Australia 282
    Of Tasmania 284
    Of Victoria 283

Natural swarming 190
    Objections raised against .. 194
Nature of honey 80
    Wax 87
Nectar of plants evaporated if
    not taken by insects 303
    Intended to attract insects .. 301
    Sometimes thrown off as
    superfluous 302
Nervous system (of honey-bee) .. 46
New South Wales, Native flora of 282
New Zealand, Native flora of 279
    Eucalypti and acacias in . 284
Nuclei forming 215
Nucleus hive 127

Objections raised against natural
    swarming 194
Observatory hives 127
Origin of the art of bee-keeping 3
Ovaries of queen bee 63
Over-stocking 296
Painting hives 129
Palestine bees 34

PAGE

Parker's comb-lever.. 167
Parthenogenesis 64
Physiology of the honey-bee .. 43
Pollen and bee-bread 96
    Artificial 98
Precautions for wintering .. 243
    To be observed against rob-
    bing 254
Preparing combs for extracting.. 152
Preserves, use of honey for .. 309
Preventing increase of colonies.. 205
Prevention of after-swarming .. 204
    Of dysentery 256
    Of swarming 203
Process of clipping the queen's
    wing 202
Profits of bee-keeping 19
Propolis 98
Providing food for winter .. 246
    Space above frames in winter 249
    Winter forage 247
Publications, Bee 23
Pure mating, Necessary distance
    apart of different races to
    secure 219
Putting on surplus boxes 234

Quantity of honey furnished by
    pasture lands 305
    Possibly consumed by stock.. 305
Queen (bee), The 41
Queen cells, Frame for raising on 212
    Frame of 214
    How to insert 216
    How to secure choice 211
    My method of inserting 218
Queen excluder honey board .. 238
Queen nurseries 220
Queen rearing 208
    Necessity of the practice .. 208
Queenless colonies, to unite .. 250
Queen's wing, process of clipping 202
Queens, Introducing.. 222
    Shipping 226
Queensland, Native flora of .. 283
Question as to grazing stock .. 304
Quieting bees 173

Recipes 315
Reducing entrance to hive .. 246
    Interior room in hive 243
Relation of bees to flowers .. 75
Remedies against bee moths .. 276
Reproductive organs of queen bee 62
Respiratory organs of honey bee 48
Reversible frame (Heddon's) .. 237
Reversing frames 236
Ripening honey 239
Robber bees 253
    How to know 253
Robbing, causes of 252
    How to stop 254
Root's roller machine 161
    Uncapping knife 152
Roumelian bees 37

PAGE

Saccharine matter derived from the atmosphero and rain water 301
    Not derived from the soil .. 300
Salicylic acid, remedy for bacillus alvei 264
Salvia officinalis 77-78
Seasons, Variability of 320
Section boxes 134
    Cramp and form for 136
    Dovetailed, To put together.. 136
    One piece 136
    Requisites of good 135
    To make 135
Section case, Heddon's .. 140
Section racks and cases 139
Securing covers 247
Senses of bearing and smelling of the honey bee .. 55
Separators 149
Shade for apiary 100
Shipping cage 227
    For two queens 227
Shipping crate for comb honey 243
Shipping queens 226
    Whole colonies 229
Silken tube of bee moth larva 273
    In comb 274
Simplicity feeder 181
Single comb extractor .. 149
Six comb extractor 149
Smelling, Sense of, in honey bee 55
Smokers 173
    Bingham's direct draught .. 174
    Clarke's cold blast 174
    Fuel for 175
Soil, can it be injured by bees? .. 299
    Saccharine matter not derived from 300
Sources, Ordinary, of bee forage 279
South Australia, Native flora of 282
Spiders 269
Spray diffuser 268
Spreading brood 234
Spring management.. 231
State aid to apiculture .. 21
Stimulative feeding 179
Sting of honey bee 59
Stings, Cure for 177
Stocking the apiary 105
Straight combs, To secure .. 170
Suitability of New Zealand and Australia for apiculture .. 16
Sun evaporator, Bliss' .. 239
Supplying old stock with fertile queens 205
Surplus boxes, Putting in 234
Surplus honey, Mode of securing and marketing 231
    Taking 238
Swarm, Issue of the 192
Swarm box 196
Swarms, Absconding 200
    Taking and hiving 196

PAGE

Swarms, Absconding, To unite .. 251
Swarming, Causes of .. 191
    Preparing for .. 192
    Symptoms of 193
Swarming season 192
Symptoms of bacillus alvei 257
Syrian bee 34
Syrup, How to feed 181
Taking and hiving swarms 196
Taking surplus honey .. 238
Tasmania, Native flora of 284
Tents (bee) 254
Timber for making hives .. 129
Tinea cereana 272
Transferring 183
    Board 184
    Wires and clasps .. 183
Uncapping can, Dadant's 153
    Knife, Bingham and Hetherington's 152
    Root's 152
Uniting 244
    Swarms .. 251
    Weak and queenless colonies 250
Use of honey, Antiquity of .. 1
    For food, drinks, manufactures, and medicines .. 308
Use of meteorological observations 320

Variability of seasons 320
Ventilation 248
Victoria, Native flora of 283

Waste of wax 90
Water for apiary 101
Wax, Analysis of 89
    Bees wasting 90
    Extractor, the Gerster 91
    Extractor, Jones's 92
    Extravagant waste of .. 91
    Method of rendering .. 91
    Nature of 87
    Smelter 166
Weak colonies, To unite .. 250
What and when to feed 180
What bees collect, and what they produce 80
Wings of the honey bee .. 55
Winters, Australasian .. 245
Winter forage, Providing.. .. 247
Winter seasons, Difference in character of .. .. , .. 244
Wintering 244
    Precautions for 245
Wire embedder, Easterday's 169
Wired foundation 167
Wiring board 169
Wiring the frames 168
Women, Adaptation of bee-keeping to 20
Worker bee, The 42
    Fertile 75
Workshop and hive store 105
Young queens, Mating 218

Mount Hobson Nursery & Seed Warehouse

(Established in Auckland and Suburbs a quarter of a century.)

The Stocks embrace almost everything required from a General Nursery and Seed Establishment, including Fruit Trees, Ornamental and Shelter Trees and Shrubs, Hedge Plants, Climbers, Flowering Plants, and Plants and Sets of all kinds. Choice Ferns and New Zealand Plants packed in Wardian Cases for all parts of the world. **Agricultural, Vegetable, Flower, and Tree Seeds in choice and extensive variety.** Seeds of Economic Plants and Novelties. Seed Potatoes, embracing the cream of the kinds exhibited at the late "European" International Potato Exhibition. Appliances and Materials for the Destruction of Insect Pests. Galvanised Wire Seed Guards and Garden Bordering. Garden Pottery and Glassware, Bee-hives and Appliances, and other articles too numerous to mention.

SPECIALITY IN BEE PASTURAGE.—"ERICA ARBOREA," or Tree Heath, a hardy, white, very floriferous evergreen shrub, 6 to 10 feet high in this climate, and can be cut or pruned to any form or purpose, as hedges, etc. This sweet-scented ornamental winter and spring flowering plant, at least fully equal to its congener—the heather of the British Isles—literally swarms with bees every fine half-hour in its season, and is invaluable to bee-keepers, coming at a season when flowers are scarce, and generally producing but little bee-feed. Safe balled plants are being produced in quantity at low rates—at present, 12 to 18 in. high, including package and shipment, 5s. per doz., 27s. 6d. per 100, £10 per 1,000, for cash on receipt of order.

Nursery and Seed Catalogue forwarded free with pleasure on receipt of address. In connection with Auckland Telephone system—Telephone, No. 209. Busses pass the Nursery frequently during the day, starting from the Greyhound Hotel, Queen Street, Auckland. The Nursery and Seed Warehouse are within 15 minutes' walk of either Newmarket or Remuera Railway Stations.

POSTAL ADDRESS—

C. T. WREN, Remuera, Auckland, N.Z.

Telegraph Address—"NEWMARKET."

A. I. ROOT,

MEDINA, OHIO,

MANUFACTURER OF AND DEALER IN

Apiarian Implements and Supplies

AND JOBBER IN

HOUSEHOLD CONVENIENCES.

A 40-PAGE PRICE LIST FREE ON APPLICATION.

Our Customers now number about 200,000, and goods are shipped to all parts of the world. To keep pace with late improvements and new inventions, our price list is kept constantly standing in type, and new editions are printed in the busy season, frequently as often as once a month. Our **A B C OF BEE CULTURE**, a book of 318 pages and 162 illustrations, is also kept standing in type in the same way. Our journal, **GLEANINGS IN BEE CULTURE**, gives reports semi-monthly of the state of bee culture in almos all regions of the globe where civilization extends.

BEE-KEEPER'S GUIDE

OR

MANUAL OF THE APIARY.

13,000 Sold since 1876.

The fourteenth thousand just out. Tenth thousand sold in just four months; 2,000 sold the past year. More than 50 pages and more than 50 costly illustrations were added in the eighth addition. It has been thoroughly revised, and contains the very latest in respect to Bee-keeping. Price by mail, $1.25 Liberal discount made to dealers and to clubs.

A. J. COOK,

AUTHOR & PUBLISHER

AGRICULTURAL COLLEGE, MICH., U.S.

BRANSTON & FORSTER

AUCKLAND.

MAKERS OF

6-Comb Honey Extractors, &c.

First Prize and specially commended at Wellington Exhibition.

THE
BEE-KEEPER'S

MAGAZINE.

A Monthly Magazine devoted exclusively to the interests of bee-keepers. You cannot do without it, because the articles it contains are written by the most experienced bee-keepers in the country.

Subscription Price, 5s. per year.

POSTAGE PREPAID.

SEND FOR CATALOGUE AND EXPORT PRICES.

ADDRESS—

KING, ASPINWALL & CO.

16 THOMAS STREET, NEW YORK, U.S.A.

GEO. McCAUL

WELLESLEY STREET EAST,

GALVANIZED IRON WORKS

AND

COLONIAL OVEN MANUFACTORY.

PLUMBERS, Builders, Contractors, and Storekeepers are informed that **GEO. McCAUL**, of Thames, has taken over the business lately carried on by J. L. KEAN, and being a direct importer and purchaser for cash, is prepared to supply all business material at lowest rates.

AWARDED THREE FIRST PRIZES

FOR

Colonial Ovens, Corrugated Tanks, Lead-headed Nails and Washers, Spouting, and Lead-edge Ridging.

GEO. McCAUL,

MANUFACTURER & IMPORTER

Wellesley Street East, Auckland,

AND

BROWN AND DAVY STREETS, THAMES.

DALY & PERRETT,
WOODSIDE APIARY, HAUTAPU,
WAIKATO, N.Z., HAVE ON SALE:—

HONEY (retail and in bulk), in 2lb., 5lb., 10lb., 25lb., and 64lb. tins; also, 1½ lb. glass pails with screw tops.

ITALIAN   **QUEENS**

SELECT TESTED (for queen rearing)   ..   £1 0 0

PURELY MATED (not specially selected) .   ..   0 17 6

YOUNG QUEENS from pure mother and laying, but with no
    guarantee as to purity of mating 0 5 0

Either of the above in three-frame nucleus, ten shillings extra. Safe arrival guaranteed to any address in New Zealand. N.B.—Orders for Queens will be filled in rotation as soon as may be possible after receipt, but it is, of course, not at all times possible to fill them immediately.

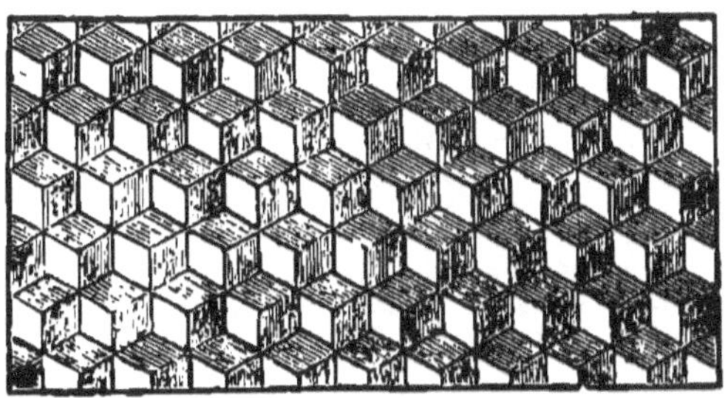

COMB   **COMB**

FOUNDATION   **FOUNDATION**

Made on the Given press, either wired or plain. Customers can have their 'own wax made up, or we will give best market price for all clean beeswax sent to us, or send cash on receipt.

Also, Bingham Honey Knives. Jones' Glass Pails, with screw tops, holding 1½lbs. Jones's Bee Entrance Guards, for excluding drones Alley's Drone Excluder, queen and drone trap combined. Wire Cloth for making Queen Cages and other Apiarian requisites.

FOR PRICES, ETC., APPLY AS ABOVE.

YATES' RELIABLE SEEDS

Yates' Choice Garden

and Agricultural Seeds

May be procured from his Warehouse, 46 Victoria Street, or from his Agents. throughout New Zealand.

Vegetable Seeds, 6d. per packet.   Flower Seeds, ordinary, 3d. per packet.

Flower Seeds, extra choice, 6d. to 2s. 6d. per packet.

Yates' Seeds are carefully saved from the BEST and PUREST Stocks only, and their growing powers are all thoroughly tested.   They are now generally acknowledged to be the best and most reliable seeds ever offered in the colonies, his largely increasing business and numerous testimonials being ample proof of this.

USE THEM AND A GOOD GARDEN IS ASSURED.

IMPORTANT.—We guarantee none genuine without name and address on packet. A. Y. has always a large stock of Agricultural Seeds on hand, including Grasses, Clovers, Turnip and other Root Seeds.   These are all of the finest quality.   Samples and prices post free on application.

ARTHUR YATES, Seedsman,

46 VICTORIA ST., AUCKLAND, N.Z.

HEAD ESTABLISHMENT:

16 & 18 OLD MILLGATE, & 97 Shudehill, MANCHESTER, ENGLAND. Estab. 1826.

YATES' GARDEN ANNUAL FREE ON APPLICATION.

TO POULTRY KEEPERS.

THE BEST BOOK

On the subject for the Colonies is

OUR DOMESTIC BIRDS

A practical Poultry Book for England and New Zealand by
ALFRED SAUNDERS, an Englishman, many years resident
in New Zealand.

PRICE 6s., or Post Free in N.Z. for 6s. 10d. in Stamps,
from

N. G. LENNOX,

BOOKSELLER, AUCKLAND.

THE NURSERY,

NEWMARKET, AUCKLAND

C. S. M'DONALD

Has for sale, a choice and comprehensive general Nursery Stock,
which includes first-class varieties of Fruit and Ornamental Trees,
Flowering Shrubs, etc., etc.

Descriptive Catalogues will be published in April, 1886, and will be
forwarded post free on application.

Opposite Railway Workshops,

NEWMARKET.

POTTIE'S
HORSE & CATTLE MEDICINES

POTTIE'S REMEDIES.—To Squatters, Breeders, Contractors Farmers, Farriers, and others.

These Medicines have the Largest Sale of any In New South Wales Victoria, and Queensland, and have been proved very rapid and safe in Curing all known Diseases. Books on Treatment forwarded Post Free on application to the Agents.

LIST OF REMEDIES.

Human) Lily Hair Oil	Hoof Ointment
Use) Highland Oil	Litzart Ointment
Glencairn Oil	Healing Ointment
Black Oils	Tumour Ointment
Physic Balls	Gall and Shoulder Salve
Embrocation	Condition Powders
Green Healing Lotion	Worm Powders
Black Healing Lotion	Urine Powders
Mange Liniment	Mange Powders
Sweating Blister	Stomach Powders
White Oils	Gripe Mixture
Cooling Oils	Cough Drink
Golden Blister	Fever Drink
Worm Balls	Blood Tonic
Fly Blister	

Pottie's Horse Medicines.

SOLE AGENTS FOR

AUCKLAND, GISBORNE, NAPIER, NEW PLYMOUTH, & WANGANUI,

HARRY H. HAYR & CO.,

HIGH STREET, AUCKLAND.

TELEPHONE No. 285.